BIOETHICS IN CULTURAL CONTEXTS

INTERNATIONAL LIBRARY OF ETHICS, LAW, AND THE NEW MEDICINE

VOLUME 28

The titles published in this series are listed at the end of this volume.

Bioethics in Cultural Contexts

Reflections on Methods and Finitude

Edited by

CHRISTOPH REHMANN-SUTTER
University of Basel, Switzerland

MARCUS DÜWELL
University of Utrecht, The Netherlands

and

DIETMAR MIETH
University of Tübingen, Germany

 Springer

A C.I.P. Catalogue record for this book is available from the Library of Congress.

ISBN-10 1-4020-4240-X (HB)
ISBN-13 978-1-4020-4240-9 (HB
ISBN-10 1-4020-4241-8 (e-book)
ISBN-13 978-1-4020-4241-6 (e-book)

Published by Springer,
P.O. Box 17, 3300 AA Dordrecht, The Netherlands.

www.springeronline.com

Printed on acid-free paper

Contents

INTRODUCTION

CHRISTOPH REHMANN-SUTTER, MARCUS DÜWELL,
DIETMAR MIETH

When we placed "finitude", "limits of human existence" as a motto over a round of discussion on biomedicine and bioethics (which led to this collection of essays) we did not know how far this would lead us into methodological quandaries. However, we felt intuitively that an interdisciplinary approach including social and cultural sciences would have an advantage over a solely disciplinary (philosophical or theological) analysis. Bioethics, if it is to have adequate discriminatory power, should include sensitivity to the cultural contexts of biomedicine, and also to the cultural contexts of bioethics itself.

Context awareness, of course, is not foreign to philosophical or theological bioethics, for the simple reason that the issues tackled in the debates (as in other fields of ethics) could not be adequately understood outside their contexts. Moral issues are always accompanied by contexts. When we try to unpack them – which is necessary to make them accessible to ethical discussion – we are regularly confronted with the fact that in removing too much of the context we do not clarify an issue, but make it less comprehensible. The context – at least some essential parts of it – is intrinsic to the issue. Unpacking in ethics is therefore a different procedure. It does not mean peeling the context off, but rather identifying which contextual elements are essential for an understanding of the key moral aspects of the issue, and explaining how they establish its particular character. Contexts, we could claim, are not a kind of packaging at all; they are not something we could reasonably try to get rid of without losing the substance of what we want to clarify. This wisdom is as old as ethics itself, and it would be surprising if it were not also applicable to bioethics.

In the current bioethical discourse, it is obvious that context dependency characterises not only the moral significance of the issue, but also the answers we produce. There are "cultural differences" in moral attitudes to questions such as disclosure of terminal diagnoses, reproductive technologies, stem cell research, therapeutic cloning, organ transplantation, brain death, assisted suicide or euthanasia. "Comparative bioethics" started

1

C. Rehmann-Sutter et al. (eds.), Bioethics in Cultural Contexts, 1–10.
© 2006 *Springer. Printed in the Netherlands.*

in the mid 1990s, following the groundbreaking study by Renée Fox and Judith Swazey (1984) comparing medical ethics in China and the United States. Different countries, with their specific mix of cultures, have different laws about these matters, and some of the official attitudes represented in a country's law might not find equal support in all sectors of society.[1] As Martha Nussbaum (2000: 49) observes, in the context of ideas of feminism, democracy and egalitarian welfarism, which now pervade every society, "the ideas of every culture turn up inside every other, through the internet and the media." The idea of homogeneous cultures apart from each other is unrealistic. Instead of homogeneity, agreement or submission, in reality there is diversity and contestation. Moral answers, as well as the methods used to produce them, are rooted in cultural traditions that are themselves diverse and dynamic. The predominant arguments have "local" histories and reveal different debates about understanding the role of the human subject in relation to life and to suffering. This holds true even for the choice of questions discussed in healthcare ethics and politics. In a world in which acute wealth differences mean that the single most important risk factor to global health is in fact malnutrition (WHO 2003), but where in some places the major risks are seen to be in the over-development of biomedical technologies, it is clear that not all ethical questions appear equally important everywhere. The choice of questions also reflects the contestation of the power structures that are responsible for these striking inequalities in the trans-cultural distribution of capabilities to live in good health.

This book is not about comparative bioethics and therefore does not contribute much to these observations. It rather starts from the evidence and opens a range of *methodological* questions. The chapters ask, for instance, how the intrinsically important elements of cultural contexts can be identified, and how they can be introduced adequately into bioethical inquiries. Can bioethics be an enterprise that does not simply isolate but also (re)contextualises? The various chapters also ask about the strengths and weaknesses of traditional and innovative modes of ethical argumentation in terms of this set of tasks. Which discipline (or method) would be capable of doing which job? And how can different approaches and their contributions be integrated without stepping into an "anthology trap" of just referring to a variety of historically developed methods (from consequentialism to deontology, virtue or discourse ethics, care or narrative approaches) one after the other, leaving people simply faced with the different results and thereby jettisoning the clarifying and guiding role of ethical investigation? What are the main differences between and limits of these diverse approaches to context? The overall question is this: if bioethics is to become context sensitive, what must be taken into consideration and how should this be done?

However, this last phrase is polemical, in that it claims the predominance of a research style in "bioethics" that is not, or is not adequately, conscious of contexts. This may still hold true for some traditions in bioethics, frequently but not exclusively in the libertarian and analytical traditions that may have become predominant in some academic subcultures. However, it is certainly not true for the whole field. Some areas are already active, and here we can hope to contribute. As a result of an "anthropological turn" in bioethics (Kleinman 1995) there is a rapidly growing body of interpretative research using methods such as qualitative interviews, grounded theory or participant observation, which explore the relationship between "cultural" norms and the understanding of the meaning of issues like autopsy, abortion, medical genetics or hard clinical decisions (Turner 2003; Scully et al. 2004). Some authors (Borry et al. 2004; Musschenga 2006) even speak of an "empirical ethics" that differs from purely descriptive approaches in that it aims to be both descriptive and normative.

Concepts need to be clarified. What is "context", for instance? Even if we restrict the meaning of the term to those surrounding elements of the situations that are interwoven with the issue under consideration, that is those that contribute to its moral significance, the concept of context still remains extremely broad. Situations typically have no clear outer limits, but extend more or less infinitely into the past and the future, into those tremendously complex spheres we call culture and society. Therefore we need to select guiding questions that can help to reduce the level of complexity. And we also need concrete empirical information about the real practice of patients, doctors and other agents, how they selectively reduce complexity and how they create situations they can cope with.

Culture and "cultural contexts" on the one hand, and bioethics on the other, are therefore related to each other in more than one way, and we should clarify to which of these relationships the contributions of this book mainly refer. Firstly, we have already mentioned cultural differences and pluralism in moral attitudes and in solutions that are held to be ethically sound within traditions and (sub)cultures. The corresponding ethical issues are organised around the relativism/universalism debate. Paradigmatic issues are women's rights for reproductive autonomy and treatment of Jehovah's Witnesses who reject blood transfusion or female genital cutting (clitoridectomy) in some African or Asian traditions (Cook, Dickens and Fathalla 2003: 262–275). "Culture", in this respect, is a synonym for particularly local traditions, particular worldviews or communities of mutual agreement and understanding. Secondly, culture can also be related to bioethics as a necessary component of its issues. Biomedical developments carry with them social and cultural meanings that must be taken into account if the accompanying bioethical dilemmas are to be understood. Biomedicine is a specific form and concept of medicine, and so are predictive medicine and regenerative medicine. In order to be sensitive to those dimensions, bioethics should include a "cultural approach" (Callahan 1999: 279; Marshall and Koenig 2004). Thirdly, culture is intrinsically related to bioethics, because bioethics itself is a social phenomenon. A reflective, culturally informed bioethics also questions bioethics' actions in relation to social institutions. Carl Elliott (2004) and Virginia Ashby Sharpe (2002) for instance have forcefully criticised the naïve behaviour of bioethicists who help to legitimise controversial practices within companies by providing them with an ethical conscience. The ways in which bioethicists communicate and interact among themselves, with their counsellees and with the public have also been investigated by sociologists (DeVries and Subedi 1998).

It is the second and third elements of the cultural contexts/bioethics relationship that stand at the centre of the following chapters. By introducing the term "finitude" in the sense of limits of human existence, limits of human knowledge, knowledge capacity and ethical judgment, a difference was set in the cultural apprehension of medicine. Medicine could also be conceptualised as an undertaking to overcome limits. "Finitude" contains a reminder of the inherent limits of such a goal. It reintroduces the cultural context on which every medicine (limits-sensitive or off-limits medicine) depends.

These concepts form a bridge between method and content in bioethics. Limits and attitudes towards the finitude of human existence seem to be at the core of the moral uncertainties of modern biomedicine. From the perspective of patients there can also be an excess of medicine, an overdose of technological "help" so to speak, which might be well-intended but fails to contribute to welfare, causing despair or at least becoming a burden for those affected by its implications. One question this raises is how to handle the heavy burden of the responsibility for deciding when to stop. The social and cultural meaning of death has been affected, even changed. It is no longer inevitably the last and limiting event imposed from somewhere outside, but has increasingly become an object of human control and decision. Some technologies already available, and even more those still in development, promise to shift limits, to expand the life- and "healthspan"[2] of human existence, even to change the human condition altogether. If reflections on limits and finitude are important for methodological clarification and improvement of ethics, the focus on limits and finitude might also provide us with some good guiding questions.

Like all medicine, biomedicine is motivated by the desire to help people relieve the burden of suffering by curing or preventing diseases that ultimately delineate the limits of human existence. Its aim must be to expand these limits, and at the same time there must be ethical reflection in order to take into consideration the *moral* limits of its limit-expanding practice. This is why we see limits as the subject matter of ethics in biomedicine. The limits of our capacity for moral judgment, or to foreseeing potential consequences, the limits of moral rationality, the limits of cultures with regard to other cultures, the limits of the explanatory power of symbols, or the limits of what should be taken into account in assessing the context of moral questions: all these represent ways in which limits appear in discussions of the methodology of bioethics. Substantive limits and methodological limits: do they relate to each other, or do they even have things in common?

We should take into consideration the fact that both biomedicine and ethics, and particularly bioethics, are products of cultural achievements. Despite the complaints about the divide between the "two cultures", both science and humanities arose from the same "occidental" or "western" traditions and have continued to interrelate. Therefore, understanding biomedicine *and* understanding bioethics presupposes a critical and interdisciplinary investigation into this cultural tradition and its current dynamics. If bioethics, as a field of active reflection, is to contribute here, it needs a research programme that is oriented towards cultural context and cultural background.

Medical ethics itself has a long tradition going back to antiquity. Until the 19th century it was essentially a professional ethics of physicians, with the task of legitimising their interventions and expressing the limits of the acceptable use of their art. *Bonum facere* and *nil nocere* were the two great moral principles of the Hippocratic tradition. In the 19th century, with the inclusion of the methodological repertoire of experimental science (Claude Bernard), an expert-centred research ethics was developed that limited the involvement of persons as research subjects. The first non-medical personnel to participate in the discourse of medical ethics were theologians. It was only in the late 1960s in the United States that bioethics was inaugurated as a new interdisciplinary field, and interdisciplinary ethics committees started to become mandatory instruments in clinical research.

Clearly, the interdisciplinary research programme oriented to cultural context and cultural background for which this book argues is not a matter of course; and even within the current field of academic bioethics, the programme is not unanimously approved. Perhaps its most radical counterpart would be the idea of a fully secular expert discipline of practical moral rationality. The problem with this approach to bioethics is that, aside from its merits in posing provocative arguments that challenge moral intuitions and producing exemplary models for rational consistency, it has a certain inherent closure that isolates it from other, non-rational aspects of human practice, and as a consequence it tends to exclude these other areas of concern from discussion. Our approach in this book is to incorporate both non-rational and rationalistic approaches into a broader interdisciplinary engagement.

Contemporary medicine, as we have said, itself is part of the distinctive cultural project of modernity. Its motivating normative programme is rooted in the idea of experimental science and technology that has developed since the 17th century. Its main aim is to master nature, and this means also to master the organic mechanisms of the human body that produce functional disorders. In other words, it aims to eliminate contingency in human existence. Limits are seen as negative. Abilities that individuals lack are classified as dysfunctions, diseases as adversaries, and death as the foe. Biomedical breakthroughs have frequently been valued as "great events of humankind" that promise to extend the limits of our physical existence (however, this begs the question of how to evaluate these limits). Therefore, moral questions that are raised by modern medicine cannot be properly understood without taking the fundamentals of its cultural construction into account.

The anthropologist Arthur Kleinman (1995: 29) has characterised Western "biomedicine", in comparison with most other medicines, by "its extreme insistence on materialism as the grounds of knowledge, and by its discomfort with dialectical modes of thought. Biomedicine also is unique because of its corresponding requirement that single causal chains must be used to specify pathogenesis in a language of structural flaws and mechanisms as the rationale for therapeutic efficacy. And particularly because of its peculiarly powerful commitment to an idea of *nature* that excludes the teleological, biomedicine stands alone." The consequence of this is that biomedicine sees disease primarily as a "biological pathology" (32) that can but does not need to be accompanied by suffering. And this again legitimises medical intervention on grounds other than the relief of suffering. Medicine can become preventive; the reparative intervention can precede actual suffering. Medicine can become spare-part medicine, detaching the individuality of the self from the integrity of its body. Medicine can also develop extracorporeal functional replacement technologies, which can then cause moral difficulty precisely because these extracorporeal functions are necessarily in the hands of powerful social institutions whose authority (over gametes, embryos, machines, resources etc.) must be justified. Medicine can even become predictive, legitimising its power by its ability to forecast of potential biological pathologies, installing new social regimes of risk reduction on the lives and bodies of persons who are not yet patients, but who are supposed to behave in anticipation of these potential diseases. Biomedicine might be a "plastic word", a place of multiple discursive signification, rather like the "life sciences" which form its background (Kettner 2004: Vorwort), however, the key aspects of its value orientation need to be clarified and exposed in order to make possible what Kleinman deplored as not yet having taken place in medicine: an open competition of paradigms.

The book's first section on fundamental aspects starts with two introductory chapters. Albert R. Jonsen describes the actual state of this originally American academic phenomenon, "bioethics", with an eye to its problems and its future. Dietmar Mieth reviews current issues in the ethics of biomedicine and biopolitics. Beat Sitter-Liver then introduces the idea of "finitude", drawing on a phenomenological and hermeneutic background of European philosophy. The section is concluded by two chapters which examine limits, one on the limits of bioethics itself and on the different ways in which bioethics is confronted with limits as a moral phenomenon (Christoph Rehmann-Sutter), the other on the limits of law as a social and political institution (Silvana Castignione).

In the second section, a series of classical approaches to bioethics are discussed, each focusing on the special strengths and weaknesses, the possibilities and restrictions that characterize it. Marcus Düwell defends a one-principle approach to bioethics, while Bettina Schöne-Seifert discusses advantages of Beauchamp and Childress' four-principles-approach. Brigitte Feuillet-Le Mintier's chapter looks at the intersection of ethics and law and criticizes too straightforward an implementation of moral ideas in new laws. Roberto Mordacci scrutinizes Kant's Categorical Imperative as an approach in bioethics. Derek Beyleveld develops a concept of practical rationality and applies it to a case of surgical separation of conjoined twins. All these approaches are sensitive in particular ways to social and cultural contexts, but they do not have developed, specialized instruments in context-analysis.

This contextualisation, however, is the aim of the chapters in the third section. Adela Cortina clarifies the role of the public in bioethics and reinterprets the role of bioethics in the context of political decisions. Sigrid Graumann critically analyses expert approaches and the idea of biopolitics. Claudia Wiesemann looks at what medical history can and should contribute to bioethics. The next chapter is a case study with a social science approach. Eva Krizova has analysed the discussions about alternative and school medicine. Pavel Tichtschenko takes a political perspective on bioethics as a discourse in the multicultural situation of modern societies.

Three further chapters build a section on the subjects of body and identity. Jean Pierre Wils defends the thesis that modern biomedicine is a radically different approach to the body. He uncovers the particular understanding of identity and of the relation of self to embodiment that this implies. Jackie Leach Scully discusses different ethical approaches from the perspective of disabilities studies. The disabled body is a challenge for the interpretation of embodiment. Walter Lesch introduces the concept of coping strategies and demonstrates its potential for a better understanding of basic bioethical limit-problems.

The concluding section is devoted to five innovative methods of analysis. Erica Haimes introduces sociology, as both theoretical and empirical research into social phenomena and structure. She clarifies what the social sciences can and should contribute to bioethics. Matthias Kettner's chapter develops a discursive method for bioethics, based on the philosophy of Karl-Otto Apel and Jürgen Habermas. Eva Feder Kittay, one of the pioneers of care ethics, reviews the current state of its methodological developments and its problems. Anthropologist Rayna Rapp shows how qualitative social sciences can improve our understanding of the thick social matrix in which bioethical issues are constructed and solved. Hille Haker completes the section with her method of narrative bioethics.

Metaphorically speaking, bioethics could be seen as a "dining table", with place settings offering different menus: different methods (descriptive, evaluative, narrative), different criteria (norms, rules, rights), different ways of questioning power relationships (participatory, communicative, political) bring their ingredients together, find suitable recipes, and set their part of the table nicely. They prepare a meal, to which a range of guests (those concerned, those involved, those affected) is invited. According to this metaphor, bioethics must be an open moral reflection, taking the plurality of cultural perspectives into account at all levels of scientific and hermeneutic methodologies. A key motif for understanding how different methodologies can work together is the *pluriperspectivity* of moral phenomena. They are social life-born, not theory-born.

The idea for this book goes back to a project of two interrelated European Research Conferences (EURESCO) funded by the European Science Foundation, under the title "Biomedicine within the limits of human existence". The first conference was held at Davos, Switzerland, in September 2001. It happened to be precisely the week of 9/11 when televised events surrounding the attacks on the World Trade Center in New York suddenly gave the limits of law and ethics and the clash of cultures a distinctive new political meaning. The participants were forced to redefine their task as intellectuals and "contemporaries" at short notice.

We would like to thank the European Science Foundation, the Swiss National Science Foundation, the Swiss Federal Office for Education and Science, and the Swiss Academy of Humanities and Social Sciences for generously supporting this series of European Research Conferences and the book project. But without the original ideas and the continuous intellectual stimulus of Beat Sitter-Liver it would not have happened at all; the credit for taking finitude as a focus in interdisciplinary bioethics is due to him. We wish to thank Irina Miletskaia, Rouven Porz, Rowena Smith and Franziska Genitsch for professional help in the editing process, Springer Verlag (formerly Kluwer Adademic) and David Weisstub for accepting the book in this series, and two anonymous referees for their helpful feedback on earlier drafts of the manuscript. Last but not least we would like to thank the participants at the 5-day conference in Davos, particularly the authors of this volume, for engaged discussions and for their continuous work during the gestation of this book.

References

Borry, Pascal; Schotsmans, Paul and Dierickx, Kris. "Editorial: Empirical Ethics – A Challenge to Bioethics." *Medicine, Health Care and Philosophy* 2004, 7, 1–3.

Callahan, Daniel. "The Social Sciences and the Task of Bioethics." *Daedalus* 1999, 128, 4, 275–294.

Carrese, J. and Rhodes, L. "Western Bioethics on the Navajo Reservation: Benefit or Harm?" *J of the American Medical Association* 1995, 274, 826–829.

de Castro, Leonardo D. et al. *Bioethics in the Asia-Pacific Religion: Issues and Concerns.* Bangkok: UNESCO Bangkok, 2004.

Cook, Rebecca J.; Dickens, Bernard M. and Fathalla, Mahmoud F. *Reproductive Health and Human Rights. Integrating Medicine, Ethics and Law.* Oxford: Oxford Univ. Press, 2003.

Coward, Harold and Ratanakul, Pinit (eds.). *A Cross-Cultural Dialogue on Health Care Ethics.* Ontario: Wilfrid Lawrier Univ. Press, 1999.

De Vries, Raymond and Subedi, Janardan (eds.). *Bioethics and Society. Constructing the Ethical Enterprise.* Upper Saddle River: Prentice Hall, 1998.

Elliott, Carl. "Pharma Goes to the Laundry: Public Relations and the Business of Medical Education." *The Hasting Center Report* 2004, 34, 5, 18–23.

Fox, R. and Swazey, J. "Medical Morality is Not Bioethics: Medical Ethics in China and the United States." *Perspectives in Biology and Medicine* 1984, 27, 336–360.

Gordijn, Bert and ten Have, Henk (eds.). *Medizinethik und Kultur. Grenzen medizinischen Handelns in Deutschland und den Niederlanden.* Stuttgart-Bad Cannstatt: Frommann-Holzboog, 2000.

Kettner, Matthias (ed.). *Biomedizin und Menschenwürde.* Frankfurt am Main: Suhrkamp, 2004.

Kleinman, Arthur. *Writing at the Margin.* Berkeley: California Univ. Press, 1995.

Koenig, Barbara and Gates-Williams, J. "Understanding Cultural Differences in Caring for Dying Patients." *Western J of Medicine* 1995, 163, 244–249.

Marshall, Patricia and Koenig, Barbara. "Accounting for Culture in a Globalized Bioethics." *Journal of Law, Medicine and Ethics* 2004, 32, 252–266.

Musschenga, Albert W. "Empirical Ethics, Context-Sensitivity and Contextualism." *Journal of Medicine and Philosophy* 2006, in press.

Nussbaum, Martha. *Women and Human Development. The Capabilities Approach.* Cambridge: Cambridge Univ. Press, 2000.

Schicktanz, Silke; Tannert, Christoph and Wiedemann, Peter (eds.). *Kulturelle Aspekte der Biomedizin. Bioethik, Religionen und Alltagsperspektiven*. Frankfurt am Main: Campus, 2003.

Scully, Jackie Leach; Rippberger, Christine and Rehmann-Sutter, Christoph. "Non professionals' evaluations of gene therapy ethics". *Social Science and Medicine* 2004, 58, 1415–1425.

Sharpe, Virginia Ashby. "Science, Bioethics, and the Public Interest: on the Need for Transparency." *The Hastings Center Report* 2002, 32, 3, 23–26.

Turner, Leigh. "Bioethics in a Multicultural World: Medicine and Morality in Pluralistic Settings." *Health Care Analysis* 2003, 11, 99–117.

World Health Organization (WHO). *The World Health Report 2003*. Geneva: WHO, 2003.

---

[1] On the disclosure of terminal diagnoses, the open discussion of death, the withdrawal and withholding of fluids and nutrition, the role of family members in protecting ill individuals from "bad news" etc. see the study of Barbara Koenig and J. Gates-Williams (1995); information disclosure to patients in a Navajo Community has been investigated by Joseph Carrese and Lorna Rhodes (1995); reflections on the different attitudes to euthanasia and end-of-life treatment in Germany and the Netherlands are published in a volume by Bert Gordijn and Henk ten Have (2000). Recent surveys on comparative cultural studies and anthropological work in bioethics include Leigh Turner (2003), Patricia Marshall and Barbara Koenig (2004), Silke Schicktanz, Christoph Tannert and Peter Wiedemann (2003), Harold Coward and Pinit Ratanakul (1999), and Leonardo D. de Castro et al. (2004).

[2] A word that appeared in the discussion because the extension of lifespan by itself would not be attractive if it did not also increase the duration of healthy life.

I. FUNDAMENTAL ASPECTS

Chapter 1

HISTORY AND FUTURE OF BIOETHICS

ALBERT R. JONSEN
San Francisco, USA

On August 9, 2001, President George Bush made his first formal media address to the American people. In this first speech, delivered from his home in Crawford, Texas, he announced his decision to allow federal funds to support research only on existing stem-cell lines, and concluded what was, in effect, a moral sermon as much as a policy statement with the words, "As we go forward, I hope we will always be guided by both intellect and heart, by our capabilities and our conscience ...".[1] This was an extraordinary moment. One commentator noted that the President chose to appear before the nation for the first time as a bioethicist! His decision came after months of public debate that had exposed the American people and its politicians to a large dose of bioethical language and argument. I intend to use this event to illustrate the nature of American bioethics. I shall suggest that its general contours follow the lines of moral argument that are deeply drawn in American history, and then comment on the role of the American bioethics community in the debates and the policy formulation. Finally, I will attempt to relate this American story to the broader theme announced in my title, the "History and Future of Bioethics".

1. PRESIDENTIAL BIOETHICS

I am aware that the debate over stem-cell research has taken place in many of the nations represented here and that law and policy have been developed in several jurisdictions. Although I have followed the European and British discussions on this contentious issue, I can hardly consider myself familiar enough with them and with the cultural values behind them to make comparisons with the American debate. I cannot, of course, claim

C. Rehmann-Sutter et al. (eds.), Bioethics in Cultural Contexts, 13–19.

that the debate over this issue is uniquely American: it has the same general form wherever it has been raised. But I will occasionally claim that some feature of the American debate is peculiarly American. European scholars who could find analogous ideas and arguments in their own moral traditions can certainly contest this claim. And, after all, we are all, Americans and Europeans, heirs of the same broad moral traditions of western Christianity and European enlightenment. Still, as a long-time participant and observer of American bioethics, I have found the stem-cell debate an intriguing paradigm for the way in which Americans deal with bioethical issues.

There has been continuing debate over the scientific use of the human embryo and fetus for over thirty years. Indeed, I have suggested in my history of American bioethics, *The Birth of Bioethics*, that the vigorous arguments between theologian Paul Ramsey and scientist Joshua Lederberg on human cloning during the 1960s initiated bioethical reflection in the United States (see Jonsen 1998: 306f.). I will not review that long history, except to say that a series of governmentally appointed commissions have recommended limited use of fetal and embryonic tissue and the Congress of the United States has persistently imposed restrictions and prohibitions. The question is obviously related to the extremely contentious matter of abortion, which has been a perpetual irritant in American politics and policy since the crucial Supreme Court decision of 1976 in the case Roe vs. Wade, which allowed legal abortion with almost no restrictions.[2] The immediate precursor of the stem-cell debate was a decision by President Clinton to permit some embryo research to be done with federal funding and a determination by the National Institutes of Health that stem-cell research could be done but that federal funds could not be expended on the derivation of the cells from embryonic sources (NBAC 1999). George Bush, during his campaign for the presidency, voiced opposition to that policy and promised to stop any research that involved destruction of the human embryo. His political support among Christian conservatives and among Roman Catholics strongly endorsed that position. In the early days of his presidency, the matter was muted and other political issues, such as tax policy, dominated the public scene. Beginning in the spring of 2001, hints that the president was considering action on his campaign promise began to appear and, with surprising rapidity, the anti-abortion constituency and the scientific community began to form sides. What might have been a rather silent administrative decision about an obscure topic took on huge dimensions and by fall, scholarly articles and media interviews with scientists, politicians, advocates, religious figures saturated public attention. The government instituted studies of the problem and consulted with experts of all sorts. The president, not known for his intellectual interests, actually immersed himself in the topic, studying documents, engaging experts in conversation and meeting with leading scholars.

One of those meetings, on Monday, July 9, was with two leading American bioethicists, Dr. Leon Kass of the University of Chicago and Dr. Dan Callahan of the Hastings Center. *The New York Times* wrote, "It was at this meeting that his thoughts began to gel", and one counselor to the President noted, "The meeting with the bioethicists was the first at which I recall pretty specific discussions about the ethics of using the existing cell lines."[3] This meeting between the president of the United States and two bioethicists was certainly a signal moment in the history of American bioethics! Although previous presidents had sought bioethical advice, they did so by creating commissions to study the issues and recommend policy. Never before, to my knowledge, had an intimate meeting taken place in which president and bioethicists explored the questions and, indeed, stimulated a resolution, which became policy within several days. Incidentally, a third bioethics scholar, Dr. Leroy Walters, met separately with the President on the same day. Dr. Walters, much more expert than the other two bioethicists in matters of genetics, has generally taken a more liberal position on stem cell research. No word of his advice has emerged from the White House.[4]

2. MORALISM AND MELIORISM

From its beginnings the debate had taken shape in a characteristically American manner. In the United States, moral arguments are often formulated in terms of two broad features of the American ethos. I have named these features Moralism and Meliorism (see Jonsen 1990). From the beginnings of our history in the seventeenth century, a powerful strain of moral righteousness has marked American culture. From the founding Puritan colonies through the continual immigrations from many parts of the world, the belief that there is a strict code of right and wrong has dominated the culture, even though there have been diverse views of what that code demanded and, unquestionably, there has been, as with all strict moral codes, widespread deviation from the true path. Despite the tolerance and the permissiveness of a pluralistic society, this moralism, at least as a cultural conception of the moral life, remains. Meliorism is the second, and contrasting, feature of American ethos. From the beginnings, as a nation founded by adventuring colonists expanded into the trackless lands, overcoming natural barriers and conquering indigenous people, until the nation spread from Atlantic to Pacific, an imperative exhorted all Americans to explore, to master and subdue nature and to improve in every respect the fortunes of the explorers and their children. This imperative to better the land and its people was preached beside the strict code of morality. Few noticed that there was an inherent tension between the two moralities, since it is not always possible to walk a narrow path of moral probity and at the same time range into undiscovered territory. The western movement, undertaken as a mission to evangelize and civilize often became the rough and cruel world of the Wild West.

Thus, in my view, the American Ethos, formed by the history of the nation and the complexion of its peoples consists of a deeply imprinted deontology, the conviction that there is a clear and distinct, even though unspecified, set of moral principles and, at the same time, a deeply imbedded teleology, the urgency to improve life for individuals and society. These two moral consciousnesses are rarely seen as incompatible and rarely perceived as paradoxical. Walking both paths seems desirable and possible, even though exploration of the unknown often disrupts settled ways of life. When conflict does appear, the response is to form sides that defend one path and repudiate the other. Moral disagreements over portentous issues such as slavery and trivial ones such as drinking liquor and dancing have become moral battles.

The stem cell debate followed this traditional pattern. One the one side, moralists announced that it was absolutely immoral to destroy even the earliest human life, regardless of the prospects of great benefit. The meliorists insisted that the possibilities of cure for many diseases not only justified but demanded that the scientific research proceed. A quote from *The New York Times* summarized the issue: "advocates argue that embryonic stem cell research can help cure an array of diseases … But abortion opponents say the research destroys embryos and, therefore, violates human life."[5] Congressional hearings, television debates and the news media rarely moved away from these two diametrically opposed assertions. There was no moral middleground. His conservative constituency, for which he has a natural affinity, to affirm the moralist position, summoned the president: thou shalt not kill. Yet voices from the community of persons afflicted with the diseases summoned him to acknowledge their pain and show, in his own phrase, compassionate conservatism. Indeed, even the icon of Republican conservatism, former President Ronald Reagan, who is a victim of Alzheimer's disease, was offered as an example favoring the pursuit of research.

The world of academic bioethics exists within this broader American ethos. It is certainly influenced by it but, in the manner of the moral disciplines of theology and philosophy, it conceives argument in a more nuanced way. From the early days of the debate, bioethicists had discussed the questions and written articles of some sophistication, exploring perennial questions, such as the attribution of moral personhood and the epistemological questions about moral certitude. American bioethicists have, from the beginning of their discipline in the 1970s, endeavored to create resolution to controversial issues that draw from the moralistic and the melioristic traditions. American bioethicists have been without theoretical or practical guilt deontologists and teleologists. It is commonly acknowledged that American bioethics has not espoused any one system of moral philosophy. Yet, in their consistent effort to reconcile deontological moralism and teleological meliorism, American bioethicists are the heirs, often unconsciously, of the one distinctive American philosophy, pragmatism.

William James whose thoughts first engendered pragmatic philosophy in the late 19th century, once wrote, " ... the essence of good is simply to satisfy demand ... and since all demands cannot be satisfied in this poor world, the guiding principle of ethical philosophy must be simply to satisfy as many demands as we can ... (thus) invent some manner of realizing your won ideals which will also satisfy the alien demands" (James 1967: 623). At the same time, he warned that creative solutions must respect the laws and usages of civilized society, which are the accumulation of creative solutions in the past. James was at the same time, moralistic and melioristic, deontological and teleological. While James and the other pragmatists left many theoretical questions unanswered, they were preeminently public philosophers, attempting to craft resolutions that would both advance life and acknowledge its limits. American bioethicists, with few exceptions, have worked in the in this pragmatic way. Although American bioethics has, strangely enough, avoided direct confrontation with the abortion question and thus has never provided a mediating position, it has done so successfully in many other areas of bioethical interest, such as human experimentation, death and dying, organ transplantation and genetics. The various governmental ethics commissions during the last three decades, have left a trail of creative solutions of this sort.

The stem-cell debate has been different. Two battlelines, the moralists and the meliorists, confronted each other relentlessly. Bioethics, in my view, contributed little and certainly did not achieve a creative solution. The President's two consulting bioethicists lined up on the moralistic side and provided the one solution that even moralists can employ, the casuistic advice that existing cell lines only be used for research, a replay of the ancient casuistic debates about whether good results can be drawn from evil causes. Even the National Bioethics Advisory Board, asked by President Clinton to advise on the matter, recommended stem cell research, using discarded and donated embryos from reproductive treatment and from early abortuses, concluded with a melioristic judgment, "A principal ethical justification for public sponsorship of research with human embryonic stem cells is that this research has the potential to produce health benefits for individuals who are suffering from serious and often fatal diseases." When it came to the contentious question of moral status, the Commission reviews various opinions and then states without further explanation, "On this issue, the Commission adopted what some have described an intermediate position ... that the embryo merits respect as a form of human life but not the same level of respect accorded persons." The Commission simply did not expose the arguments that would persuade anyone of the reasonableness of their "intermediate position". Indeed, they admit, with an almost audible sigh, "It is unlikely that by sheer force of argument, those with particularly strong beliefs ... will be persuaded to change their positions" (NBAC 1999: 50). Thus, in the one public forum where bioethics could have been creative, it avoided the central question of the moral status of the embryo in favor of a simplistic melioristic claim.

In my opinion, the general contribution of bioethicists was disappointing. A few, but very few, bioethicists wrote serious analyses of the issue. Some bioethicists did, however, make brief statements in the media, from which it could be discerned on which side they stood. Although this was an unprecedented moment for bioethics in America, both as a discipline and as a stimulant for public discourse, the bioethics community, in my view, failed at what it had done best in the past, that is, formulate through vigorous debate and in a pragmatist spirit, a coherent analysis of the issue. It did not contribute to a public policy resolution that would, in William James' words, satisfy as many demands as possible.

3.   BIOETHICS IN ADOLESCENCE

In this episode, I see the future of bioethics. The stem-cell debate is not over in the United States. The president's solution simply opens a new stage of the discussion. This opens the way for a more serious and sustained bioethical study of the questions. First, its practitioners must become much more erudite in the biosciences. Despite the name, "bioethics", the field has, in the United States, become a modern form of medical ethics in which the moral aspects of the relationship between patient and physician, and its central value of personal autonomy, has become the center of attention. Stem cell research, genetics, and many other problems on the horizon, such as the implications of the neurosciences for conceptions of human freedom and responsibility, require a much deeper understanding of science than many bioethicists possess. Second, American bioethicists have viewed ethics so exclusively as a function of interpersonal relationships that we have forgotten the Aristotelian admonition that ethics is a part of politics. Debates over moral principles and values are but a part of larger debates about power and authority within forms of policy and culture. In the future, the politics of a democratic republic must be factored into our understanding of ethical questions and into the resolution of the dilemmas. Bioethicists will have to develop explicit doctrines about how ethical arguments fit into public policy and politics. Finally, bioethics must find a public and popular voice. Its words have been heard, of course, in governmental commissions but its serious reflections remain hidden in their own journals and meetings. A few bioethicists have become media figures but too often only to state conclusions rather than justifications. They must learn how to express complex arguments in a way that is comprehensible to a broad public.

These tasks have been partly undertaken in the history of American bioethics but the stem cell debate has revealed that they remain as yet unresolved problems for the field. The stem cell debate is only the early intimation of the radical changes that molecular biology will bring to research and to medicine. As this future opens, bioethicists must address these problems. I entitled my book on the history of bioethics, *The Birth of Bioethics*. The stem cell debate and what will follow is the critical moment that should compel bioethics from its adolescence into a mature discipline. If it fails to do so, it will die.

References

James, William. "The moral philosopher and the moral life." In *The Writings of William James*, John J. McDermott (ed.). New York: Random House, 1967.

Jonsen, Albert R. "American Moralism and the Origin of Bioethics in the United States." *Journal of Medicine and Philosophy* 1990, 16, 113–130.

Jonsen, Albert R. *The Birth of Bioethics*. New York: Oxford University Press, 1998.

NBAC (National Bioethics Advisory Commission). *Ethical Issues in Human Stem Cell Research*. Washington D.C.: U.S. Government Printing Office, 1999, vol. 1.

---

[1] New York Times, 10.8.2001, A17.

[2] United States Supreme Court, 410 U.S. 113, 93 S. Ct. 705. January 22, 1973 (reprinted in many textbooks, e.g. Munson, Ronald. *Intervention and Reflection. Basic Issues in Medical Ethics*. Belmont, CA: Wadsworth, 1993, 61–67).

[3] New York Times, 11.8.2001, A17.

[4] It must be noted that after this lecture was given, the President's Commission on Bioethics, headed by Professor Kass, issued *Human Cloning and Human Dignity: An Ethical Inquiry*. Washington, D.C.: US Government Printing Office, 2002. The Commission recommended a ban on cloning for reproductive purposes and a four year moratorium on cloning for purposes of biomedical research. In this writer's opinion, the report did not add anything substantial to the argument proposed in this essay.

[5] New York Times, 21.7.2001, A9.

Chapter 2

THE NEED FOR ETHICAL EVALUATION IN BIOMEDICINE AND BIOPOLITICS

DIETMAR MIETH
Tübingen, Germany

1.   INTRODUCTION

The revolutionary developments taking place in genetic research as regards pharmacology, plant and animal breeding, and human medicine, make a revision of long-established views on human nature a concern for anyone involved with this matter. The success in mapping and sequencing the human genome, thanks to methodological achievements no-one thought possible a few decades ago, seems to speak for itself. Ethical evaluations are either the task of helpless philosophers or theologians who are not easily persuaded of the positive results of the genetic revolution, or of professionals who aim for positive feedback from the public. Scientists and ethicists seem to come from different planets, to speak a different language, and to regard each other as having rather reductionist views on humans and humanity. This paper aims to reopen the discussion by stressing the need for ethical evaluation when social changes of this magnitude are possible or even contemplated.[1]

Basic research on the human genome cannot be, and is not meant to be, separated from its possible applications in various fields of medicine, biology or pharmacology, for basic research and its applications are dependent on each other. Nevertheless it is important to differentiate ethically relevant questions that arise in one or the other area. Therefore, the first issue to be discussed will be the general implications of the Human Genome Project itself, which was

C. Rehmann-Sutter et al. (eds.), Bioethics in Cultural Contexts, 21–43.

primarily engaged with the question of basic research into human genetic material.

To consider its implications from an ethical perspective means addressing questions of the rights and duties of moral subjects, questions of which institutional structures would enable persons to act morally correctly, or background questions behind research policies. But first I will list some general ethical evaluations, which I will not discuss in detail later.

a) It is unclear whether international research priorities are set correctly with respect to the duty of distributive justice. Here issues surrounding the international challenge to improve medical standards, especially in countries in the Third World must be considered. Whether the effort to analyse and sequence the human genome – a goal that in itself does not raise questions – can be integrated into this challenge or not, is obviously of significance for ethical evaluation. If the final result of basic research is only the improvement of therapies for so-called industrial diseases, such as cancer, heart disease, Parkinson's or Alzheimer's disease, it may not be justifiable in the light of the urgent need for basic medical care on a global level, where much more financial support is necessary than is available today.

b) The increase in genetic knowledge without diagnostic or therapeutic effects, apart from for some rare monogenetic diseases, is a problem that has not been adequately discussed. Research into disposition to certain illnesses could also cause problems if the increased knowledge about an individual's potential susceptibility to diseases cannot be paralleled with adequate medical therapies.

c) Commercialisation of knowledge and patenting rights are problems that have emerged since the first "inventions" or "discoveries" (see Dworkin, in EHGA: 175f) of genes and monogenetic diseases. That human genetic material should be open to such commercialisation depends on a specific – materialistic and consequently dualistic – view of human life that is not at all evident. We will come back to this when we examine the issue of biopatenting.

d) Another aspect of the general commercialisation problem concerns questions of data protection in the context of health and life insurance. It is conceivable that the health and life insurance industry may become interested in data on the genetic makeup of their clients. Recent reports of the situation in the Netherlands, for example, demonstrate that this is not a scenario in the distant future (see de Wachter and van Luijk, in EHGA: 165f). The balance of principles of fairness (how much solidarity must a community of insurance holders accept?) with the principle of the individual's right to privacy is ethically relevant in this case.

e) At a more theoretical level one might ask whether the definition of what counts as disease could bring about possible discrimination by selection. It is not clear whether genetic researchers, or even members of the medical profession, should set the standards on their own. Since the complex terms "health" and "disease" cannot be described empirically – although this is a very important indication – the hermeneutic background has to be considered in every step that is taken to define a certain genetic anomaly as disease. Otherwise a questionable reductionist view would cause inadequate conceptions of health (see Beck-Gernsheim, in EHGA: 199f).

f) The right or duty to know and the right not to know are relevant even at the level of basic research. The individual's right to information and the question of whether there could even be a duty to obtain information has to be balanced against the definite right not to know everything about one's own genetic makeup. This individual right must be seen in the context of a socially influenced definition and interpretation of health and disease. The more knowledge of human genetic structure is gained, the more it concerns the individual, who is confronted with demands to know even in contexts where no direct connection to the individual's health exists. This is the case in possible screenings in the workplace, or in screening programmes for all pregnant women in a specific area or within a specific age range.

While some of these problems of evaluation, such as questions of data protection and the spread of commercialisation, can be seen as problems that require a juridical answer, others appear to be strongly dependent on moral attitudes that differ immensely from one country to another. This is a matter of general ethical relevance, that is theoretically discussed as the problem of moral pluralism leading to a social and political theory of modern societies, and which must be applied to the general implications of high technology research.

Having examined these general ethical questions I will now turn to a number of fields of application for research on the human genome, in order to give an idea of the complexity of the ethical issues that arise at this second level.

2.   APPLICATIONS OF THE HUMAN GENOME PROJECT: GENETIC TESTING

2.1   The difference between conventional and genetic diagnostic tests

The most dramatic difference between conventional and genetic diagnostic tests is the number of persons who may be affected by the result. If a genetic disease or disorder is diagnosed, other family members or relatives may also be affected – in the present or the future. Therefore, the nature of decision-making also changes: how can we take the interests of other family members into consideration? How is their right to privacy to be weighed against their possible interest in knowing their genetic status, and especially in finding out about preventive or prophylactic measures (where applicable)? Until now ethicists and medical professionals have been quite vague in their formulations:

"Confidentiality of the results of the test is an ethical imperative. Genetic data should not be released, except with the free and informed consent of the woman or couple. If the genetic data are relevant to the interests of other family members, the woman or couple should be strongly recommended by the genetic counsellor to allow the release of such data to these family members."[2]

2.2   Genetic diagnosis of adults

With respect to genetic diagnosis we face the problem of what to do with genetic information when no therapy can be offered to a client or patient. Knowledge alone may have an effect on attitude and cause a person to be regarded as ill when no symptoms would otherwise have been recognised. Additionally it can cause a period of desperation about the therapeutic "cul-de-sac" that must be faced on a psychological level.

In general when the person's own life is concerned, ethicists agree on the priority of the right not to know over the duty to know. Most ethicists also defend this position where future children are concerned, but the consensus here is not as broad as it is in the first case. However, ethicists do agree on the necessity of genetic counselling as a precondition of diagnosis, and demand that more institutions should be established to implement this.

2.3   Screenings

Another field of application concerns screenings that could improve medical or security standards. Genetic screenings are considered to be continuous progress in the course of public health improvement. Three specific types have been discussed during the last few years: carrier screenings of adults who might carry a genetic risk of developing a disease and who, for example in the case of cystic fibrosis, might otherwise refrain from reproduction: prenatal screening that addresses pregnant women but which also diagnoses the genetic makeup of foetuses; and genetic screenings in the workplace. Each is in the specific interest of the people concerned.

Although the development of screening programs may be considered advantageous in certain cases, discussion in recent years has shown that strong reservations exist, not only in public opinion – reminding the protagonists of screenings that social pressure is an unavoidable side effect – but also among carrier interest groups, such as carriers of Chorea Huntington, who fear discrimination.

2.4   The hermeneutic problem of justified interests

It has become common to define utilitarianism as the ethical approach that should be used in the determination of justified interests. This can be implemented through the determination of the ranking of the envisioned goods or, as in the typical point of departure in utilitarianism, through the greatest sum total of interests. In any event, it is characteristic of the concept of interest that it proceeds from a conception of the human being as an individual with specific competencies. This concept determines the holder of interests on the basis of individual capabilities: self-movement, self-reflection, self-decision. Whoever does not have these capabilities at his or her disposal, cannot hold an interest. Also significant here is the fact that interests appear to be bound to expectations for the future.

On the one hand, the capability of having interests constitutes a characteristic that enables people to draw distinctions among human beings. "Speciesism" is countered with what I would like to call "personicism" (as distinct from "personalism"): not all human beings are persons. On the other hand, holders of interests, whether they represent their own interests or those of others, are to be understood in the sense of the concept of the individual confronted with the interests of other individuals in a supplemental and external way.

Interests can neither be objectified (for instance, in the sense of needs in an objective theory of needs), nor can they be formulated inter-subjectively. Naturally, as the interests of individuals they are calculable once they have been articulated. But two factors remain characteristic of a concept that deals with justified interests: individualism and perfectionism. That is why Wolfgang Huber, the former ethicist at the Faculty of Protestant Theology in Heidelberg (now bishop of Berlin), distinguishes an ethics of interests (for those who are capable of expressing individual interests) from an ethics of dignity (which is independent of developed capabilities).[3]

The problem is: do we not have to understand the human being differently if we take his or her social dependency seriously, and if we take studies of identity in solidarity into consideration? Personalism, inter-subjectivity, contextualism, and other theories see "You/Thou", respectively the "other", as constitutive to being a subject, and to identity. The individual interest as such does not become superfluous; it simply acquires a different ranking. Theories based on need become more important than theories of interest; solidarity is ranked equally with self-realisation.[4]

2.5   The eugenics argument

The subject of eugenics is often raised in connection with the methodology of human genetics – genetic diagnosis, gene-transfer, and cloning. Here, we are obviously dealing with an important point of reference in a moral context. Of course, the content of the word eugenics is fundamentally descriptive rather than ethical. According to Francis Galton (1883) it deals with the re-conceptualisation of inheritance as a statistical condition of populations and future generations. This re-conceptualisation can serve the idea of selection, negatively of the "bad", positively of the "good". The extent of its misuse is far reaching, from the sterilisation of the mentally ill (Indiana, 1907) to the extermination of "those unworthy to live" for racist or public health reasons as in Nazi-Germany. Only eugenics deals with the types of influences that instrumentalise the individual. "Eugenics should concern itself with all conditions which can contribute to the cultivation of genetic characteristics in future generations and to the development of those genetic assets in the lives of individuals which can contribute to the benefit of the whole."[5]

Let us hold on to the phrase "to the benefit of the whole", also known as the "gene-pool" argument, which assumes that the human genome as the "inheritance of Humanity" (UNESCO-Declaration of

1997) can be influenced through such interventions. One is counting here on two possibilities: genetic degeneration (for example because of human influences through environmental degradation, misuse, or accidents) and genetic improvement (also because of human influences). On a scientific level there is disagreement over whether it is possible to strip populations' influences on their genetics of their arbitrariness, and guide them. On the whole this is hardly feasible, but in individual cases the reparation of burdens (genetic diseases) may become possible. In bioethical documents the distinction is drawn between "health purposes" and "enhancement". Even if this line is continually redefined, it still has the potential to restrict the "benefit of the whole" to a reduction of health burdens. Of course, this opens up an ethical responsibility according to which, for instance in the sense of Robert Edwards, a general genetic obligation on parents should exist to bring the healthiest possible children into the world.[6] This naturally also introduces the evidence of illness as a justification for selection of embryos and foetuses, and involves either the idea that being put to death is better for a human being than illness, or that the burden of a sick individual who is dependent on others is to be regarded as unreasonable. Against the background of such an argument, a plausible morality of exceptions in borderline situations easily turns into the moral command that competent individuals should bring about an indirect benefit to the whole through their decisions. Furthermore, interests can be influenced: through an appropriate offer, through social expectations, and through the reduction of the dignity argument regarding early human life forms into a piety argument.

What can one make of the eugenics argument under these circumstances? It can be summed up morally as an argument against a collectivism of population-genetics ("benefit of the whole"), and against the misuse of self-determination at the cost of others ("indirect eugenics"). In the first case the argument is directed at the preservation of individual human dignity, in the second, at the consideration of the anthropological status of not-yet-competent human beings or those humans who no longer seem to be competent (the euthanasia problem). Because of this, the eugenics argument is dependent on those parameters that differentiate individual well-being and individual rights from general welfare options. Here, the morality of autonomy (self-determination and self-obligation) is a good antidote. In the case of indirect eugenics, of course, the moral discussion shifts to the moral obligation to protect embryos even against the allied interest of parents only to accept and love their offspring if they are perfect.

3.   ETHICAL ASPECTS OF GENE THERAPY

3.1   Gene therapy

"Human gene therapy is a revolutionary new approach to the treatment of incurable disease. The approach is to insert a functional gene into the cells of a patient in such a way that the disease is corrected. The initial applications will probably be to cancer, AIDS and genetic diseases (such as ADA-deficiency, sickle cell anaemia, thalassemia and haemophilia)" (French Anderson, NIH Washington, Poster for his project, 1989).

A somewhat broader description of experimentation on the human genome is offered by R. Wimmer, who takes into account the fact that there are also experiments "that do not directly serve a therapeutic purpose directed towards a specific individual and/or his or her future descendants but indirectly, insofar as experiments are undertaken on healthy human embryos for the benefit of 'sick' individuals or experimentation under the auspices of basic research which are not primarily oriented towards application and therapy" (Wimmer 1991: 186).

It is also necessary to differentiate between experiments on human cells (somatic gene therapy) and those that take place in the germlines of zygotes or embryos. In the first case the effects are limited to the individual, in the second they are passed on with the genotype.

3.2   Germline therapy

Closer consideration indicates that the general global rejection of germ line therapy has gradually been replaced by a highly differentiated bioethical discussion. The former categorical rejection assumes that we may one day conceivably refrain from applying possible therapies for widespread genetic/hereditary diseases because it is not permitted to change the status quo of genetic information, seen as 'nature'. Conversely, other arguments assert our responsibility to develop therapy and our responsibility to future generations (Wimmer 1991; Agius 1989).

If we move to the evaluative level, the arguments for a legal ban clearly predominate. However, the presuppositions under which this juridical restraint is formulated could change, even though this is somewhat improbable at present. The following arguments against the authorisation of germ line therapy are often presented.

The method of researching and making germ line therapy operational requires exhaustive and highly consumptive experimentation on early embryos, experiments that according to this opinion are ethically unacceptable. However, should the achievement of so-called "higher ranking goals" become feasible, the debate will undoubtedly be reopened, even in a country as restrictive as Germany.

The isolation of the therapeutic experiment, and thus of the effects, is so difficult that ultimately unacceptable side effects are to be expected.

The slippery slope argument cannot be denied. Experimentation on the human genotype will lead to a complicated debate over a new definition of limits. What is disease? What constitutes a diseased or 'unhealthy' gene? Does the individual interest, as is generally the case at the moment, determine the need for therapy? It is also possible to formulate this argument as a principle: a narrow definition of limits is preferable to a broader one if the broader one is unclear and difficult to implement.

These considerations/objections will be seen as a pragmatic barrier, repeatedly calling for renewed reflection. At present, arguments of a more principal nature are hardly possible. What would be immensely useful, however, is a serious scholarly debate on the *physicalness* of personhood: what protects the phenotype, what protects the genotype?[7]

3.3 Somatic gene therapy

It is a widely held opinion that somatic gene therapy, once it has gone beyond the experimental and preliminary phase (for instance, retardation of the development of metastases in cancer) is a therapy like any other, or more precisely, a therapy that falls under the same criteria of medical ethics as every other therapy. However, it cannot be denied that the application of these criteria deserves particular attention in this case.

The following questions must be directed at the possible implementation of somatic gene therapy:

Are the goals formulated realistic, or are we nurturing hopes that cannot be fulfilled, but which nevertheless serve to psychologically motivate individuals to participate in such experiments?

Is the indirect method of attaining a 'free and informed consent' adequately taken into account in the various series of experiments?

Has the possible toxic side effect of vectors for the supply of genetically engineered substances been thoroughly analysed and considered (for example the gene therapy accidents in the USA)?

Do the necessary control mechanisms and institutional structures exist?

These questions, which concern experimentation on somatic cells that is limited to the human phenotype and can be clearly defined as therapeutic, are not applicable only to human genetics. They are questions about modern medical techniques in general. The prominence of human genetic engineering in public discussion should not lead us to overlook the fact that the accelerated development of any highly sophisticated technique in medicine confronts us with the need for responsibility that has an institutional basis.

4.   AN ETHICAL AND CHRISTIAN APPROACH TO BIOPATENTING

When patent applications involving living organisms began to be filed on a regular basis, a turning point had been reached. Although there was already a predisposition to regard patenting biological resources as no different from patenting anything else, the decisions of the US Supreme Court in the landmark Chakrabarty case established a principle that "the relevant distinction was not between living and inanimate" things; rather the question was whether living products could be seen as "human made inventions". This was part of a major but invisible cultural shift, expressed by a senior UK patent expert, R. S. Crespi:

"Historically, the patent system came to birth to meet industrial needs. Industry was perceived as activities carried on inside factories ... Manufacture was the key word. Agriculture was felt to be outside the realm of patent law. Living things were also assumed to be excluded as being products of nature rather than products of manufacture. ... This restricted view no longer persists in most industrialised countries. Thus the European Patent Convention of 1973 declares agriculture to be a kind of industry."[8]

In almost all ethical systems, however, a vital distinction is made between how we treat what is living and what is not.

The EU Biopatenting Directive strongly affirmed the principle of patenting almost everything biological. It merely added a set of arbitrary exclusion clauses for applications such as human cloning, known to be politically sensitive to the European Parliament. It did not respond appropriately to the full range of relevant ethical concerns, and made it clear that its prime concern was European economic growth and competitiveness. This drive to patent everything biological turns the commercial paradigm into an idol.

The second distinction concerns what has been invented. With regard to genetic modifications to an animal or plant, the addition of two or three genes to an animal with perhaps 100 000 genes does not turn the animal into a human invention. The inventive step is to add the new gene construct to the animal. The novel construct, or the inventive use of a modified animal for a specific purpose might be rewarded by a patent.

The same applies to a gene. It may take great intellectual effort to decipher a gene and identify its function, but the gene is just as much a product of nature as the animal. Despite the considerable investment involved, the identification of a gene's function is not an ethical basis for claiming inclusive rights. Although intellectual effort has been made, it is of the nature of discovery not of invention. However, the EU wanted to find a premise to patent human genes. The Biopatenting Directive states that genes are patentable inventions because they have to be copied, using bacteria or chemicals, in order to be isolated and identified.

The Church of Scotland expressed its opposition to patenting living organisms as follows:

"Living organisms themselves should therefore not be patentable, whether genetically modified or not. It is wrong in principle. An animal, plant or micro-organism owes its creation ultimately to God, not human endeavour. It cannot be interpreted as an invention or a process, in the normal sense of either word. It has a life of its own, which inanimate matter does not. In genetic engineering, moreover, only a tiny fraction of the makeup of the organism can be said to be a product of the scientists. The organism is still essentially a living entity, not an invention."[9]

I am sure that a scientist in favour of patenting would answer that it is the inventive step in the processing of a product that provides the justification for patenting. Genes, for example, are only "products by process". And so they are not considered as a creation, they are only included in the term "invention".

This may be a demonstration of why a top-down argument in theological ethics does not tackle the real complexity of the problem. The decisive issue is whether patents on process and patents on living resources can be separated: and they can.

If this is the case, the living resources in themselves are the basis for financial gain. Taking into account the recognised principle of non-commercialisation of the human body and its parts, this is ethically unacceptable. It is a question of respect for human dignity and it is the real justification for rejecting biopatents on human genes. This argument is reinforced by, but not dependent on our belief in creation; the ethical concept of human dignity in itself is sufficient.

5.   ETHICS IN HUMAN PROCREATION: AN ANALYSIS OF SOME DILEMMAS

5.1   The scientific dilemma

On the one hand, science wants to serve knowledge in itself. On the other, it tries to be socially useful and has a covenant with society. Modern (biological) science needs intervention in nature, and instruments with which to conduct this intervention. An ever more sophisticated technology is not only a result but also a presupposition for the practice of modern science. Therefore the freedom of knowledge has its limits in the social acceptance of the instruments, and of the theoretical aims and practical application of knowledge. Perhaps the scientific dilemma between the insistence on freedom of knowledge and the need for social control is not the central question. In the promotion of the *technicalisation* of Human Procreation there are always individual needs and social advantages that are given as reasons for its promotion. These include health purposes, but are not limited to them. This means that in the case of Biomedicine and in the special case of Human Procreation, science never is only science in itself. The promotion of science often has political implications, and involves moral options and individual and social preferences. The dilemma is that science is not neutral and it is not above the given interests. But all scientists are perceived as experts, even if they are only experts in their own field of research and its application. They are not experts in contexts, assessments, ethics, education and so on. But in our society, which has an irreversible contract with science, technology, economy and their development, the scientific expert is often accepted as an expert in many things, and this may prove a temptation for scientific lobbyists. It is easy to imagine that a group of scientists might claim a societal development, but it is harder to imagine that a group of sociologists would propose a scientific method. Scientists are members of the scientific community and high-ranking experts, but they are also distinguished citizens. Therefore it is absolutely necessary to bring societal and ethical dialogue into the scientific community, and not only to allow scientific lobbyism into society.

5.2 The societal dilemma: increasing individual options, pluralism, tolerance and the lack of restrictive consensus

Modern and especially post-modern societies are based on individual rights and their protection by institutions. The liberal state corresponds with a pluralistic society comprising very different options. However, the choices that are made are often not authentic, but follow and conform to social trends. Conformity in the leading figure (concept, *Leitfigur*) of "authenticity" is one example. Here, paradoxically, the more an atomistic concept of individualism is promoted, the greater the conformity in concepts of individualisation. People want to be authentic and original, but they take the same thing (clothes, travel, behaviour) as an expression of this authenticity. Perhaps paradoxically, pluralism in society is its own worst enemy. A similar paradox exists for tolerance. Post-modern tolerance even has problems distinguishing intolerance from tolerance. Because solidarities are founded on a pluralistic concept, a societal solidarity that integrates some differences seems to be possible only during times of great counter-experiences and pressure by negative factors (e.g. "Tschernobyl"). But in most cases the distinction between good and evil depends on experience and on individual (or societal) options, which can be quite different. Therefore solidarities beyond pluralism can only be reached by transparency of interests and argument, and by a common understanding reached through narration and memory, which form convictions that need to be promoted or preserved.

5.3 The political dilemma between promotion and restriction

In the same simplified manner in which P. Snow spoke of the two cultures of science and humanities (*Geisteswissenschaften*), we can also speak in a heuristic sense of two mentalities with regard to questions of biomedicine. The biomedical mentality, typical of those at the interface between experimentation and clinical application, is a mentality characterised by hope of promotion and acceleration. Limits are seen to be connected with self-understanding – for example, it is forbidden to knowingly create human monsters – or to be confined to the present social context, or to be individual limitations assumed by specific options, for example those of religious groups. As experts who feel responsible for promoting health interests or other aims within a constantly developing society, those with a biomedical mentality often see political opposition that opts for excluding restrictions as a mixture of ignorance, conservatism and fundamentalism ("stupid and crazy"). On the other hand, for example in Germany, the "bioethics" of scientific experts and their philosophical servants, are considered as a kind of a conspiracy ("*Verschwörung*") against the needs and values of the people.

But it is not necessary to refer specifically to the German debate. As an example of the hermeneutics of suspicion I will refer to a lecture by Ivan Illich, given in Chicago in 1989 to the Evangelical Lutheran Church in America. The lecture was entitled "The Institutional Construction of a New Fetish: Human Life" (cf. Ivan Illich, *In the Mirror of the Past*, New York/London 1992), and Illich started with the following thesis:

"'Human Life' is a recent social construct, something which we now take so much for granted that we dare not seriously question it. I propose that the Church exorcise references to the new substantive life from its own discourse."

For Illich, the new notion of "life", so essential for modern ecological, medical and ethical discourses in the Western tradition is "the result of a perversion of the Christian message". In this message "life" (bios, zoe, vita) means something moved by an internal teleology of "soul" (vegetative, sensitive or intellective). The notion of "life" in today's society does not belong to the world of such sacred and contemplative feelings. It is a word that belongs to the field of modern management, to the language of planning so-called "human resources". Following the cartesian dualism, "life" is an objectification and an area of experimental intervention and manipulation with the intention of ameliorating. The context of epistemic presupposition that is part of an unquestioned progress turns "better life" into a new fetish. This has created a struggle between two contradictory options, between "pro life" and "pro choice". "Pro life" means "pro vita". "Pro choice" favours the use of biological material for a better quality of self-determined "life" for individuals. The "life" of individuals means "vita", but in a restricted, anthropocentric way. This discourse neglects the connotation of non-objectification, which was primarily given with the gospel of life. The gospel of life is a personal "I am life", from Moses to Jesus. The modern notion is a "value", a good, which must be preserved or destroyed according to social options.

In his lecture, Illich made five observations on the history of life, which we should not forget.

"First, life, as a substantive notion, makes its appearance around 1801." Instead of the religious and philosophical tradition of psyche, bios and zoe, the term "biology" came to mean "a science of life" (Jean-Baptiste Lamarck). From then on, "life" became a construct of organic phenomena like reproduction, genetic development and so on.

"Second, the loss of contingency, the death of nature and the appearance of life are but distinct aspects of the same consciousness." The loss of contingency here is the loss of dependent and actual connection with the breath of creation. The mechanistic model replaces the creative-processual model.

"Third, the ideology of possessive individualism has shaped the way life could be talked about as a property." It can easily be demonstrated in the patenting-life discourse that life is being discussed in its "elements", not as a matter of discovery but as a matter of human invention, even if it is identical to its natural state. The instrumentalisation of human life, as we will demonstrate in the section where we discuss cloning, is clearly a result of this "possessive individualism".

"Fourth, the fetitious nature of life appears with special poignancy in ecological discourse."[10] To think of life as a system of correlations between living forms and their habitat is a reduction of imagination and a subjection of life to all kinds of empirical and virtual objectification.

"Fifth", the pop-science fetish for life tends to void the legal notion of person. "The *distinction* between "human life" and "human person" has created the notion of a "human non-person", who is not a member of the "moral community". Illich concludes that "The new discipline of bioethics ... mediates between pop science and law by creating the semblance of a moral discourse that roots personhood in the qualitative evaluation of the fetish, life."

I refer to these critical observations by Illich as an important perspective on "bioethical" questions, because we are seldom aware that, before beginning an ethical discourse, we have to exorcise "a language and a language politics which does not allow a real ethical approach, only the semblance of it". Bioethics is an invention of the scientific language of biology and its derivations. This paradigm dominates ethics, and legal and social responsibility. It is clear that there is also an interdisciplinary approach to ethics, and that this makes sense in ethics that are concerned with sciences and new technologies. Nevertheless, the perception is not inadequate that the concept of the life fetish as a fetish for a scientific paradigm is strongly present in this kind of discourse.

5.4   The language dilemma

Perhaps some of Illich's pretensions are a caricature of the more pluralistic world of biology and medicine. However this experimental approach to the basis of political power and the paradigm of scientific promotion may be helpful when we consider motifs and presuppositions that are also present when we speak, for example, about the need to help someone who is infertile or who has ambiguous expectations about the health of a future child.

There is a biomedical lobby that provides a background to biomedical and bioethical committees. There are other kinds of lobbies and politically relevant pressure too, but the biomedical lobby is defining the language, before the ethical and political debate begins. One example is a distinction made in human cloning,[11] which will be important for the future of in-vitro-options for experimentation on embryos. Here the ethical debate will have nearly the same frame of reference as it does in other in-vitro-options, such as PGD (preimplantation genetic diagnosis).

In this case, we have to differentiate between cloning techniques that target in-vivo development (copying and raising humans), and cloning techniques that are limited to the in-vitro phase (to the embryos before implantation). Particular goals are pursued when transforming embryos into cell-cultures, for example in connection with therapies to prevent immune-rejection in transplantation, and in connection with the development of early forms of human organs.

Here, at present, we are basically discussing science fiction. But each day it becomes less science fiction and more scientific reality. However, we are dealing with interesting situations in which a society, with the help of ethical reflection, can declare its position before the technique as such exists. Various advisory groups have been commissioned in order to present experts' opinions and so define ethical standards. These include: the Clinton Commission in the USA, the GAEIB, and the above-mentioned advisory group to the European Commission, which completed its Opinion on May 20, 1997.[12] With regard to the cloning of human beings, a distinction was introduced in the EU and the US Commission advisory groups between reproductive and non-reproductive cloning. I have already mentioned that we have to differentiate between the in-vivo and in-vitro situation. Our language is not capable of grasping all new phenomena immediately. Any concept of a distinction, by definition may be misunderstood, and this also applies to the differentiation between reproductive and non-reproductive cloning. Reproductive means that the cloned embryo is implanted, and that a human adult person develops from it. In-vitro cloning does not – yet – determine whether an early embryo that has been manipulated accordingly will be implanted or not. If it is not implanted, but experiments with particular long-term therapeutic goals are carried out on it, it is classed as non-reproductive cloning.

The EU advisory group concluded that reproductive cloning must be forbidden. The Clinton Commission, using the same terminology, was more liberal insofar as it demanded a moratorium at present. The formula "at the present time and the present social context" did not enter into the conclusion of the European advisory group. In my opinion, a limitation to the present assumes that the prohibition of reproductive cloning presupposes a future revision. It is self-evident that every rule that we formulate under present circumstances can be revised in the future. Yet whoever wants to specifically articulate this has ulterior motives that must be made clear. Thus it can be said that the Clinton Commission declared itself in favour of a moratorium, whereas the European advisory group was in favour of a strict ban, but only in the so-called reproductive area. In the non-reproductive area, that is, in-vitro cultivation of cloned embryos in cell cultures, the EU advisory group stated that in those countries which allow experimentation on embryos (such as Belgium or Great Britain) in-vitro cloning should not be forbidden on the following conditions: that it is for high-priority therapeutic purposes; that a licensing body (i.e. an ethics commission) is consulted; and that manipulated embryos will not be implanted and become independent human beings. This is consistent with the recent British report of the Human Fertilisation and Embryology Authority (HFEA) and the Human Genetics Advisory Commission (HGAC) in December 1998, preparing the British law promoting embryo research in vitro for so-called therapeutic cloning.

I did not sign the declaration of the EU advisory group, because I am of the opinion that the differentiation between reproductive and non-reproductive cloning was not decided on a factual basis. Rather, it intended thereby to reach a differentiation between the treatment of embryos and humans. In a 1994 document I found a statement by the ministerial council of the European Council regarding tissue-banks in which reproductive and non-reproductive were defined in this context. In this document, egg cells, sperm cells and embryos were considered to be reproductive. Their use as tissue was to be forbidden exactly because they are reproductive, that is because human beings could develop from them. The expression reproductive was intentionally narrowed down to implanted embryos by the advisory group in the interest of practicability. I call this the politics of language. It is not without precedent. This difference between reproductive and non-reproductive had previously been introduced by an American advisory group on in-vitro fertilisation.

Non-reproductive cloning is promoted in the case of therapeutic cloning. In my opinion, this is also politics of language. The concept of therapy or the concept of human health plays a central role in the politics of language. I would like to elucidate this with an example from the Bioethics Convention of the Council of Europe. Article 18.2 of the Convention states that the production "of embryos for research purposes" is forbidden. Although Article 12 is explicitly concerned with the possibility of sex selection, it also deals implicitly with research on embryos. It states that sex selection is possible only if it serves health purposes, and mentions sex selection in order to prevent hereditary disease as an example. The text does not clarify to which method it is referring, and it could mean abortion after prenatal diagnosis, embryo selection before implantation, and possibly also sex selection through semen centrifuging. At the same time, Article 12 states that these health purposes also include "research on health purposes". Of course, at this point one wonders what is meant by the prohibition of embryo production for research purposes in Article 18.2. Are research purposes for health purposes, which are always allowed elsewhere, excluded? If this were the case, then the research purposes of Article 18.2 would mean little, because one could declare any research purpose in this area to have a therapeutic purpose.

With this example I want to make it clear that with such politics of language the Bioethics Convention delays but does not solve certain problems. We could illustrate this with further examples. In its supplementary report on cloning, signed in January 1998, the Bioethics Convention leaves the definition of a *human being* to the nation-states. In an interview, the president of the EU advisory group, Noëlle Lenoir, explained: "when human life begins is determined by the nation-states." One can imagine that, in the attempt to reach a valid European consensus, the openness of this question resulted in any consensus remaining unclear in the specific case. According to Noëlle Lenoir, Germans consider an early embryo to be a human being, other countries only recognise individual human beings after birth. The Human Rights Convention on Biomedicine does not determine what a human being is.

This has particular significance for the cloning of human beings. The supplementary report (January 1998), which was so highly praised in the European press, stated that it is forbidden to create a human being with the intention of making him or her identical to an already living or deceased person. Someone reading this without bias assumes that this means that any human cloning is forbidden. Yet things are not so simple. The misleading political language is further intensified by the fact that in the second paragraph it is stated that this prohibition is so strict that there are no exceptions. However in the official explanation things are rather different.

Here, it is stated that one has to distinguish between three levels of cloning. Firstly, the cloning of cells in general, which is morally unproblematic – an opinion with which I concur. Cloning is not morally problematic when dealing with a living being without an independent destiny. Secondly, in-vitro cloning; at this point there is a careful examination of embryo cells, but behind *embryo cells* in the totipotent state embryos are hidden. Thirdly, the explanation discusses the cloning of *human beings*. This means that the cloning of human beings is viewed as distinct from the cloning of embryo cells, and it is expressly indicated that this supplementary report applies only to the cloning of human beings. Thus, we return to the politics of language: although the expression non-reproductive cloning is not used, it is clear that the technical differentiation which it implies – reproductive *non-reproductive* cloning, no, and in vitro, yes – was adopted by the Human Rights Convention for Biomedicine. Some interpreters of the Human Rights Convention on Biomedicine claim that this is not that serious since it is stated in the aforementioned Article 18.2 that one may not produce embryos for research purposes, and that this therefore effectively prohibits the use of germ cells to create embryos. Yet I have already noted that the concept of research purposes is not yet clear. Is research for health purposes really excluded by this? In the future the courts will seek to clarify this, since at present it is not clear from the text. Thus, one cannot assume that there is a total prohibition of human cloning in the Human Rights Convention on Biomedicine and its supplementary report. Such a prohibition does not exist at all in the framework of the UNESCO Declaration, since in the UNESCO Declaration only reproductive cloning is forbidden. The UNESCO Declaration adopted this expression from the EU advisory group.

It seems to me a political dilemma that we often accept within the political debate a pseudo-ethical discourse, in which we speak of the ethical task of weighing risks and benefits. In most cases the question of what is a risk, what is a chance, and what is a benefit, is a question of ethical reflection on the right criteria. And therefore this discourse is pseudo-ethical as long as questions regarding the criteria and their foundation are not involved.

5.5   Pluralism, tolerance, mutual respect –not clarified

In the Opinion of the European Group on Ethics (EGE) no, 12, on Ethical Aspects of Embryo Research, published on November 23, 1998, we can read about the diversity of ethical views.

1.23 "The diversity of views regarding the question whether or not research on human embryos in vitro is morally acceptable, depends on differences in ethical approaches, theories and traditions, which are deeply rooted in European culture ..."

1.25 "The diversity in policies and regulations concerning embryo research in the Member States of the EU reflects fundamentally differing views ... and it is difficult to see how, at these extremes (cf. embryo as human life or as human being), the differences can be reconciled."

The introduction of this kind of opinion often leads to a substantial restriction being unacceptable.

Mutual respect is also mentioned in the EGE Opinion:

1.27 "Pluralism may be seen as a characteristic of the European Union, mirroring the richness of its tradition and asking for mutual respect and tolerance." (cf. 2.5)

In other papers mutual respect is precisely focussed on moral choices. Therefore more liberal positions always have a political advantage. They cannot be dominated, because if they are respect for different approaches and moral choices is not upheld. This is what in critical ethics we call "repressive tolerance". You always can suppress substantial restrictions but not a substantial liberalisation.

The ethical discourse on pluralism, tolerance and compromise seems to be very underdeveloped. Most members of bioethical committees speak of these attitudes, but the words remain without clarification. If pluralism is not *laisser faire*, as EGE implied, what is it? If pluralism is not the same as the lowest restrictive level, how can it be precisely defined? If pluralism has a tendency to compromise, what distinction can be made between a practical compromise and an ethical judgement? I am speaking from a concrete experience, which I had as a member of a European Project on Pluralism. The ethical paradox of the result was that if you take pluralism as the *norma normans*, you need no more ethics because all arguments can be silenced by the norm of pluralism. And if there are no limits to pluralism, then the result is nothing other than a kind of fundamentalism. If we try to have a moral discourse that aims for consensus, we must begin, for example, with doubt about our own moral position and with reflection on the conditions that are necessary to avoid the dominance of the power of definition, the politics of language, or repressive tolerance. We must also understand that a moral conflict is not contrary to the respect for persons. If ethicists must learn something about scientific specialities and the scientific use of language, the same spirit is necessary for scientists in the public debate. In both cases there is a danger of there being too little education of public opinion for the conditions of a moral discourse to occur.

6.   A FINAL REMARK

The problem of human genetic engineering is not only about respect for human beings as regards their genotypes and phenotypes. It also concerns the issue of human contingency, which may not be ignored in science and modern technology (human fallibility and finitude). We should be aware that by reducing fundamental problems to technical ones and by isolating them, we might solve technical problems by technical means. However, this could cause new problems with the adequacy of problem-solving strategies. In addition, other dimensions related to the technical problems can be easily obscured: for instance social problems arising from individual decisions, ecological problems, anthropological problems etc. Therefore, in continuing our research we have to keep in mind that progress in one dimension will not guarantee progress for human destiny in general.

As an ethicist, when dealing with the high hopes and great promises of biomedical advances, I cannot help wondering about the breakthrough mentality that refuses to take into account essential factors of the human constitution. Humans are prone to error. We often attempt to rationalise our motives and ignore the danger of instrumentalising others. We can postpone, but we cannot abolish our abiding finitude.

References

Agius, E. "Keimbahntherapie. Unsere Verantwortung für künftige Generationen." *Concilium* 1989, 25, 259–266.

Haker, H.; Hearn, R. and Steigleder, K. (eds.). *Ethics of Human Genome Analysis* (EHGA). *European Perspectives*. Tübingen, 1993.

Graumann, S. *Die somatische Gentherapie. Entwicklung und Anwendung aus ethischer Sicht.* Ethik in den Wissenschaften, Bd. 12. Tübingen, 2000.

Haker, H. *Ethik der genetischen Frühdiagnostik. Sozialethische Reflexionen zur Verantwortung am Beginn des menschlichen Lebens.* Paderborn, 2002.

Hildt, E. and Mieth, D. (eds.). *In vitro Fertilisation in the 1990s. Towards a medical, social and ethical evaluation.* Aldershot GB, 1998.

Junker-Kenny, M. (ed.). *Designing Life? Genetics, Procreation and Ethics.* Aldershot, 1999.

Mieth, D. "The Ethical Relevance of 'Justified interests' as an Hermeneutical Problem in Genome Analysis." In *EHGA, Ethics of Human Genome Analysis, European Perspectives.* H. Haker, R. Hearn and K. Steigleder (eds.). Tübingen 1993, 272–289.

Mieth, D. *Was wollen wir können? Ethik im Zeitalter der Biotechnik.* Freiburg i. Brsg. 2002, 135–213.

Wimmer, R. "Kategorische Argumente gegen die Keimbahn-Gentherapie." In *Ethik ohne Chance?* J.P. Wils, D. Mieth (eds.). Tübingen: Attempto, 1991, 182–209.

---

[1]   See for a broader discussion: H. Haker, R. Hearn and K. Steigleder: *Ethics of Human Genome Analysis. European Perspectives.* Tübingen, 1993 (in the following quoted as EHGA); Bioethics 7, 1993, Special Issue: *Inaugural Congress of the International Association of Bioethics*; P.J.M. van Tongeren. "Ethical Manipulations. An Ethical Evaluation of the Debate Surrounding Genetic Engineering." In *Human Gene Therapy* 1991, 2, 71–75.

[2]   Quoted from the opinion of the Group of Advisors on the Ethical Implications of Biotechnology to the European Commission (GAEIB) No. 6 on "Ethical Aspects of Prenatal Diagnosis", Sécrétariat Général, Bruessels. This Advisory Group, of which I was a member from1994 until 2000, is now called the "European Group of Ethics" (EGE). It is foreseen in some European Directives and has an important influence.

[3]   Compare this with: Huber, W. "Grenzen des medizinischen Fortschritts in christlicher Sicht." In *Möglichkeiten und Grenzen der Medizin.* C. Herfarth and H. J. Buhr (eds.). Berlin, 1994, 140–152.

[4] Scholars involved in feminist bioethics debates have been arguing for a reevaluation of these aspects by theorizing autonomy in terms of relational autonomy, reflecting on ethics of care, and reframing the question of justice in the context of the necessary embeddedness of human life in asymmetrical relationships. For a systematic reflection on feminist bioethics and its contextualization within the general debate on biomedical ethics and for further references see H. Haker: "Feministische Bioethik." In: *Einführung in die Bioethik*. M. Düwell and K. Steigleder (eds.). Frankfurt am Main: Suhrkamp, 2002.

[5] This definition is given by C.H. Hörz et al. in the *Wörterbuch zu philosophischen Fragen der Naturwissenschaften*. Berlin, 1991 (Dictionary to philosophical questions of natural sciences). Compare also: Graumann, S. *Die somatische Gentherapie*. Tübingen: Francke 2000, 25–30.

[6] Robert Edwards, the co-inventor of the first successful in-vitro fertilization (Baby Louise 1978), pretended in a conference of the European Network on Biomedical Ethics (1997), of which I was the Director (1996-1999), that he had insisted on this aim of IVF since 1971. Compare his contribution to the first volume of this Network: *In vitro fertilization in the 1990s*. E. Hildt and D. Mieth (eds.). Aldershot GB/Brookfield USA, 1998, 3–18.

[7] This question was significant for my first study on genetic engineering: "Moraltheologische Aspekte der Genetischen Technologie." *Wort und Wahrheit* 24, 1969, 557–561.

[8] Compare with: Crespi, R.S. "Patents in Biotechnology: the legal background." In: *Proceedings of International Conference on Patenting Life Forms in Europe*, Brussels, 7.-8. February 1989, unpublished. Compare also: *Biopatenting and the Threat of Food security – A Christian and Development Perspective*. Ed. by the International Cooperation for Development and Solidarity (CIDSE), Brussels 1999, 16.

[9] Cf. *Biopatenting and the Threat of Food security – A Christian and Development Perspective*. Ed. by the International Cooperation for Development and Solidarity (CIDSE), Brussels 1999, 12.

[10] The word „fetitious" introduced by Ivan Illich means to treat something as a fetish.

[11] Compare this with: Mieth, D. "Cloning: Ethics, Morality and Religion." In: *Cloning*. A. McLaren (ed.). Council of Europe Publishing 2002, 119–140.

[12] See note 2.

Chapter 3

FINITUDE –A NEGLECTED PERSPECTIVE IN BIOETHICS

BEAT SITTER-LIVER
Bern, Switzerland

1. INTRODUCTION

"The energies and possibilities of medicine must be given direction." And this, as H. Tristram Engelhardt stresses, in a situation where "unexpected possibilities are becoming real", and in which it seems impossible to elaborate a common fundamental intellectual response to that challenge (1991: XVI).

In this paper I am going to argue two points:

1) That by explicitly reconsidering the existential finitude of human beings, we may gain at least some common ground on which to become moral relatives (even if not necessarily moral friends);

2) That by analysing the practical implications of existential finitude we may be able to elaborate some criteria that will allow us to reasonably tackle some of the moral dilemmas in biomedical practice.

In biomedical ethics, the concept of the finitude of human existence is rarely discussed. When it is, it is used more as a declaration than made transparent in its practical consequences. There are exceptions, and I shall come back to some of them. It is astonishing that the implications of human finitude for biomedical ethics have not been generally analysed. Enhancement of human existence and well being, their prerequisites, are key concerns of biomedicine. And when we strive to diminish and possibly abolish the limitations of human existence, such as disease, suffering and mortality, finitude is the existential and factual presupposition. Perhaps this fact is too evident, and this is why it escapes in-depth investigation. In

C. Rehmann-Sutter et al. (eds.), Bioethics in Cultural Contexts, 45–57.
© 2006 *Springer. Printed in the Netherlands.*

contrast, I hold that anthropological reflection which stresses human finitude provides a perspective from which we may develop criteria and solutions when engaging in moral and ethical controversies. It may allow us to do so without having recourse to religious or metaphysical anchoring, remaining within the philosophical framework that H. Tristram Engelhardt has named "secular humanism" (1991: XI, XV, passim). Yet while Engelhardt stresses finitude as an ontological limitation of bioethics, I believe that it could be analysed as an opportunity for a secular understanding of what might be considered meaningful aims and goals for biomedicine.

2.   CLARIFYING THE CONCEPT OF FINITUDE

But let us first clarify the notion of finitude, since this rarely used term may create irritation when introduced into ethical debate. I consider this fact in itself worthy of ideological critique, since "finitude" refers to something we experience personally daily and often painfully. Yet this is not the place for such a critique; I will content myself with stating that the neglect of existential finitude belongs to the realm of psychological repression. It is part of that attitude which Martin Heidegger so aptly described as a lack of existential authenticity (1963 [1927]: §§ 26 f.).

It is indeed by referring to Martin Heidegger's existential analysis of human being (1927) that I wish to clarify the notion of finitude. This notion does not directly relate to any particular limits human beings encounter inside or outside themselves, and it must be well distinguished from scarcity. The term signifies a constitutive trait of human existence. Today Engelhardt speaks of our "ontological infirmity" (1991: XI) or of "an inescapable element of our ontological condition" (ibid. XIII); Heidegger calls it existential. I shall briefly dwell upon two aspects of that existential constitution (for more details see Sitter-Liver 1975: 31–36, 64f., 150f., 236f).

1) Human being exists as being-in-the-world with, or better, within possibilities. Whenever it chooses one, it discards others. While being the source of certain concrete realizations, it is the cause of non-existence for others. There could always be more than what it is in a position to achieve. Moreover, human being is never self-sufficient in so far as its mere existence is not its own doing: it has been thrown into the world – destined not to be a self-supporting entity and to be the cause of lost realizations. It is thus thoroughly finite, and there is no remedy for that ontological fact.

2) The second aspect concerns the personal finiteness of human existence. Death is not just something that unfortunately happens to humans; much rather their being-in-the-world is constituted as being to its proper end. Dying and death may be neglected or repressed, but this is the result of a

lack of existential authenticity. We cannot choose the fact of our death, for we exist as dying entities. The seal of our finitude is our existing as beings with and to their end ("Sein-zum-Tode", cf. 1927: §§ 49–53). There is no sense in trying to change this ontological condition, and any attempt to do so is impregnated with existential finitude. Paul Kurtz presents it as "the brute finitude of existence, the contingent and precarious, often tragic, character of human life" (Kurtz 1983, cit. by Engelhardt). Let me add that the various methods of searching for immortality about which we have been hearing do not escape the ontological constitution of human existence. They correspond to the tenacious refusal to confront one's own or someone else's death, as frequently displayed by those cherishing faith in a transcendent life or state.

It is this existential meaning that I have in mind when using the term "finitude". Limitations in intellectual power, moral capacity, strength of will, empathy and sympathy, and so on are grounded in the existential human constitution and thus are not to be overcome. This does not mean that they are not to be tackled, quite to the contrary: we only live up to possible authenticity and true autonomy when, driven by pressing reasons, we work towards overcoming controversies and tensions while keeping in mind that our findings will remain finite, and are therefore provisional and ever open to modification. Findings are the steps of a staircase that does not end until death.

3. FINITUDE, FREEDOM, AND RESPONSIBILITY

Markus Zimmermann-Acklin, one of the authors in the field of bioethics who explicitly considers human finitude (2000: 30–32), distinguishes three levels of discussion in today's bioethics: practical decisions, theoretical work in ethics, and the level on which the meaning and interpretation of expressions of human finitude are at stake (16–32). On the third level, fundamental intentions, moral convictions, images or visions of human being, of world, and God attempt to make sense of painful experiences such as disease, suffering, dying and death. Zimmermann-Acklin holds that third level issues will be intensely studied in the years to come (43) with 'Baconian' optimism concerning the power of reason, material progress, and the moral enhancement of humanity giving way to a more adequate and comprehensive understanding of human existence. He comments on the idea put forward by Gerald P. McKenny (1997) that medicine becomes inhumane when it turns a deaf ear to human border experiences (op. cit.: 30f.), recalling that the topics at stake are central to theologies and religions, a fact that explains to a certain extent the ongoing process of re-theologizing within the bioethical debate (Zimmermann-Acklin 2000: 32).

While being far from underestimating this approach, I wish to emphasize that when talking about existential finitude we do not raise the question of the meaning of particular human limitations. Our approach is ontological, and thus remains formal or secular. But for all that, it does not lack practical relevance. Existential finitude is the transcendental as well as the factual key to freedom. If the realization of possibilities were in no way restricted, choice would not exist, freedom would be of no relevance, nor could we witness the phenomenon of responsibility. However, in order to fully understand the phenomenon of responsibility we have to consider another condition of human existence, which I would call the existential fact of being related to the distinctions between true and false, good and bad. Under normal conditions, humans cannot act without reference to what they consider true or false and evaluate as good or bad. This existential disposition does not imply that they factually realize existential transparency and truth, or that they follow what they perceive to be good after thorough consideration. But it is the prerequisite that enables them to consider themselves free and to understand what responsibility is: they personally have to grasp their own existential possibility in order to turn it into moral and ethical reality (Sitter-Liver 1975: 208, 116–131).

In my insistence on the formal character of existential finitude I find myself in the company of Dietmar Mieth, another bioethical expert tackling human finitude. His tune reminds us of what we have heard from McKenny, but it was played several years earlier and on different instruments. Progress in genetic engineering, according to Mieth, teaches us that the sciences' proposed solutions to problems are solutions found by neglecting contexts (1999: 196). In this way, the knowledge of human finitude as elaborated by philosophical and Christian theological traditions has simply been forgotten (147). But the human being is "a finite, limited, socially dependent being capable of mistakes. Finitude, limitations, and proneness to mistakes or the openness to failure lie beyond human control and power. Even though one tries hard to act in the best possible way applying all one's knowledge and consciousness, a life project may simply fail and break down" (204). The corollary is evident: whatever we do – biomedical research and clinical application included – will turn into an abortive and inhumane undertaking if we refuse to recapture existential finitude and its manifold impacts. We ought to make "New Finitude" a practical slogan as Mieth suggested (209). This slogan might set us free from our all-too-hectic efforts to optimise the human being – efforts which tend to become the sheer instrumentalization of those we pretend to serve.

4.   EXISTENTIAL FINITUDE RECONSIDERED

4.1   H. Tristram Engelhardt and secular humanism

As I have said, although existential finitude is not a general concern in bioethics, there are a number of authors who have examined it to a greater or lesser extent. I have mentioned Engelhardt, Mieth and Zimmermann-Acklin, and I might add Lazare Benaroyo, Alberto Bondolfi, Michel Doucet, Eve-Marie Engels, Florianne Koechlin and others. But as I see it, it is Engelhardt who made finitude a key concept when he used it to support his comprehensive study on "Bioethics and Secular Humanism". While he attaches great importance to the historical reconstruction of post-modern secularism – an expression in which the epithet "post-modern" signals the unamendable failure of the project of modernity with its belief in the power of reason and human moral capacity – he presents finitude as an ontological trait of human being. He draws from it the few normative principles that remain universal – at any rate for those willing to accept the moral point of view and to resolve controversies without recourse to force. The grammar of controversy resolution for such ethically engaging moral strangers contains the principle of mutual respect and the unconditional prohibition to use others without their consent. Although Engelhardt firmly maintains that it "is not possible rationally to discover a canonical moral account with content" (1991: 110) – an outflow of the ontologically rooted finitude of knowledge, evaluation, and wisdom (119) – he aptly shows that involuntary euthanasia, use in research without consent, treatment without permission, and a few more norms can be justified starting from the gained platform. Such norms are of course common in routine bioethical discourse. However, Engelhardt thinks that in secular humanism they do not involve any supreme value or particular moral content (119–122). I think this is correct, for the will to engage in ethics may be a purely prudential act where respect for others is merely functional, and not grounded in an acknowledgement of their inherent worth or dignity. One ultimate value remains, of course, and that is the value of individual existence proper to every moral stranger. But this value has no moral significance, since it remains irrelevant beyond the individual human being. As such it may become the foundational element of egoism as an ethical theory. We may therefore conclude that although existential finitude might extinguish moral reasoning, it does not affect the possibility of ethical reflection and construction. The conclusions Engelhardt reaches, both here and in the field of political ethics, are viable and open to meaningful public debate. Let me give one example. If in a given political community there persist unresolved controversies as to the ethical acceptability of particular biomedical research projects or therapeutic practices, public goods provided by that community may not be engaged in supporting either of them (133). Existential finitude expressing itself in moral limitations will always lead to inequalities and presumably also to inequalities through multiple systems of health care (132 f.).

For all that, there is also a positive side to the "finitude of human moral arguments" (133). It leads to tolerance and to the acceptance of "collateral, competing, or supplementary health care systems, supported from private funds" (ibid.). Tolerance is indeed promoted, but it is doubtful whether it will persist if it is not complemented by solidarity. Obvious and pressing discrimination might cause social uproar. It is doubtful whether a prudential approach in securing solidarity would suffice. Yet this is a matter to be discussed on another occasion.

4.2   Limits of the duty to help and heal

The duty to help those in need and to heal disease and suffering has lain at the heart of medical ethics from its beginning. The obligation is precious, and yet it can be misused. If it is made absolute, it risks turning into absurdity. Considering existential finitude may amend its misplaced use and reveal that the physician's vow to heal does have its limits. By this consideration we affirm that only in critical situations can the principle be correctly interpreted.

This has consequences for the sometimes undifferentiated appeal to the Hippocratic oath that seeks to provide ethical justification for whatever research project or clinical experiment is at hand. Existential finitude, taken seriously, obliges the different actors to scrutinize and evaluate the goals and intentions that guide their practice. Only if these aims prove reasonable in the light of finitude can they be ethically sufficient. I shall come back to this issue when discussing some examples in the next section. Let me add that finitude made practical in this way provides us with an antidote to the normative power of routine and medical paternalism. For patients it establishes the moral right not to surrender to social constraints and pressures, or to scientific and therapeutic expertise and interest. It withstands the conception and financing of hopeless efforts to produce the perfect human being. In contrast, it fosters the readiness to accept and support those who suffer from more significant imperfections than the average members of the communities in which we live.

5.   FINITUDE AND AUTHENTICITY: THE EXAMPLE OF GENE TECHNOLOGY

In this section I wish to illustrate the potential role for the consideration of existential finitude by applying it to the field of gene technology. This technology, particularly when related to human beings, is, as we know, at the forefront of the biomedical debate.

Today, gene technology is widely praised and propagated by those with visions and promises with respect to serious or deadly diseases. Truthfulness and authenticity demand that legitimate hope for new and effective therapies does not veil what fashions human existence: feebleness, sickness and death, as well as frailty and lack of knowledge, skills, and risk assessment techniques. When anticipating and evaluating the hoped-for achievements of gene technology, finitude should serve as a criterion by which to examine the meaning – and if necessary to establish the absurdity – of the objectives and methods in research and medical practice. Difficulties and problems should be made public; for example, the fact that research is still far from understanding the complex functioning of genes, particularly their dependence on environmental factors beyond the human genome. The uncertain success of gene therapies should be explained, their social range analysed, and the costs of their development and application discussed. Authenticity demands respect for societal sectors and the political community as fully-grown partners and not as wards that must be given time to adapt to the trends and routines introduced by gene technology. Animating public debate is a corollary to truthfulness. It includes being prepared to renounce research or diagnostic and therapeutic activities, should the outcome of the debate – which is not simply a scientific one – request it. Yet even before that, authenticity fashioned through the perspective of finitude might oblige us to refrain from particular methods of research and application. This is the case when bona fide and proven scientists oppose, on serious grounds, the development and use of certain methods under consideration (Spaemann 1980: 204–206). This is where the notion of moratorium comes in, and authenticity requests that it should not be rejected on the basis of arguments like: you can't teach someone to swim without putting him into water.

Finally, looking at existential finitude may help us to cope with another and widespread argument, which stresses that not only performing an action but also renouncing it, entails responsibility. Formally correct, the argument leaves finitude out and hastily identifies something quite different. When acting in an ethically sound way, we do our best to anticipate and evaluate certain probable and possible consequences, and for that we may be held to account. Should we renounce an action on the same premise, our responsibility would not be questioned. But this is not the situation in which we find ourselves, for if we abstain from concrete action or plead for a moratorium, we do it precisely because consequences of whatever sort cannot be clearly established or satisfactorily evaluated.

When existential finitude expresses itself in such a dilemma, we are ethically free not to act. More than that: we are ethically relieved from responsibility. It would be morally wrong to force a proactive decision in the context of the argument I have just discussed.

6.   FIVE EXAMPLES

Markus Zimmermann-Acklin holds that debates about phenomena such as suffering, dying and death, as well as hermeneutical inquiries into the images of nature and of human beings are significant, but will not help us to solve practical controversies. He restricts their function to the disclosure of the backgrounds to moral positions with a view to facilitating a process of mutual understanding (2000: 34). I would maintain that the potential of ontological or existential consideration in ethics reaches farther. To argue for this position, and to descend from what have been somewhat general and abstract considerations, I am going to test their ethical potential on five roughly sketched examples.

6.1   Karen Ann Quinlan

You may remember the case of Karen Ann Quinlan, a young lady who, for whatever reasons, took tranquillisers together with alcoholic beverages and then fell into persistent unconsciousness, in a chronic and persistant vegetative state. She was taken into intensive care. Her state deteriorated continuously, but feeble brain activity was still observed. She obviously suffered from apallic syndrome. After three months her adoptive father requested that she should be taken from the respirator, arguing that her life depended entirely on complex artificial devices and that she should be granted a dignified death. His demand was rejected. A first lawsuit resulted in the verdict that professional ethics and competence opposed the father's will. However, the court of appeal judged in his favour. The respirator was removed. Karen continued to breathe by herself; she was taken to another hospital. Ten years later Karen left this world, without ever having regained consciousness (Dulitz and Kattmann 1990: 45f.; The Matter of Quinlan 1976). Finitude had been neglected at least twice in the process, both by those responsible at the first hospital, and by the first judge. A narrow deontological reasoning (and the physicians' professional ethics) was given priority over the consideration of an inescapable existential condition. The principle not to kill – which is not identical with the principle not to do any harm – was made absolute and turned into existential absurdity. So did the physicians' reference to the Harvard definition of cerebral death. Due consideration of existential finitude would have lead to the disconnection of the respirator at the father's request.

6.2   The seventy-one year old renal patient

The second example concerns a seventy-one year old patient suffering from a progressive bilateral renal sclerosis. He was refused both haemodialysis and a position on the waiting list for an allotransplant. Budgetary reasons imposing other priorities were used as arguments to legitimise the decision. The example is cited in a collection of cases published by the European network "Medicine and human rights" (1996: 381ff.). The ethical comment provided by the editor is sharp: the case is interpreted as an example of euthanasia for economic reasons. Since such an act discriminates against the poor, it is judged as entirely contrary to ethics. In so far as the verdict is rooted in principles that firstly forbid unjustified discrimination, and secondly secure equitable access to medical treatment and, moreover, refer to international covenants supporting fundamental personal rights, it sounds reasonable. Yet once again the perspective of finitude is lacking, and formal juridical and ethical arguments prevail. A critical stance permits us to inquire whether the meaning of equity or the criterion of justification in the case of discrimination are as clear as they appear. Of course they are not. Taking existential finitude into account might help to sharpen both issues on a case-by-case basis, even before consideration of economic and other external constraints. Later, such constraints might render the concept of finitude even more significant in the process of weighing the goods and interests at stake. Although I have not yet studied the matter in detail, I would not exclude the possibility of constructing an ethically sound obligation to renounce treatment. Appealing to fundamental rights without considering fundamental obligations is one-sided. For the time being, I would start further investigation from the heuristic basis that the decision concerning the seventy-one year old renal patient might be ethically defended.

6.3   Xenografts

Xenotransplantation of tissues and organs is still a lively project with researchers, medical experts, politicians, and pharmaceutical firms. Intensive research has shown that it is accompanied by the hardly rateable risk of serious, if not deadly infections, particularly by porcine endogenous retroviruses (PERV). The risk not only affects the patient, but also care personnel and all those with whom he or she comes into close physical contact. There has even been speculation regarding the danger of a pandemic. Arguments taken from animal ethics also speak against the production of xenografts, among them the large number of animals used to produce the transgenic individuals needed, and the pain, suffering and death inflicted on animals by xenograft research, and by breeding and living conditions. Thus arguments drawn from human as well as from animal ethics oppose xenotransplantation. The project is meant to alleviate the lack of human organs; it should serve those persons certain to die if not grafted. The obligation to help and heal is invoked to justify the continuation of the project. Yet one may ask whether the exposure of many people to serious risks and the infliction of suffering and death on numerous animals are outweighed by the possible benefit for comparatively few patients who are near the end of their physical existence. The aspect of finitude might relativize the physician's vow to heal and help, and give more weight to another obligation that burdens him: the obligation to protect (cf. Sitter-Liver 2000).

6.4   Embryonic stem cells

The next example refers to the currently much discussed use of human embryonic stem cells, either taken from surplus embryos or specifically produced. It will serve to demonstrate that the relevance of existential finitude is limited, too. A decision concerning the ethical acceptability of the use of embryonic stem cells depends of course on the moral status we award embryos. But even if these are defined as humans in the biological process of their existence, one might ask how it is possible to explain to a fully-grown and seriously affected person that his wish to survive is outweighed by the claim to life assigned to embryos that must be destroyed in any case. Could we not have recourse to the aspect of finitude, and release the embryos for research and applications, supporting our decision by saying that while dying, they will serve to help and heal others?

The argument is tempting, but inadequate. For with respect to the patient, we do speak of dying; as to the embryos, we correctly speak of killing, independent of the certainty of their death. And it is quantitatively important killing in the interest of as yet uncertain and unknown third parties. If we consider embryos not just as a heap of cells, but as human beings in their development and therefore as entities endowed with human dignity (a controversial and not easily articulated standpoint [cf. Spaemann and Merkel, both 2002] but one which we may assume in the present context without any further inquiry), the perspective of existential finitude will not help us clear the situation. When embryos are defined simply as a cluster of cells the argument is, of course, utterly irrelevant.

6.5   Eternal brains

My last example might appear to belong to science-fiction, yet it touches actual reality. In a recent article published by "Le Monde diplomatique" (Vol. 7, No 8, August 2001: p.19), Mariano Sigman, a neuroscientist at Rochester University in New York, commented on the general programme of the Japanese Riken Institute, "the rising sun of neurobiology". The Institute has defined its overall aim as 1) understanding, 2) protecting, and 3) creating the human brain. The third goal includes the construction of robots endowed with the intellectual and emotional capacities necessary for participating in human intellectual controversies and discourse. By the year 2020, the programme aims to have led us to the disclosure of the mystery of human thinking; it will have rendered the brain immortal, by providing it with the necessary technical devices to continue its existence indefinitely. The consequences of the scientific and technological progress will reach far beyond the implications of genetic engineering. "They have launched", says Mariano Sigmann, "the most important attack on human identity that has ever been attempted. And they threaten to make human beings vanish in a 'post-human' epoch."

We may leave that prognosis open and content ourselves with the fact that the Institute's goals and programmes are supported by public as well as private funds. The goal of creating external human brains independent of human beings as we know them is, to my mind, clearly a product of the eternal longing for redemption and the ensuing constantly renewed struggle to overcome human finitude. It is rooted in a truly religious motive (Mutschler 1998–99, particularly pp. 72–74). If, from a bioethical position that takes existential finitude seriously, we can approve the Institute's first and second goals as endeavours to enhance human well-being, we must judge the third as goal ethically wrong. For by trying to overcome human finiteness it works towards abolishing human existence. It is bioethically evident that public funds must not be invested in such an enterprise; and one may raise the question of whether private funding should also be legally prohibited. Of course, this conclusion has a normative premise. It is valid only in so far as we deny the third goal ethical legitimacy on the grounds that we oppose scientific and technological developments that tend to put an end to humane existence in this world.

We may close this section by acknowledging that some of the examples are controversial. I have simply aimed to illustrate that the perspective of human finitude may be helpful in tackling moral dilemmas. But if it is, then decisions on biomedical issues that claim to be taken in an ethically responsible way must not forget existential finitude.

7.   CONCLUSION

I hope to have shown that consideration of human finitude may help to clear and resolve controversial issues in biomedical ethics. H. Tristram Engelhardt's effort to construct a platform for moral strangers that allows them to meet in a rudimentary common moral world has been very instructive indeed (1991: 42). I would, however, maintain that making human finitude a concern of bioethics might enrich the instruments offered by his secular analysis of human nature. He does aim, as we remember, to find interests that all humans share with a view to constructing "a basis for secular cooperation excellence" (ibid.).

This aspiration is essential. The range Engelhardt offers may yet be extended, without having recourse to religious or metaphysical premises. Introducing the perspective of human finitude into bioethical discourse constitutes an effort to draw from an ontological (or existential, or anthropological) study conclusions that help to tackle moral dilemmas, and to guide scientific and content-oriented political practice – without falling into the trap of the naturalistic fallacy.

References

Bondolfi, Alberto. "De la bonne conjugaison des arguments anthropocentriques et zoocentriques dans le contexte d'une légistation autour des xénogreffes." In *La dignité de l'animal*. D. Müller and H. Poltier (eds.). Geneva: Editions Labor et fides, 2000, 363–375.

Dulitz, Barbara and Kattmann, Ulrich. *Bioethik. Fallstudien für den Unterricht.* Stuttgart: Metzler, 1990.

Engelhardt, H. Tristram. *Bioethics and Secular Humanism. The Search for a Common Morality.* London/Philadelphia, 1991.

Engels, Eve-Marie. "Le statut moral des animaux dans la discussion sur les xénotransplantations." In *La dignité de l'animal*. D. Müller and H. Poltier (eds.). Geneva: Editions Labor et fides, 2000, 317–361.

Heidegger, Martin. *Sein und Zeit*. Tübingen: Niemeyer, 1963 (first published Tübingen 1927).

Kurtz, Paul. *In Defense of Secular Humanism*. Buffalo N.Y., 1983 (cit. by Engelhardt 1991, 117, 187 [note 28]).

McKenny, Gerald P. *To Relieve the Human Condition. Bioethics, Technology, and the Body*. Albany, 1997.

Merkel, Reinhard. "Rechte für Embryonen?" *Die Zeit. Zeitdokument* 2002, 1, 75–79.

Mieth, Dietmar. "Die Krise des Fortschritts und die vergessene Endlichkeit des Menschen." In *Moral und Erfahrung – Grundlagen einer theologisch-ethischen Hermeneutik*. Dietmar Mieth (ed.). Freiburg, 1999, 193–210.

Müller, Denis and Poltier, Hugues (eds.). *La dignité de l'animal*. Geneva : Editions Labor et fides, 2000.

Mutschler, Hans-Dieter. "Technik als Religionsersatz." *Scheidewege. Jahresschrift für skeptisches Denken* 1998–99, 28, 53–79.

Sigman, Mariano. *Le Monde Diplomatique*. August 2001, Vol. 7, No 8.

Sitter-Liver, Beat. *Dasein und Ethik. Zu einer ethischen Theorie der Existenz*. Freiburg/München, 1975.

Sitter-Liver, Beat. "Un défi et un parcours ardu. Vers l'atténuation de la conception utilariste des animaux." In *La dignité de l'animal*. D. Müller and H. Poltier (eds.). Geneva: Editions Labor et fides, 2000, 427–450.

Sitter-Liver, Beat. *Xenotransplantation aus der Sicht der Tierethik*. Folia Bioethica Geneva 2000, 27.

Spaemann, Robert. "Technische Eingriffe in die Natur als Problem der politischen Ethik." In *Ökologie und Ethik*. D. Birnbacher (ed.). Stuttgart, 1980, 180–206.

Spaemann, Robert. "Gezeugt, nicht gemacht." *Die Zeit. Zeitdokument* 2002, 1, 71–74.

The Matter of Quinlan. Excerpts from the State Supreme Court's decision, including the basic case history 1976. Cf. http://www.csulb.edu/~jvancamp/452_r6.html. (For more information see http://www.who2.com/karenannquinlan.html.)

Zimmermann-Acklin, Markus. *Perspektiven der biomedizinischen Ethik: Eine Standortbestimmung aus theologisch-ethischer Sicht*. Folia Bioethica, Geneva 2000, 26.

Further literature

Enzensberger, Hans Magnus. "Putschisten im Labor." *Der Spiegel* 2.6. 2001, 23, 216–222.

Gould, Stephen J. "Die große Asymmetrie: Vergehen ist leichter als Entstehen." *du. Die Zeitschrift der Kultur* 2001, 718, 55–57.

Günter, Joachim. "Menschenwürde und Leidverminderung. Die Bioethik im deutschen Kulturkampf." *Neue Zürcher Zeitung* 18./19.8.2001, 190, 62.

"La santé face aux droits de l'homme, à l'ethique et aux morales", published by le Réseau européen de coopération scientifique "Médecine et droits de l'homme". Edition du Conseil de l'Europe, Strasbourg, 1996.

Lutz-Bachmann, Matthias. "Menschen und Personen. Über einen Grundsatz der praktischen Vernunft." *Information Philosophie* 2001, 3, 16–19.

Meister, Martina. "Der Terror der Normalität." *Frankfurter Rundschau* 9.7.2001, 156, 9.

Meyer, Verena. "Was würde Kant zum Klonen sagen?" *Information Philosophie* 2001, 3, 20–22.

Schaber, Peter. "Genetik, Ethik und Biopolitik. Eine Diskussion oder ihre Unterdrückung?" *du. Die Zeitschrift der Kultur* 2001, 718, 65–67.

Venter, Craig. "Revolutionen werden nicht von Schaffern gemacht." *Frankfurter Allgemeine Zeitung* 1.6.2001, 126, 49.

von der Weiden, Silvia. "Die persönliche Pille." *Die Zeit* 5.7.2001, 28, 22.

Weisshaupt, Brigitte. "Ethik und die Technologie am Lebendigen". In *Grenzen der Moral. Ansätze feministischer Vernunftkritik*. U. Konnertz (ed.). Tübingen 1991, 75–92.

Wetzel, Sylvia. "Glaube nicht, es sei zu wenig." *Gen-ethischer Informationsdienst* Dezember 1999, 137, 14–17.

Chapter 4

LIMITS OF BIOETHICS

CHRISTOPH REHMANN-SUTTER
Basel, Switzerland

> Our actions are like ships which we
> may watch set out to sea, and not
> know when or with what cargo they
> will return to port.
> *Iris Murdoch: The Bell*[1]

If ethics is to say anything helpful concerning human practice it must develop a vision of what is good in human life. This is the job of ethics. But in attempting this we are to a certain extent afflicted by the epistemological limitations of our vision and knowledge of practice. An ethics that does not reflect on these constraints would not be eligible as a good guide.

This statement may have some intuitive appeal, but of course it contains an inherent assumption that ethics is or can be a guide to practice at all. It assumes that ethics is not content-free. In this paper, I will start with this assumption and give some reasons for it. However, the main topic will be the strengths and limitations of moral vision, and of ethics as a 'fabric of practical rationality'. An important question in this context is how limits appear in morality and in ethical discourses. Ethics not only deals with limits (reflects on them, struggles with them, defends them, tries to localize them), it also relies on respect for boundaries and is in itself intrinsically limited. My thesis is that limits are a necessary constituent of ethics and need to be appreciated positively if we are to seek and hopefully find at least glimpses of practical wisdom. However, this is a general thesis about ethics, and in this paper I shall discuss these ideas more concretely with reference to bioethics.

C. Rehmann-Sutter et al. (eds.), Bioethics in Cultural Contexts, 59–79.

What is bioethics? Bioethics, as I understand the term, is a specialized field of ethics, concerned with moral issues of the life sciences, including genetics, biology, medicine and ecology. In these sectors, where moral dilemmas and existential conflicts abound, the otherness that can only be approached by respecting limits is particular and concrete: the life of living creatures, the health and well-being of patients, the self-organization and creativity of nature. It is the relation of boundaries to these special kinds of alterity that gives bioethics an identity.

My account of the limits in and of ethics draws extensively from metaphors of seeing. Several of the influential convictions shaping ethical discussions are commonly expressed through visual metaphors: the 'moral point of view', ethical 'aspects', 'perspectives', the 'ideal spectator', the 'view from nowhere', the 'observation of the observer', to name but a few. Instead of rejecting this visual imagery, a critical perspective on ways of seeing in ethics can be attained by reflecting on the visual regime those concepts represent. A critical evaluation of moral vision can lead us towards a reassessment of the norms of visions in morality – which are moral norms themselves.

Every disciplined way of thinking has a perspective, and so does every kind of serious ethical thought. In order to understand what we see in such a perspective, we need to know what we are capable of seeing within this perspective, and also where our flaws and limits lie. On the other hand, moral vision implies engagement both in the realization of a better life, and against negligence, injustice, oppression and violence. For this we need to trust our eyes and to ensure that we are reliable partners in communication. What I have in mind when starting my account of the limitations of ethics/bioethics is not a weakening of the ethical project. The real strength of ethical thought lies in a reflected moral vision that is aware of other possible or actual perspectives and tries whenever possible to integrate its arguments with other ways of seeing. A critical inquiry into moral perception is therefore a necessary part of serious ethical methodology.

1.   THE AIM OF BIOETHICS

From its beginnings in Antiquity, 'ethics' was meant to be a philosophical investigation into practical wisdom. In Plato's *Republic* (352 D) Socrates says, the question we are talking about is *how one should live*. This is of course not trivial at all. The question was understood to be solvable by philosophy, in an inquiry that is bound to an attitude of reflective generality and tries to gain clarity on the basis of rationally persuasive arguments (Williams 1985: 2). In the sixth book of his *Nichomachean Ethics*, Aristotle explained practical wisdom *(phronesis)*, particularly the capacity to reflect about what are good practices, as one of the means by which our soul can reach truth. 'Good' was not to be understood in a narrow sense of what is good for health or what gives force and power, but in the sense of a good and happy life (1140a 25–28, cf. Urmson 1989, Broadie 1991). But this investigation should not lead away from concreteness, because practical wisdom concerns not only the general but also the particular. Practical wisdom in Aristotle's conception belongs to practice, or is itself part of practice; acting necessarily concerns the particular goals, needs and circumstances of the situation (1141b 14–16).

The idea that ethics is a purely formal moral theory without being a guide to practice is much more recent and was never uncontroversial within philosophy. In some ways, the two great ethical conceptions of eighteenth century – Kantian deontology and Bentham's consequentialism – can be seen as formal constructions of a practical rationality. Both conceive a method of evaluating plans for activities without basing the validity of this evaluation on the ethical values of the individual concerned. The plans for activities (plans for singular actions or plans for rules) are tested against the criterion of universalizability (Kant), or compared with the alternative options under the criterion of maximization of happiness (Bentham). However, both these constructions of a formalized theory, which can be utilized in everyday practice and harmonized with different moralities, are based on ethically relevant conceptions. The picture of ethical rationality they support contains assumptions about the configuration of the moral self. In Kant's philosophy, only the will can be a true subject of good or bad (Grundlegung, AA 402), and the will is a capacity to control one's behaviour. In social life, this will has to do with Others who are perceived as others of the same sort, other subjects of a behaviour-controlling will. From a critical perspective we can recognize that Kant's conception of the will organized the human self in nontrivial and ethically controversial ways. If the categorical imperative is the answer, the corresponding question must have been clear beforehand. This question was, in a simplified form: is there a universally valid criterion to distinguish between a good and a bad will (which can materialize in actions and decisions)? This question is not ethically neutral but rather it is the expression (or foundation) of an ethically significant concept of the human mind and its relationships with alterity (body, environment, other living beings, other persons). Related observations about a presupposed identity as moral selves could be made about Bentham's moral philosophy. Charles Taylor (1989) has analysed how the configuration of modern identity has been influenced by and can be understood from the conceptions of moral philosophy. Hence, despite the aims of the main approaches to moral philosophy in the eighteenth century

– to find an Archimedean point outside material moral positions in order to fix a scientific system of morality – they did not have the intention of keeping moral philosophy from guiding practice.

Analytical moral philosophy or "metaethics" in the twentieth century came closer to pure formality by recognizing the meaning of the words of our moral language as the basis for philosophical inquiry, or by clarifying the meaning of questions before trying to find answers to them (Moore 1903; cf. Scarano 2002). But the value of this field of inquiry has also been seen as a "prolegomena" for an ethics that could indeed guide practice (G.E. Moore in the preface to *Principia Ethica*). It was the need for clarification of a series of difficulties that caused misunderstandings in moral communication, not the intention to keep ethics from its guiding task.

H. Tristram Engelhardt, in his *Foundations of Bioethics* (1986), argued for a "content-poor" conception of "secular bioethics" that would be capable of securing "moral authority and purpose for common action across moral communities" (53). Engelhardt's aim was to reconcile intersubjective and universalistic rationality with the obvious and insurmountable fact that modern societies are conglomerates of many different religions and diverse material conceptions of the good life. Pluralist societies consist of "moral strangers". Secular bioethics, in his understanding, "is an enterprise in policy making" (ibid.) and has therefore to bridge the polytheism of pluralist society with a non-violent procedure that could only be a purely formal and content-poor procedural rationality. In this sense Engelhardt reinterprets the principles of autonomy and beneficence. The first principle, autonomy, "justifies the process for generating content" (71) in holding everybody from using others merely as means: their own decisions and value judgements matter, and must be respected. The second principle, beneficence, stands in a fundamental tension with autonomy and says: "Do to others their good" (76). It expresses a general concern for providing others with the goods of life without imposing on them a particular value system beyond those principles that are compatible with their own.

It has been argued against this conception of bioethics that the two principles offered cannot actually reach the goal of content-emptiness (cf. Tishtchenko: this volume; Ach and Runtenberg 2002: 67). Indeed, the concept of autonomous individuals is based on a particular understanding of the human self and represents the anthropological ideal of self-governance, whereas the principle of 'doing others their good' is – however admirable – not free from material morality. I would say that a good morality is contained in both of these principles. The theory, however, can be improved. Engelhardt himself later replaced his 1986 account of autonomy with a more restricted interpretation of autonomy as the principle of permission for and respect for privacy (Engelhardt 1996; Ach and

Runtenberg 2002: 70). I fully support Engelhardt's recognition of the plurality of material moralities in modern societies. Only an attitude of respect for other visions of the good life can form a basis for a mutual understanding of those differences. However, although I see, in the openness, fairness and transparency of procedures, the key to acceptable consensus in public policies about bioethical issues (laws, guidelines etc.), I am hesitant to restrict bioethics to "an enterprise of policy making", and to restrict it *within* the field of policy making to discussion about content-free (or at least mutually acceptable) procedures. I would argue that the account of secular bioethics that Engelhardt defends draws limits that are too narrow.[2] I will now explain briefly the main reasons I see for these two hesitations.

(1) *Bioethics also has a role outside policy making*, for example in discussions about difficult counselling issues, in situations of conflict in hospitals, in decisions by patients, potential patients or participants in research (Scully, Rippberger and Rehmann-Sutter 2004). In these cases, the implications of individual and social practices are relevant from the perspective of individuals. I cannot see why the traditional field of 'individual ethics' (often distinguished from 'social ethics') should not be included in the scope of bioethical research. Engelhardt certainly agrees with this, but he would probably say that it needed to be outside 'secular' bioethics, because individuals can also draw from their deep moral traditions, from the religious views and values they share with some others (but perhaps not with everybody in society). My reply to this is that religious and moral views often differ even within small groups and relationships, for example within the team of doctors, nurses and relatives that tries to find out what should be done next when the patient is not able to decide for him- or herself. In trying to find a common understanding and a morally acceptable decision in consideration of their different religious and moral views, the participants in a way, *practise* bioethics.

Theoretical (and professional) research in bioethics can contribute by helping to clarify the issues and concerns, by discussing what is at stake, and by reflecting on the consequences and implications of different ways of proceeding. But theoretical research in bioethics can never replace the moral reasoning of participants with theory-deduced prescriptions. In practice, bioethics as a philosophical and professional investigation does not decide the issues 'for' those involved. This would be moral paternalism (see Tishchenko, this volume). The participants themselves have the burden of finding the right decisions, probably in conjunction with what they understand from bioethics literature and/or with the help of what they hear from ethical counsellors. It is they who will have to live with the outcomes. Their own reflections are bioethics in practice.

(2) *Bioethics has a role both in issues of procedure and in matters of content,* even within the field of policy making. In order to make this point I will refer to the example of stem cell research. From the start of the discussion with the publication of Thomson et al. (1998), in my country (Switzerland) and in many others the issue was predominantly seen as one of regulation. Should it be permissible to use spare embryos from IVF for harvesting stem cells under special circumstances? Or should it be permissible to purposefully create research embryos? Should the use of somatic cell nuclear transfer ('cloning') technologies be allowed for this purpose? These are some of the questions under debate. Other questions that are ethically at least as important have been much less discussed: how can regenerative medicine change the meaning of medicine and what impact will it have on our basic understanding of life, disease, death and healing? How will it affect our culture of coping with limits and how might a society be changed when egg- or embryo-donation becomes an institutionalised routine? Which criteria should be respected to safeguard the participants (and non-participants) from injustices?

One of the most important functions of ethics in public discourse is to raise those questions that tend to be overlooked. Bioethics should always look behind the questions that have already arisen and been debated, and attempt to find language for them, rather than simply stepping on board already rolling wagons. A bioethics that restricts itself to procedural suggestions would not be capable of doing this.

However, even regulatory issues require an engagement by moral philosophy in the matter itself. How else can the central regulatory questions of the stem cell debate be ethically clarified? It is, for example, very difficult to determine whether respect for the moral status of human embryos created for IVF is consistent with harvesting stem cells (and thereby ending the life of the embryo as an organism) in a situation where the embryo cannot be transferred to the womb and would otherwise have to be abandoned for legal reasons (the use of 'spare embryos' in the sense of Swiss law; NEK-CNE 2002). Here, it is necessary to decide whether the instrumentalization of an embryo as a source of stem cells in a situation where for other independent reasons its life would not continue, is ethically acceptable. If bioethics restricted itself to formal and procedural issues, this question could only be decided by counting votes. Perhaps in the end we may indeed count votes, but preferably after a broad, precise and clarifying discussion where those who will vote (in a parliament or a public referendum)[3] have a chance to see the ethical implications of each option. For this, an open ethical discourse with the participation of the best thinkers is crucial. If bioethicists, intellectuals who specialize in such issues, refrain from participating materially (with the argument that rationality cannot contribute to the issue), they cannot meet what they view as their social responsibility.

The role and power of philosophical thinking in bioethics should not be underestimated. Bioethics would underestimate itself if it did not think that philosophy is capable of understanding problems in the cultural project of modernity and criticizing it with a guiding intention. Philosophers will probably not reach a consensus on all the points, but a well-argued dissent can also serve to clarify. One danger of bioethics is the underestimation of its challenge, the result, as Eva Feder Kittay aptly pointed out at the conference in Davos, would be "irrelevance". The other danger – "moral priesthood" in Kittay's terms – would be its overestimation as a fabric of rational, quasi-scientific solutions for those nontrivial moral issues of the biopolitics debate, solutions that rational citizens would only have to understand, accept and apply.

2.   LIMITS IN MORALITY

The landscape of morality is structured by different sorts of limits and boundaries. They not only happen to be there as something we regularly encounter in our moral practice (behaviour, thought, perception and talk), they are in many ways constitutive of morality. Without limits, morality would not be possible, and many of the limits are crucial aspects of morality. I will start with a phenomenological tour through the landscape of limits, with an eye to those phenomena of limits that might be particularly relevant for bioethics.

(1) One of the central problems of environmental bioethics is the moral considerability of nonhuman entities: animals, living beings in general, natural habitats etc. (Attfield and Belsey 1994). The debate about anthropocentrism is about the relation of humans to nonhuman others. Are they other beings with intrinsic value, even with their own moral rights that must be respected by us? If they are morally considerable in themselves, beyond their utility for humans, they possess at least some of the same kind of Otherness we see (much less controversially) in other human beings. If they are not, they still matter for us morally, but only as things, as resources or means that are to be protected because (and in so far as) the life of humans relies on them. Our relation to them in the latter case would not be a moral relationship; they would only appear as necessary elements in moral relationships that we have to other human persons. In human bioethics this kind of problem appears at the boundaries of life: experimentation with embryos, with totipotent embryonic cells, in the context of the definition of brain death or the dignity of a patient in a persistent vegetative state. The

meaning of 'moral consideration' is something that needs to be clarified where considerability is controversial: at the limits of normal moral consideration, in other words, in the borderland of the moral community.

Otherness, in the sense of moral considerability however, has its place right at the centre of a moral relationship: it constitutes 'the ethical' in this relationship. Otherness means (i) that the other is in his or her own 'sphere of sense' and (ii) that this sphere of sense matters for me. If my actions cause harm to him, her or it, I am affected myself. The other's humiliation diminishes me. In one sense this moral otherness is based on a moral community, and in my being part of the community I am similar to the other. This kind of similarity is not the same as equality of rights, but it makes possible a community of equals whose well-being matters. It is the basic moral relationship of Self and Other, where Other depends on the awareness of the Self about the other's boundaries. Otherness of the other is constantly endangered by the arrogance of the self that claims to know better, to have more relevant desires. These limits that constitute moral sensibilities are the very heart of morality. Otherness, as the basic sphere of the ethical, is made by the self through the sensible membranes that constitute its limits. Moral responsibility – the capability and readiness to respond to the other – is achieved by those limits of the self toward the other (see Benhabib 1987; Ricoeur 1990; Kellenberger 1995). In order to keep this first set of limits separate from others, I will refer to them as *relational limits*. When I claim to know all of the other, I eradicate his, her or its otherness and replace it with sameness. An other remains 'other' when he, she or it can still surprise me. Hence, being attentive to relational limits creates the relational space for responsibility.

(2) A second set of limits is what most questioning in practice-oriented ethics (like bioethics) is *about*: what limits define an 'acceptable' course of action? Which biomedical interventions are unacceptable? What should be allowed or forbidden? Bioethical debates are typically about setting barriers in the right places: on this side of or beyond therapeutic cloning? Just beyond active indirect euthanasia but before direct euthanasia? The bioethical debate very often remains within this frame, that is, within the assumption that the debate should be concerned with questions of the limits to 'moral' conduct. I will call these limits therefore *moral limits*. Connected to them, like the other side of a coin, is the common perception of biomedical progress as an enterprise that, whether we like it or not, has the inherent and sometimes uncontrolled desire to transgress the limits of our human existence. James Whale's film *Frankenstein* (Columbia Pictures 1931) is a perfect illustration of this. The following conversation between Dr. Waldman and Henry Frankenstein follows the creation scene (the monster is – still – captured in the cellar, the two scientists sit in the laboratory):

> *Dr. Waldman:* This creature of yours should be kept under guard! Mark my words, he will prove dangerous!
> *Frankenstein:* Dangerous! Poor old Waldman. Have you never wanted to do anything that was dangerous? – Where should we be if nobody tried to find out what lies beyond?
> (from Anobile 1974)

This 'Frankensteinian' picture of science (Rehmann-Sutter 1996) seems to be regularly confirmed by experiments that seem to contribute to Huxley's Brave New World, like growing fertile oocytes from embryonic stem cells in a petri dish (Hübner et al. 2003). Those who most explicitly try to bolster the transgression pattern of scientific progress with a 'light' version of bioethics include the so-called Transhumanists. They enthusiastically (and naively) embrace literally every biotechnology that promises to lengthen the human life span and to enhance human capabilities – such as body strength, health, intelligence or emotional control – from cryonics, to nanotechnology, to uploading the self onto computers, right up to enhancement genetic engineering.[4]

Limiting the realization of human desires and making a peaceful human coexistence possible is one of the key functions of social norms (Merton 1968). Ethics is a reflection on the right content, and on the implications and foundations of norms. Hence, it is always concerned with setting the right limits to what individuals might want and try to do. Moral limits, however, take on a slightly different and more positive significance if they are investigated in the light of relational limits: at least some of them are necessary for respecting and protecting others. Limits, materialized as social norms, are also expressions of an attention to others who otherwise would not be considered in certain interest based life plans.

As Bernard Williams has pointed out, an unlimited life is not even desirable: "The fact that there are restrictions on what [an agent] can do is what requires him to be a rational agent […], it is also the condition of his being some particular person, of living a life at all. We may think sometimes that we are dismally constrained to be rational agents, and that in a happier world it would not be necessary. But this is a fantasy (indeed it is *the* fantasy)" (Williams 1985: 57).

(3) I agree with this, but restrictions are sometimes hard to accept. There is 'akrasia' – weakness of the will: the gap between the best judgment about what to do, the formulation, and the actual realization of an intention. People can be driven by other forces that also have their origin within their self (passions, desires, fear etc.), which lead to a behaviour that is contrary to reason. All our reason can do in such circumstances is to think 'around' those forces: given they are present, what should we do with them?

There are other weaknesses of the self: weakness of judgment and weakness of self-understanding. Weakness of judgment, for instance, has come very forcefully under consideration as the limit of foresight in the discourse about ecological and technological risks. We know only some of the consequences, probably a very small fraction. The 'risk' we typically take into account has as its essential component a large number of unknowns. If we know that a certain technological use (e.g. deliberate release of certain GMOs) *can* theoretically have dangerous consequences, but we do not (and cannot) know whether those possibilities are mere theoretical constructs or real possibilities, then our practical reasoning is hampered, or at least the ideal of a thorough rational guidance of practice has serious limits. Practical reasoning, however, can be improved when the limits are considered and included in the ethical and political deliberations about GMOs. The knowledge gaps are themselves a feature of the actions we plan and should be made transparent (see van Dommelen 1996).

Another weakness is that we are not transparent *ourselves*; we know only parts of ourselves. '*Gnoti se auton!*' (Know yourself!), the inscription at the oracle of Delphi is and will remain an imperative. Every attempt to give an account of oneself has its limits. Foucault in his later writings puts this limit of self-understanding into relation with regimes of truth. The recognition of the self is limited from its very beginning by what counts as a recognizable form of being and what does not (Butler 2003). I would like to call this third set of limits within morality the *limits of the self* (see Table 1).

A further component of these limits of the self is *dependency*. In many circumstances we are very far from the ideal of the autonomous self that has independent control and perception of its actions. The Kantian anthropology of enlightenment has been fundamentally criticized in that its picture of the independent controller is a highly idealized (and perhaps also gendered) self-image with an underside: all the rest of ourselves (the body, the emotions, the 'female' nature, 'the other') is subject to control and possibly even to oppression (Böhme and Böhme 1983). But in everyday situations we are very often dependent on others with whom we live in relationships. And there are persons who are always dependent on the care of others. Eva Feder Kittay defines 'dependency work' as follows: "the work of caring for those who are inevitably dependent" (Kittay 1999: ix). Others are dependent on us and we are more or less dependent on them to varying degrees. If ethics does not take those practices that inevitably happen within situations of dependency into consideration, a very large part of practice that is relevant in biomedical contexts would be omitted. The relationship between medical personnel and patients can be one of dependency, and patients' medical decisions are sometimes far from the ideal of autonomy as independence. One woman, who was asked by her severely sick brother whether she would be ready to donate the larger half of her liver for a transplant, told me that she never felt that this was a situation where she weighed risks and chances, values and duties. She knew immediately that she had to do it – even if this constituted a high risk for herself – because her brother's life was dependent on her. In her 'snap decision' she was not 'free' in the sense that Kantian ethics would have demanded. Diana Meyers (1989) has developed a relational approach to autonomy that is not based on the ideal of independence but on harmony with one's own true self. Also Alasdair MacIntyre wrote on "the virtues of acknowledged dependence" (MacIntyre 1999: 119 ff.).

(4) Very closely connected to the limits of the self, self-understanding and self-control is a fourth set of limits that is essential in communicative practice. Communication implies the goal of mutual understanding. Many practitioners and ethicists find that the more they face the particularities of a situation and the more they try to base their arguments on the concrete context, the better their concerns are understood by those actually involved in the decisions concerned. This is not an argument against universalization as a method of moral philosophy, and it does not inevitably lead to relativism. But whatever theory we believe expresses the true foundations of ethics, anyone who wants to make an ethical argument in the context of any practice is dependent on an understanding of the concrete social and cultural contexts of those she or he is listening and speaking to. In ethics there are no formulae like the numbers and symbols in mathematics. Ethics starts in the middle of practice and its reflection and guidance can only succeed if it speaks in an understandable language that takes the particularities of that practice into account. In this sense, a practically guiding ethics has to be grounded in an understanding of what Arthur Kleinman has called "local processes" (Kleinman 1998: 88), and therefore ethics needs an ethnographic, sociological and historical moment. This fourth set of limits I will call *communicative limits*.

Table 1: Limits in morality: a tentative map

I	Relational limits	- boundaries with the Other
II	Moral limits	- norms
		- transgression by science/technology
III	Limits of the self	- weakness of the will
		- weakness of foresight
		- limits of self-understanding
		- dependency
IV	Communicative limits	- basis in life worlds

3.   TRANSFORMATION OF MORAL VISION:
SEEING FROM INSIDE

It was my decision at the beginning of this paper to reflect on the possibilities of bioethics from the perspective of the limits that both afflict and enable perception. It is perhaps natural that philosophers who start their reflection not from perception but from the language of morality, from the meaning of the words like 'good' or 'just', obtain a different result. Richard M. Hare (1981) said that the only kind of moral judgment that moral language can be used to express is universally prescriptive, and this implies a hypothetically unlimited perspective (that of his famous archangel) able to consider the question from all points of view.[5] But perception cannot be considered as anything other than subjective. It was an assumption I would have to defend elsewhere, that language is a tool, a method of perception, and not vice versa. Moral evidence, as I see it, is based on a special kind of perception and so is practical wisdom. If our moral language implies universal prescription then it poses a special set of questions about the situations in which we inevitably find ourselves, questions that make us consider the possibility of all other points of view. Which parts of a moral judgement rest on our accidental subjective involvement and which can probably stand for everybody who might be in the same situation? These are productive questions indeed. As Thomas Nagel put it: "Objectivity is a method of understanding" (Nagel 1986: 4). We could elaborate on this and say: it is a visual strategy, a strategy that produces a natural tension with the undisturbed view from the inside. However, it is a strategy that does not really lead us outside. We remain subjectively engaged participant observers within the world, involved in relationships and entangled in histories; but while unchangeably placed like this we can reflect on the visual regimes we adopt and we can work towards transparency and a refinement of the norms constituting these regimes.

Now I will look at the four sets of limits once again, with these questions in mind, and give a brief commentary on each.

(1) The problem of moral considerability, of membership or non-membership in the community of morally relevant beings for whom we are intrinsically responsible, may be much more tightly connected to issues of perception than most theories normally assume. I do not expect these boundary issues of the moral universe to be solved via deduction from a moral principle, because every principle that implicitly carries the necessary information, would itself need a justification in its competition with other principles. So it is much more natural to admit that the principle we rely upon in deciding about the existence or non-existence of moral considerability, or about the assignment of a moral status, itself establishes

those boundaries and scales of status. If this is true, it is futile to bind the assignment of considerability or status to a particular kind of entity with reference to the principle. This would secure consistency within one regime but it would confer no justification for it. By saying, for example, that moral considerability is a function of rational capacities, or of sensibility, or of life per se, we establish different regimes of moral vision. To know more about the capacities of our moral sensitivity we need to actively test how far we are able to recognize and consider within moral relationships. The ethical recognition of the dignity of the human embryo, of animals etc. requires a change in perception. Ethical theories interpret the setting and are behind this change.

So, we can describe the visual regime as a configuration of the boundary between self and other. The position of the boundary of the moral universe (what belongs to it and what does not) seems to be dependent on this configuration, and not vice versa. 'Otherness' is established by those sensible membranes that constitute the limits between self and other. The ethical qualities of a relationship depend on their sensitivity.

(2) With regard to 'moral limits' I want to emphasize the limitedness of an approach that views bioethical issues primarily as issues of transgression into morally forbidden zones, and sees bioethics as a guardian of moral limits, or in religious language: a guardian of the taboo. This presupposes a kind of authority that bioethics cannot possess ('moral priesthood', as Eva Kittay called it). Some stakeholder groups may attempt to force such a role on bioethicists. However, here I see the real problem of religious versus secular bioethics: if ethics is charged with a role it cannot fulfil, it should give an appropriate answer without denying the problem and without accepting the role.

The role of ethical expertise is under discussion. What can it offer? Ethics can help to clarify what is at stake. Situations must be understood, and this needs a hermeneutic competence in complex practical contexts. Ethics should help us to understand better what kind of situations we are in. This necessarily includes religious dimensions if they are involved – and where ethicists are hired as 'guardians of the limits' such dimensions certainly are involved and await clarification. When bioethics offers 'points to consider' in practical situations instead of 'rules to be obeyed if the agent wants to be rational', the limitedness of its power of judgment is no obstacle, and the autonomous responsibility of other members of society is probably better served. The problem of religious arguments versus secular ethics is often debated in a way that questions whether 'religious arguments' can be regarded as respectable in philosophical ethics or not. Here I see much less of a real problem. Many of the issues discussed in bioethics touch fundamental existential questions. For instance, the ethical discussion of euthanasia in any of its possible forms is inevitably connected to the understanding of the role of medicine with regard to death and what might come after. Such questions naturally involve religious feelings and thought. Any effort to understand these situations and make them more transparent for those involved must take this dimension into consideration. If ethics tries to understand and clarify, it cannot stop when it approaches religion.

The role of 'moral limits', of the boundaries between what is acceptable practice and what is not, is different from what is often assumed by those who see the task of ethics in setting the right moral limits. Moral limits can only be a necessary external limitation of practice, and not the central object of ethical reflection. We will not have tackled visions of 'good' (or 'evil') if we only think about positions of moral limits. Moral limits decide at the most about the right or wrong of a planned action or rule, but they do not reveal the good in it.

(3) The limits of the self, or the constitutional weaknesses that afflict our moral epistemology and self-control, are relevant for assessing the opportunities and possibilities for communicative interaction in ethical discourses. A rational contributor to moral discourse might also be affected by a certain weakness of will, and this could change his or her attitude towards others whose actions seem to be guided to a greater extent by intuitions and feelings. A discussion that takes account of the weakness of foresight will be more modest, but perhaps it will also be more sensitive to actual knowledge and knowledge gaps. As humans who try to understand who we are, we are seekers of truth about ourselves. A common understanding among seekers who know that they are still seeking the truth about themselves is often much easier to establish than between 'knowers' who believe (or bluff) that they already know. A similar point can be made about dependency. We are all dependent, and the degree of our capabilities of independence is rather an accomplishment of dependency relations, not vice versa. If discourse remains sensitive to dependencies that play a constitutive role within it, it can perhaps be more perceptive.

In fact, ethicists who discuss any material issue in bioethics, mostly disagree: it is debates that characterize bioethical research, not consensus – or the latter just occasionally. There is disagreement about the status of the human embryo, about the sanctity of the therapy-enhancement distinction, about the meaning of ethical principles, even about the proper method of reasoning. In another field of expertise this situation would be rather devastating, and would signify a very poor performance of the expert disciplines. If there were to be a true understanding of a moral issue and a right set of methods to resolve it, the sign of reliability or robustness of the solutions would be consensus. In a Habermasian 'ideal speech situation' a consensus even counts as the ultimate measure of the right judgment. But if ethics is grounded in an awareness of the limits of judgment, it is probable that consensus cannot play this role. A divergence of opinions should not be seen as something that discredits ethical methodology. There is a productive role for disagreement: it makes different possibilities of seeing explicit and debatable. Dissent is not a sign of an absence of understanding, and dissenting views that make their reasons transparent can sometimes be of greater help to those concretely involved in a decision making project than an opaque consensus paper. Each view argues and explains its reasons and tries to show why other views are flawed, and it necessarily does that with the aim of convincing. In some cases people can really be convinced for their own good. However, practical problems do have different perspectives that cannot always be unified. Awareness of the epistemological limits of the self supports a mutual understanding between different views, because each party must assume that probably not all aspects of the situation can be seen. Different ethical approaches throw light on the structures and complexities of the situation from different angles, but each ray of light also casts shadows.

The criterion of universal validity (explained as 'the rules of morality are exceptionless') is claimed for several different reasons. The strongest is perhaps that oppression cannot be criticized as 'morally wrong' until those persons or groups who are oppressed by others can argue that oppression is a violation of a general moral rule: in that case, it is not only wrong because they (understandably) would *like* to be free from oppression. Despite the best arguments however, history shows that oppressors rarely give way to argumentative force. They must experience oppression themselves, or realize that those whom they marginalize are in fact people like themselves. This seeing means that they must 'come into the presence of those others' (Kellenberger 1995), and a relationship of trust must develop. Therefore, I would not try to bind the quality of moral insights to universal validity, but rather to the fairness and transparency of the processes where they are developed, and to the quality of the relationships in which the insights arise. There is a need for philosophical investigation into real communicative processes (in non-ideal speech situations) in the field of bioscience, and into the qualities of different types of possible communicative relationships.

(4) Lastly, let us briefly consider those limits we have grouped together as communicative limits. They are connected with the situatedness of the other with whom we have moral and communicative relationships. Here, the issue of relativism arises. David Wong has suggested that the critical claim of moral relativism is: "There is no single true morality" (Wong 1984; cf. Kellenberger 2001: 28). There is long-running discussion among philosophers about whether this claim is correct or not. Here, I cannot argue for either side, although my inclination is to think that the claim in this form

is probably correct, whether we believe it or not. But even if it proved to be false and there is in fact one single true morality, then all the inquiries regarding practical problems would still have to integrate perspectives into this true view of morality. However if relativism is correct, the problem of how to integrate different evaluations into a common understanding (which might be relative) also arises. Therefore, the relativism problem primarily needs not a theoretical but a *practical* solution based on dialogue. Research is needed into ethical communications that can mediate between different situations, 'local processes', contexts and perspectives. In order that perspectives can be better understood, one essential element is the support of those groups (in medicine, primarily patients) that are most often excluded by research from ethical discourse (Scully et al. 2003, cf. also Haimes this vol.). Empirical social research into ethical issues and normative reflection in ethics are naturally related to each other. Normative ethics is dependent on empirical information, firstly about the moral questions and concerns of those who have experience of the situations from inside; secondly about the details of the different opportunities to act or decide in the situations concerned and about the complications that may arise with each of them in real life histories; thirdly about the needs of the people who live in the situations and what they really wish; and fourthly about the ethical thought of those persons to whom ethicists want to explain their concerns and arguments.

4.   CONCLUSIONS

We have seen that limits have a positive role for bioethics. A reflection about the different limits in morality can shift the point of departure for ethical and bioethical methodology. Biomedicine can offer extremely instructive case examples for ethics in general where ethical debates touch existential issues and key cultural values. What does that change for moral vision? First of all, it seems to me that ethics can offer stronger guidance in practice if it does not claim a monopoly in investigating morality. There are other approaches to morality that can contribute to and are needed for a deeper understanding of practical issues. Social sciences, theology, psychology, history, natural sciences, medicine, art, literature and cultural studies – to name but a few – have their own ways of seeing and perceive other aspects of morality than moral philosophy. The more concrete and explicit an ethical investigation is – in bioethics we are confronted both with very concrete particular cases and with highly controversial background issues – the easier it is to establish fruitful collaboration between ethical approaches and other disciplines. Together they will produce a richer description and lead to a more informed reflection about the issues concerned. This is a plea for an integrative interdisciplinary approach to bioethics.

This of course needs further clarification. Each discipline has to develop an awareness of its own attitudes and of its norms of seeing. This is a prerequisite for cooperation. When an approach acknowledges what it can see, what it cannot see, and how it participates in the reconstruction of the things seen, it can collaborate with other approaches that see different things or the same things differently.

A large part of moral philosophy has been about norms, rules, rights, duties, values and moral principles. This frame setting imposes a particular visual regime on morality. Moral issues are seen through the lens of these concepts and morality appears to consist essentially of norms, rules, values and principles. If I contest the exclusivity of this paradigm, I do not deny that studies within it are helpful. But morality also includes singular, case-directed decisions after an intuitive exposure to concrete local processes of attributing meaning to situations, of story finding and telling, of correlating issues with narratives and interpreting them (Lindemann Nelson 1997). As Jonsen and Toulmin (1988) have brilliantly demonstrated, casuistry is an ethical method of basing decisions in new cases, via analogy, upon the moral evidence available in other singular paradigm cases. Responsibility can be an ethical concept that is open to the moral perception of the singular individual in the singularity of a situation because it is directed to caring for concrete others. The singular individual can use rules, norms, rights, duties, values and principles, the words of a generalizing moral language, in a non-deductive way as instruments to see relevant aspects of the singularity of the case. Moral evidence then comes from these aspects (which still belong to the situation), not from the truth of the rules.[6] This is a reflective paradigm for ethical methodology. As I have shown elsewhere, the basic distinction in a contextual approach to bioethics is between conditional and relational contexts (Rehmann-Sutter 1999). The conditional context describes a situation with a set of narrowing conditions that are essential descriptors of the case. For instance cheating is morally wrong in general, but might be acceptable and even praiseworthy *if* somebody can save the life of a judge who is prosecuted by the Mafia. This would be a conditional approach. A relational approach to contextuality would see the opportunity of saving the life of the judge as a direct responsibility that calls this person into an appropriate course of action. He would not live up to his responsibility to the endangered judge if he told the Mafiosi the truth. For him, the question is not under which rule the case could be subsumed, but on which side he prefers to stand. The conditional account of context is still oriented to the deductive paradigm of ethical reasoning: the conditions narrow the field of

applications of a principle or combine it with another rule that is valid
within this subclass of cases. A relational account of context perceives a
situation literally not as 'a case' (which is always seen as one of a class) but
as a *constellation*, which is established essentially by the dynamic texture of
relationships with its history and possible future. In bioethics, this opens
new ways of approaching difficult decisions, such as the renunciation of
multiple surgery on some newborns with multiple disorders, and the
decision for palliative care. Life perhaps could have been prolonged for
some weeks longer, but the life quality for a child suffering from the
immediate consequences of such surgery would be extremely negative. Such
hard decisions can only be responsible if they are singular, and ultimately
oriented to the baby concerned and not to any generaliseable rule (cf. Fischer
2002: 30 ff.). A bioethics that is inspired by a reflective paradigm of
reasoning and a direct vision of the particularities and relationships has
fewer difficulties with situations where non-generalization is the ethical
approach.

Ethics, as I see it, is a reflective and evaluative communicative practice
that takes place in and is interwoven with other practices. It collaborates
with different approaches and methods that can be integrated, even if
tensions remain. But morality – the topics and issues that ethics evaluates
and reflects on – is broader than the scope of ethical methods. The social,
cultural and political contexts of ethical issues that provide them with a
definite meaning, need to be investigated by other sciences as well, some of
them hermeneutical others empirical, and ethics relies upon their work.
Where ethics ends you do not necessarily find immorality, but with a range
of other approaches that aim to understand the complex phenomena of
morality. Ethics can help to facilitate communication across boarders: across
the divide between science and the humanities, and that between
subcultures and religious traditions. But it can contribute better by helping
to understand and interpret the concerns of different people than by
deducing rational solutions to the issues from moral principles. Concerns
can never be false. They do not call for a defence against all possible
reasonable counterarguments. Concerns can sometimes be groundless, but
even this is no reason not to take them seriously or communicate
respectfully with those who express them. An ethics that starts with the
concerns of the people who live in the situations under discussion does not
need to deal exclusively with those concerns (it can add its own), but it is
better equipped to deal with communication across boarders or cultural
divides.

This has one key consequence for the style of ethical thought that we
may find appropriate. Practising ethics means participating in a moral
communication; ethics is not a theory of morality located outside the field of

communicative practice. Ethics is always a communication, it participates in social practice as a special kind of action. Practising ethics is therefore different from practising science, because scientific and moral epistemologies differ fundamentally. In practical ethics, even in ethical theory, the "I", the subjectivity, the perception of the author(s) must be visible. In science, subjectivity must methodically be eliminated, and truth is impersonal, objective. In ethics, practical wisdom is necessarily subjective. It can become trans-subjective, sometimes even partially intersubjective. But in such a case of socially shared evidence, the moral truth remains subjective and rests on the authenticity of the subjects. If an ethical discourse wishes to be oriented toward impersonal, objective truth it will not only hide the subjectivity of the enterprise but it will also lose the primary reference: the "Thou" would be hidden together with the "I".

As Tristram Engelhardt noted, ethics is necessarily a peaceful and cooperative approach to moral differences. Ethics is "an alternative to the resolution of disputes through force" (Engelhardt 1986: 53). Beyond the boundary of ethics lies the danger of overpowering, even violence, and this boundary is far too often crossed.

Acknowledgment: This paper contributes to the methodology part of the research project "Time as a contextual element in ethical decision-making in the field of genetic diagnostics" (Swiss NSF grant 1114-064956). I am grateful to Jackie Leach Scully and Rouven Porz for inspiring discussions, to Katrin Bentele for clarifying questions and to Rowena Smith for English revision.

References

Ach, Johann S. and Runtenberg Christa. *Bioethik: Disziplin und Diskurs. Zur Selbstaufklärung angewandter Ethik.* Frankfurt a.M./New York: Campus, 2002.
Anobile, Richard J. (ed.). *Frankenstein.* London: Picador, 1974.
Aristoteles. *Nikomachische Ethik VI.* Herausgegeben und übersetzt von Hans-Georg Gadamer. Frankfurt a.M.: Klostermann, 1998.
Attfield, Robin and Belsey, Andrew (eds.). Philosophy and the Natural Environment. [Royal Institute of Philosophy Supplement: 36] Cambridge: Cambridge Univ. Press, 1994.
Benhabib Seyla. "The Generalized and the Concrete Other. The Kohlberg-Gilligan Controversy and Moral Theory." In *Women and Moral Theory.* Eva Feder Kittay, Diana T. Meyers eds. Savage MD: Rowman & Littlefield, 1987, 154–177.
Böhme, Hartmut and Böhme, Gernot. *Das Andere der Vernunft. Zur Entwicklung von Rationalitätsstrukturen am Beispiel Kants.* Frankfurt a.M.: Suhrkamp, 1983.
Broadie, Sarah. *Ethics with Aristotle.* New York/Oxford: Oxford Univ. Press, 1991.
Butler, Judith. *Kritik der ethischen Gewalt. Adorno-Vorlesungen 2002.* Frankfurt a.M.: Suhrkamp, 2003.
Caputo, John. "The End of Ethics." In *The Blackwell Guide to Ethical Theory.* Oxford: Blackwell, 2000, 111–128.

van Dommelen Ad. (ed.). *Coping with Deliberate Release. The Limits of Risk Assessment.* Tilburg/Buenos Aires: International Centre for Human and Public Affairs, 1996.

Engelhardt H. Tristram. *The Foundations of Bioethics.* New York/Oxford: Oxford Univ. Press, 1986.

Engelhardt H. Tristram. "Konsens: Auf wieviel können wir hoffen?" In *Moralischer Konsens. Technische Eingriffe in die menschliche Fortpflanzung als Modellfall.* Frankfurt a.M.: Suhrkamp, 1996, 30–59.

Fischer, Johannes. *Medizin- und bioethische Perspektiven. Beiträge zur Urteilsbildung im Bereich von Medizin und Biologie.* Zürich: Theologischer Verlag, 2002.

Hare, Richard M. *Moral Thinking. Its Levels, Methods and Point.* Oxford: Clarendon, 1981.

Hübner, Karin et al. "Derivation of Oocytes from Mouse Embryonic Stem Cells." *Science* 2003, 300, 1251–1256.

Jonsen, Albert R. and Toulmin, Stephen. *The Abuse of Casuistry. A History of Moral Reasoning.* Berkeley: Univ. of California Press, 1988.

Jonsen, Albert R. *The Birth of Bioethics.* New York/Oxford: 1998.

Kant, Immanuel. *Grundlegung zur Metaphysik der Sitten.* Akademie-Ausgabe, Bd. 4, 387–463.

Kellenberger, James. *Relationship Morality.* University Park PA: Penn State Press, 1995.

Kellenberger, James. *Moral Relativism, Moral Diversity and Human Relationships.* University Park, PA: Penn State Press, 2001.

Kittay, Eva Feder. *Love'Labor. Essays on Women, Equality, and Dependency.* New York/London: Routledge, 1999.

Kleinman, Arthur. "Moral Experience and Ethical Reflection." *Daedalus Fall* 1999, 69–97.

MacIntyre, Alasdair. *Dependent Rational Animals. Why Human Beings Need the Virtues.* London: Duckworth, 1999.

Merton, Robert King. "Social Structures and Anomie." In *Social Theory and Social Structure.* Robert King Merton (ed.). New York: The Free Press, 1968, 185–248.

Meyers, Diana T. *Self, Society and Personal Choice.* New York: Columbia Univ. Pr., 1989.

Moore, George Edward. *Principia Ethica.* (1903) Übers. B. Wisser, Stuttgart: Reclam, 1970.

Nagel, Thomas. *The View from Nowhere.* New York/Oxford: Oxford Univ. Pr., 1986.

NEK-CNE, Nationale Ethikkommission im Bereich Humanmedizin. Zur Forschung an embryonalen Stammzellen. Stellungnahme 3/2002. Bern, Juni 2002 (see www.nek-cne.ch).

Lindemann Nelson, Hilde (ed.). *Stories and their Limits. Narrative Approaches to Bioethics.* New York/London: Routledge, 1997.

Rehmann-Sutter, Christoph. "Frankensteinian Knowledge?" *The Monist* 1996, 79, 264–279.

Rehmann-Sutter, Christoph. "Contextual Bioethics." *Perspektiven der Philosophie* 1999, 25, 315–338.

Rehmann-Sutter, Christoph. "Bioethik." In *Handbuch Ethik.* Marcus Düwell et al. (eds.). Stuttgart: Metzler, 2002, 247–252.

Ricoeur, Paul. *Soi-même comme un autre.* Paris: Sevuil, 1990.

Scarano, Nico. "Metaethik – ein systematischer Überblick". In *Handbuch Ethik.* Marcus Düwell et al. (eds.). Stuttgart: Metzler, 2002, 25–35.

Scully, Jackie Leach; Rippberger, Christine and Rehmann-Sutter, Christoph. "Additional Ethical Issues in Genetic Medicine Perceived by the Potential Patients." In *Populations and Genetics. Legal and Socio-Ethical Perspectives.* Leiden/Boston: Nijhoff, 2003, 623–638.

Scully, Jackie Leach; Rippberger, Christine and Rehmann-Sutter, Christoph. "Non-professionals' evaluations of gene therapy ethics." *Social Science and Medicine* 2004, 58, 1415–1425.

Taylor, Charles. *Sources of the Self. The Making of Modern Identity.* Cambridge, Mass.: Harvard Univ. Press, 1986.

Thomson, James A. et al. "Embryonic stem cell lines derived from human blastocysts." *Science* 282, 1998, 1145–1147.

Urmson, J.O. *Aristotle's Ethics.* Oxford: Blackwell, 1988.

Williams, Bernhard. *Ethics and the Limits of Philosophy.* Cambridge, Mass.: Harvard Univ. Press, 1985.

Wong, David. *Moral Relativity.* Berkeley: Univ. of California Pr., 1984.

---

[1] London: Vintage 1999, p. 165 (orig. 1973).

[2] I am much indebted to lively and fruitful conversations with Engelhardt during an ESF conference organized by John Brooke at Harris Manchester College in Oxford in autumn 2003.

[3] In November 2004, Switzerland had a public referendum on the law on stem cell research that was previously accepted by both houses in the parliament. The law has finally been adopted by 2/3 of the votes.

[4] See the informative website of the Transhumanist movement: www.transhumansim.org

[5] See the critical discussion of Hare's position by Thomas Nagel (1986: 162 f.), under the headline of "overobjectification".

[6] On this point I disagree with Caputo (2000) who sees moral reflection by individuals about the singularity of a situation as an impossibility for ethics, beyond its end. But it depends on the selection of ethical methods, or calls for the development of other ethical approaches.

Chapter 5

THE PROBLEM OF LIMITS OF LAW IN BIOETHICAL ISSUES

SILVANA CASTIGNONE
Genoa, Italy

1.   INTRODUCTION

Scientific and technical progress has led to changes, unimaginable up to a few years ago, in our way of being and in our confrontation with reality, including that of our own body; (changes which, perhaps, we still have difficulty in understanding completely). Genetic manipulation has permitted us to modify the human biological structure and, in part, psychological structure too. The sexual frontiers are becoming less defined; the phenomenon of trans-sexuality is just one example. Assisted reproduction has multiplied the parental figure and if, at one time, the Latin maxim "mater semper certa, pater incertus" had significance, paradoxically the opposite is now almost becoming true, with the donation of the oocyte and use of surrogate mothers.

The frontiers between life and death, too, have become more uncertain and confused, and definitions of "death" have multiplied. These and other situations are causing a crisis not only in traditional morality but also in the legal field where there is increasing difficulty in finding concepts and legal categories to apply to the constantly expanding new realities.

The problem of legal limits in biomedicine and biotechnology can be viewed from at least two angles. The first one involves asking ourselves if these activities can be disciplined by law, and what limits can be imposed. The second questions the internal limits of the legal system, the inadequacy of its categories and its concepts, and the necessity of formulating new ones.

C. Rehmann-Sutter et al. (eds.), Bioethics in Cultural Contexts, 81–90.
© 2006 *Springer. Printed in the Netherlands.*

2.   MORAL AND LEGAL ATTITUDES IN BIOETHICS

On examining the first question, we can see immediately that the relationship between the law on the one hand and biomedicine and biotechnology on the other feels the full effect of the different attitudes and moral principles that all these new scientific discoveries and activities provoke. Generally speaking we can distinguish two main attitudes: refusal and acceptance. Both attitudes include and produce convictions and precepts that are moral and subsequently also juridical, and that vary to a great extent from each other.

2.1   Attitudes of refusal

Total or partial refusal offers at least two different positions as motives. On one side, there is refusal due to fear of the transformation in progress and of the possible consequences.

We could call this a consequentialist form of refusal: transgenic products are a danger to health, genetic manipulation will lead to an Orwellian society, the destruction of nature is inevitable, and so on. In other words, there is a predominant fear of what the modifications introduced by science might lead to.

This is the attitude of the various Green and environmentalist movements, concerned not only with the protection of the ecosystem but also for health in general and the social injustice that could result from genetic engineering. In the case of assisted procreation there is concern about the difficulties and psychological problems of so-called test tube babies with parental figures that are distinct from the natural parents. In the case of euthanasia, we are warned to guard against the dangers of establishing a practice that could be used to free ourselves of sick people or unwelcome relatives. In short, what prevails is fear of the consequences and of an immense technocratic power lying in wait, capable of expropriating human freedom.

Then there is the second type of refusal, and I would say that it is the form with a major influence not only on public opinion but also on the cultural and philosophical world. This form justifies its opposition to much biotechnology by considerations of the ontological nature of humankind. We could call this a deontological attitude of refusal. Human activities are not carried out in an ontological void: valid ethical standards that derive from the natural order exist and are easily recognisable, and these standards must be respected. This position is tied, more or less directly, to options of a metaphysical and religious type, within a 'jusnaturalistic' horizon and strictly dependent on ethical knowledge. The most prominent representation is constituted by the doctrine of the Roman Catholic Church, at least in Italy. Here the concept of the sacredness of life leads to the condemnation of abortion and euthanasia. Only a minority of genetic interventions are accepted and only for therapeutic reasons. Medically assisted reproduction is considered admissible only where it takes place between couples united in marriage.

2.2 Attitudes of acceptance

Moving to the attitude of total or partial acceptance, we can see how this too can have various motivations: (a) the conviction that the freedom of science and research should be safeguarded; (b) trust in science and scientists; (c) separation of research and application of the results: limits can be set solely on the second stage and not on the first; (d) without scientific progress humanity would still be living in the Stone Age, and therefore it is worthwhile running a certain degree of risk.

This attitude of acceptance implies, on almost every occasion, abandoning the positions of essentialism and ethical cognitivism, that is rejecting the possibility of attaining a rational knowledge of ethical standards. There is no "nature" to which one can make reference, one cannot impose certain behaviour in the name of non-existent "natural laws". Ultimate importance is given to the autonomy and free choices of the single individual, with respect for the decisions and ethical standards of others, and the sole limit is represented by Mill's principle whereby damage to third parties must be avoided. Autonomy, responsibility and tolerance are the moral standards at the centre of this prospective.

2.3 Consequences on legal plane

Now taking a step forward from the moral to the legal plane, in what way do the different attitudes outlined above influence the creation of laws tied to bio-research?

The positions of refusal, whether of a consequentialist or deontological type, usually converge in a request for a "firm" normative intervention to fix limits and prohibitions, whether through fear of possible disasters or out of respect for natural laws, the ontological structure of the phenomena.

The positions of acceptance can instead lead in two directions: either to oppose any form of regulation whatever, putting hope in the so-called "legislative void" in certain matters, in the name of scientific liberty and of conscience; or to accept a "soft" law, a moderate regulation that tends to permit the interested parties to make use of the new technologies with the maximum autonomy compatible with the prevention of abuse and harm to third parties. According to this formulation, which appears to me the most suitable for a pluralistic and multi-ethical society as is ours, the law should have the following characteristics: (a) be a sort of "limited law" that intervenes as little as possible; (b) be open, which means respectful of the plurality of ethical standards; (c) be flexible, that is the least rigid possible, because bio-medical and bio-technical science is in constant development and space must be left for the progressive adaptation of legal regulations to such development.

The latter is not easy, as it implies continual modification and constant work of mediation between the different opinions and exigencies. At this point we are faced with the question of what normative intervention is preferable, in terms of not the substance but the method, the system to be employed. Many have affirmed that there are already adequate legal instruments to resolve bioethical questions without having to resort to the creation of a law "ad hoc". There are forms of auto-regulation that are expressed through deontological codes, and there are also the views of the various groups of ethicists, and international declarations. Above all, there is the constant work of the body of jurisprudence itself that applies the legal dispositions and principles already in existence within the national and international systems, mainly using analogical argumentation to adapt them to new demands.

In effect all these normative or para-normative activities have been and are carried out daily by judges who up to now, at least in Italy, have been entrusted with the task of producing legislation in specific cases and filling the normative voids.

However it is fairly easy to point out the limits of these interventions. The opinions of ethical groups, the deontological codes and so on have little efficacy, in that the former depend on the willingness of people to follow them voluntarily and the latter, although furnished with disciplinary sanctions, mainly regard the duty of doctors rather than the needs and rights of the sick, apart from being still imbued with principles that derive from the Hippocratic tradition and that are no longer universally shared. Another discourse is that of jurisprudence which has assumed a front-line role. However, two points should be noted: first, there are areas of bioethics in which we are unable to find laws and principles that are specific enough to be applied in the appropriate manner, and so judges are obliged to resort to very generalized principles that can be interpreted in various ways, ending up by becoming legislators to an extent going far beyond that which usually occurs during interpretative operations. It follows, and this is the second point, that cases which are similar may be decided in very different ways dependent on the ideology of the judge in question. Naturally, this can be the outcome of any interpretative activity, due to the inevitable power of discretion of the interpreter who, within set limits and depending on the case in question, can choose the meaning to attribute to normative statements. But for bioethics, there are too many solutions in conflict with each other as to render some form of legislation indispensable. As already suggested, this should be the simplest and least invasive possible in order to respect the plurality of opinions and the liberty of choice of the interested parties.

3.   NECESSITY OF REDEFINING SOME JURIDICAL CONCEPTS

At the beginning I said that the "limits of law laid down in respect of bioethics" can also be taken as referring to internal limits of the law itself and its concepts, born in scientific and social contexts very different from those of the present day. First of all, however, we must remember that the juridical categories and qualifications such as "person", "subject", "state of death" and now even "parent", "father", "mother", "right of corporal ownership" and so on, are never the simple transposition into juridical language of biological data: science will never be able to say with certainty when the embryo becomes a "person" or when a human being can be considered truly defunct. Or, to put it in a better way, scientific and legal definitions may not coincide because the latter are the result of a choice that morality and law exert, passing over and above the scientific data and interpreting it in the light of set principles prevalent in a certain historical context.

3.1   Death

Let us take, for example, the concept of death. According to Engelhardt (1991: 233ff.) science proposes three different definitions of death. The first, which is the oldest and is tied to easily verifiable empirical events, defines death as the irreversible cessation of the circulatory and respiratory functions. The second, following the introduction of various techniques of reanimation and the prolongation of the heart beat and respiration by auxiliary external methods, defines it as the irreversible cessation of cerebral activity in its entirety. Lastly, in the third case, what is important is the cessation of activity in the upper brain centres, even though the patient continues breathing and the heart functions without mechanical aid. The adoption of one of these definitions is certainly dependent on the level of medical and scientific knowledge, but not only this. The religious ideals prevalent in a given society play a major role, through convictions as to what "to have life" signifies: whether it simply signifies to be alive in a biological sense of the word, or whether is also implies a certain level of awareness and autonomy. In addition, the choice of one definition or another has consequences in the case of transplants. To this end we tend to privilege the third definition (death of the cortex) as this permits the removal of organs that are still functioning, at the same time creating a series of problems regarding the ascertainment of death and the psychological reactions of the various persons involved.

3.2   Person

Another very ticklish point is that of determining the beginning of human life, or rather the beginning of the human person. It is a well known fact that there is absolutely no agreement on this question. For some, personhood begins at the moment of conception, for others at the fourteenth day, for others still when the foetus begins to manifest certain reactions, or at the moment of birth or even later. Here, too, there is a very strong entwinement between biological life and conscious and/or subconscious life, between diverse religious points of view and ideological convictions. All of which lead to somewhat different answers to the legality or not of abortion. Again, as Engelhardt writes: "The problem is where to draw the line. Possibly it is not clear what level of consciousness is required for there to be life in a person," and this is valid both for the beginning and the end (1991: 242). Science alone cannot give us a precise answer, from a biological point of view there is more than one plausible solution: but the concept of the human person goes beyond this sphere.

3.3   Parenthood

In the field of artificial conception, too, we are witnesses to a continual confrontation between conflicting cultural models. The concepts of maternity and paternity have broken down into a constellation of diverse figures: a biological, or natural, mother, a surrogate mother, a mother donor of the oocyte, a biological father, a father donor of the sperm, a social father, couples united in marriage, couples co-habiting, homosexual couples and so forth. The traditional moral and legal categories no longer exist and judgement as to who is the real mother or father are at times in deep conflict according to whether weight is put upon one characteristic or another, on one phase or another in the process of assisted reproduction. In 1994, for example, an Italian judgement granted a request for annulment of paternity presented by a husband in respect of the son borne of his legitimate wife by means of heterological assisted reproduction (insemination by donor), notwithstanding the fact that he had given his consent to it, using the argument that heterological fecundation is equivalent to adultery: a fact that would render a husband's consent devoid of legal effect in that the right to conjugal fidelity is a binding law. Numerous subsequent sentences have, however, rejected requests for annulment in the presence of consent.

3.4 'One's own body"

A final point that I would like to take into consideration is that of the relationship between the individual and the body. The traditional conception considers them as one; the body is considered indistinguishable and inseparable from the person, from the self, so much so that the inviolability of the body is asserted (with the exception of a few renewable parts such as hair, milk in the case of wet-nursing, and blood).

Kant has written in a categorical and total manner that man cannot utilize himself and that he is not allowed to sell a tooth or any other part of himself. Up to a few decades ago, there was no motive whatsoever to deprive oneself of a kidney, the cornea, or any other part of our body for the simple reason that no-one would have known what to do with it. It would have merely signified an auto-mutilation, to be explained, perhaps, from a psychiatric point of view. The law, forbidding transplants, therefore, followed the natural structure and situation. But with new techniques in transplantation and with the progressive increase in the number of body parts that can be utilized, natural limits no longer exist (transplantation from a corpse obviously creates problems of a very different kind.)

Legal and moral codes can follow all the new paths opened up by science, but they can also diverge, setting limits and imposing stops. Regarding the body and the laws for its administration, we are witnessing an attenuation in the distinction between persons and objects, which has always been one of the fundamental legal distinctions. There is no doubt that although the organs of our body constitute an integral part of our person, of our self, from the moment in which they are removed, either as a donation or for payment, they are being treated as objects.

Must the relationship between a person and his or her body, therefore, be conceived in the same way as a property right, analogous to that which we can have over an external possession? Many considerations of human dignity, equality and health are opposed to such a solution. In some cases an attempt has been made to follow the path that considers the organs given up as abandoned, in other words as *res nullius* or, better, as *res derelictae*: but this, too, does not appear an easy route to follow. In fact, there arise numerous questions of the type: "who can take possession of them, and to what ends?"

It is evident, to repeat a phrase used by Stefano Rodotà (3), that the body now appears as "a new legal possession" (1995: 204). It is, therefore, necessary to invent new legal classifications, surpassing those which I have called "the internal limits of law" in respect of bioethics. This kind of problem will increasingly emerge from the field of genetic engineering, where the conflict between the two different attitudes (of refusal on one side and of acceptance on the other) is becoming very strong. Conservatives ask for rigid limitations, fearing dangerous modifications of human nature, while liberals appeal not only to the freedom of science but also to the improvements that could be introduced for the benefit of humankind. New concepts appear and are used, for instance the idea of a right to genetic welfare, held by liberals (Castignone 1987), and on the opposite side the right to an unmodified genetic patrimony (unless for strictly therapeutic reasons), or the idea of the "inviolability of biological foundations of personal identity" (Habermas 2002: 29).

4.   CONCLUSION

What I would like to underline is precisely the following: the challenge that bioethics is now making to the legal system consists not only in the opportunity, or sometimes the necessity, of creating laws to face the emerging problems that science presents to us; to tell us, officially, how we must or must not behave. It also consists in creating concepts and categories that did not previously exist being fully aware that, in many cases, those inherited from the past are no longer tenable and that it is useless to cling on to them, forcing them this way and that in an attempt to match realities that are completely different. And it is also necessary for us to realize that such concepts and categories are not '*wertfrei*' or neutral in terms of human values but in a more or less hidden and unconscious manner, they imply valuations and choices.

In view of the conflicts and the variety of religious, ethical and ideological points of view which characterise our present day society, and the resulting lack of generally shared ethical principles, it appears evident that there is also a necessity that in constructing new categories and legal concepts, they should present the characteristics of elasticity and lightness, in order to permit their use without endangering the autonomy and liberty of the single individual.

References

Castignone, S. "Per un welfare genetico." In Biblioteca della Liberta, 99, 33–39, 1987.

Engelhardt, H. T. *Manuale di bioetica*. Milano: Mondadori, 1991 (*The Foundations of Bioethics*. Oxford: Oxford Univ. Press, 1986).

Habermas, J. *Il futuro della natura umana. I rischi di una genetica liberale*. Torino: Einaudi, 2002 (*Die Zukunft der menschlichen Natur. Auf dem Weg zu einen liberalen Eugenik?* Frankfurt am Main: Suhrkamp, 2001).

Rodotà, S. *Tecnologie e diritti*. Bologna: Il Mulino, 1995.

Further literature

Alpa, G. "I limiti dell'intervento giuridico." In *Questioni di Bioetica*. S. Rodotà (ed.). Roma-Bari: Laterza, 1997, 57–62.

Atienza, M. "Giustificare la bioetica. Una proposta metodologica." *Ragion pratica* 1996, IV, 6, 123–143.

Baud, J. P. *L'affaire de la main volèe. Une histoire juridique du corp*. Paris: Ed. du Seuil, 1993 (trad. it. *Il caso della mano rubata: una storia giuridica del corpo*. Milano: Giuffrè, 2003).

Beauchamp, T. L. and Childress, J. F. *Principles of Biomedical Ethics*. Oxford: Oxford Univ. Pr., 1989 (3rd edition).

Bompiani, A. *La bioetica in Italia. Lineamenti e tendenze*. Bologna: Dehoniane, 1992.

Borsellino, P. *Bioetica tra autonomia e diritto*. Firenze: Zadig, 1999.

Braibant, G. "Pour une grande loi." *Pouvoirs* 1991, 56, 109–119.

Brunetta d'Usseaux, F. *Esistere per il diritto. La tutela giuridica del neonato*. Milano: Giuffrè, 2000.

Ciliberti, R. *Medicina, etica e diritto nella rivoluzione tecnologica*. Torino: Edizioni Medico-Scientifiche, 2001.

Cattorini, P. *Etica e giustizia in sanità: questioni filosofiche, principi operativi, assetti organizzativi*. Milano: Angeli, 1998.

D'Agostino, F. *Bioetica*. Torino: Giappichelli, 1996.

Della Torre, G. *Bioetica e diritto*. Torino: Giappichelli, 1993.

Ferrando, G. (ed.). *La procreazione artificiale tra etica e diritto*. Padova: Cedam, 1989.

Flis-Treves, M.; Mehl, D. and Pisier, E. "Contre l'acharnement législatif." *Pouvoirs* 1991, 56, 121–134.

Flamini, C.; Massarenti, A.; Mori, M. and Petroni, A. "Manifesto di bioetica laica." *Il Sole* 24 ore, 9 giugno 1996.

Harris, J. "Regolamentazione bioetica e legge." In *Questioni di bioetica*. S. Rodotà (ed.). Roma-Bari: Laterza, 1997, 331–341.

Lecaldano, E. "La bioetica e i limiti del diritto." *Democrazia e diritto*, luglio-ottobre 1988.

Lecaldano, E. *Bioetica. Le scelte morali*. Roma-Bari: Laterza, 1999.

Marzano, M. M. *Penser le corp*. Paris: Puf, 2002.

Mori, M. *La bioetica: questioni morali e politiche per il futuro dell'uomo*. Milano: Bibliotechne, 1990.

Mazzoni, M. C. *Una norma giuridica per la bioetica*. Bologna: Il Mulino, 1989.

Nielsen, L. "Dalla bioetica alla biolegislazione." In *Una norma giuridica per la bioetica*, C. M. Mazzoni (ed.). Bologna: Il Mulino, 1998, 45–62.

Perlingeri, P. *La personalità umana nell'ordinamento giuridico*. Napoli: ESI, 1982.

Rodota, S. "Per un nuovo statuto del corpo umano." *Bioetica*. A. di Meo, C. Mancina (eds.). Bari: Laterza, 1989, 41–68.

Scarpelli, U. *Bioetica laica*. Milano: Baldini and Castoldi, 1998.

Vasquez R. (ed.). *Bioética y derecho*. ITAM, 1999.

Violante, L. "Bio-Jus. I problemi di una normativa giuridica nel campo della biologia umana." *Bioetica*. A. Di Meo, C. Mancina (eds.). Bari: Laterza, 1989, 259–270.

Zanetti, G. *Elementi di etica pratica*. Roma: Carocci, 2003.

Zatti, P. "Verso un diritto per la bioetica." In *Una norma giuridica per la bioetica*, C. M. Mazzoni (ed.). Bologna: Il Mulino, 1998, 63–76.

II. CLASSICAL APPROACHES

Chapter 6

ONE MORAL PRINCIPLE OR MANY?

MARCUS DÜWELL
Utrecht, The Netherlands

1. INTRODUCTION

From the very beginning of the bioethical debate, this new (sub-) discipline emphasized its independence from the big normative theories, like utilitarianism and Kantianism. Of course, there have been very influential bioethicists, like Engelhardt or Singer, who have explicated their normative ethical theories systematically and have shown which moral principle formed the basis for their moral judgements. But the mainstream in bioethics wanted to evade an explicit normative framework (Jonsen 1998: 325–351). The most popular approaches in bioethics tried to avoid the impression that their normative judgments are dependent on only *one* normative ethical theory. Instead, approaches became popular which could hope to deal with moral problems without needing a philosophical foundation for their normative basic assumptions. In that context we could mention a casuistic approach (Jonsen and Toulmin 1988), a common morality approach (Gert 1998 and 2004; Gert, Culver and Clouser 1997) or the very popular four principles approach of Beauchamp and Childress. These approaches hope to find the normative basis for moral evaluations in well established practices or in widely shared moral standards. Respect for autonomy, informed consent or the duty to avoid harm, seem to be moral principles that are morally acceptable by everybody, independent of other convictions concerning morality, religion, politics or metaphysics. The normative force of such principles seems to be easily defensible and no great effort to provide a foundation for those principles seems to be necessary. The meta-ethical presuppositions behind such an approach often form a

C. Rehmann-Sutter et al. (eds.), Bioethics in Cultural Contexts, 93–108.

coherentist epistemology (Nida-Rümelin 1996), sometimes with explicit reference to Rawls' reflective equilibrium (Beauchamp and Childress 2001: 397–401; Burg and Willigenburg 1998; Daniels 1996), sometimes on the basis of (a moderate version) an intuitionist epistemology (Audi 2004). But very often these presuppositions are not clear, e.g. one can only assume that casuistry could be interpreted against the background of a particularistic meta-ethical position (Dancy 2004). In any case most of these bioethical approaches will avoid a clearly deontological or utilitarian theory. Particularly strong allergic reactions can be seen when bioethicists talk about one principle approaches, like the greatest happiness principle or the categorical imperative. The idea that applied ethics could be a 'deduction' from *one* basic principle to concrete moral judgments, a so-called top-down approach, is often criticized in this debate (Beauchamp and Childress 2001: 385–391).

Although the use of moral principles in bioethics is controversial, defenders and critics share the assumption that only mid-level principles like 'not harming' or 'respect for autonomy' can be serious candidates as bioethical principles. The defenders assume that those principles are valid justificatory reasons *because* they are elements of a common morality. They assume that the bioethicist has to explicate these widely shared moral principles of the common morality and to transfer them to challenging moral problems of life and death. In doing so the bioethicist has the task of making moral consensus possible. The role of normative ethical theories in the work of bioethics is very limited, in fact even problematic, because there is no generally accepted normative ethical theory (Timmons 1999). An exclusive relation between justificatory moral principles and one normative theory even seems to endanger the possibility of a moral consensus. Therefore mid-level principles are introduced that are evident in the context of different normative ethical theory. This understanding of 'principle' is criticized as confusing and unclear (Clouser and Gert 1990) but it has become dominant in the bioethical debate. Even the critics of the use of 'principles' in bioethics presuppose this notion. For example, in the plea for context-sensitive approaches, for virtue- and care-ethics, for a particularistic alternative as an 'Ethics Without Principles' (Dancy 2004), the same notion of moral principle is used. However, the presupposed concept of moral principle does not seem to me to be self-evident at all. There is good reason to discuss the presuppositions of the use of moral principles in more detail. We need to discuss the *nature and content* of moral principles, the *use of moral principles* for the justification of moral judgments, and the question of whether the principles themselves are in *need of a justification or foundation*. This whole discussion is of course closely interrelated with the *task of bioethics* as a discipline, which I will discuss at the end. But I want to start with a short outline of the place of moral principles within the broader context of moral philosophy.

2.   MODERN MORAL PHILOSOPHY

The introduction of moral principles is closely interrelated with the idea that this is what ethics is talking about. Before we discuss whether a top-down, a bottom-up or a coherentist approach is appropriate, we should also discuss the overall goal of ethical research. What is the question that ethics wants to answer? In general two kinds of questions for ethical investigation can be distinguished: questions concerning the *perfection of our lives* and questions concerning *our obligations towards each other*. In the first perspective we are asking the perfectionist question of what kind of people we want to be (Hurka 1993). In the second we are asking whether and to what extent we are obliged to take the interests of others into account. While the perfectionist question was central to premodern ethical thinking (Annas 1993), in modern ethics the second question is of prior importance. Although the relation between those two questions can be disputed and can be further differentiated (van den Brink 2000), the distinction is of fundamental importance for the understanding of the whole project of ethical investigation (Habermas 1991; Krämer 1992; Steigleder 1999).

Perfectionist ethics often uses narratives or examples to illustrate the ideas of a good and successful life. Perfectionist ethics teaches us to think about desires and needs. If such ethics deals with rules, these rules are the result of common experience of life and luck. In the perfectionist perspective rules are the condensation of an everyday knowledge about the ways in which, in general, people should deal with the preconditions of human existence in order to reach the goal of a successful life. Those rules and principles are only rules of thumb with a very limited scope of application and dependent on specific cultural circumstances. This particularist and contextualist approach is internally related to the goal of perfectionist ethics: to help us find ways to a successful and fruitful existence.

That moral principles are introduced in modern ethics is not an accidental event in the history of ethics. Modern ethics asks whether we have moral obligations. Moral duties demand to guide all our actions, but in the first place the relation between our interests and those of others are of fundamental importance. While in perfectionism ethical advice is more concerned with the means of reaching a good life, modern ethics introduces an independent 'moral point of view', a sphere of moral evaluation sui generis. From the moral point of view agents are confronted with demands to limit their scope of action to show respect for others, even if these limitations may result in restrictions affecting their own happiness. At this point we do not have to decide whether or not the reference to a 'moral point of view' implies some specific normative content (such as impartiality or the like). Even if there exist very different ideas about the content of these moral demands, all of them conceptualize morality as an evaluative standard of its own. And I want to add that this concept of a moral point of view can be interpreted within very different philosophical frameworks. Kantians see moral obligations as a result of our rational capacity, contractualists see them as guidance to ensure the necessary precondition for a secure life for all of us, and Humean philosophers see them as a cultivation of our moral sentiment.

In comparison to perfectionism the reference to a genuine moral point of view introduces a different role for *moral principles* and *moral justification*. While in a perfectionist perspective it is sufficient that a rule or a narrative has some evidence for the agent that it really shows him a way to reach his life goals, the normative demands of morality confront us with an obligation that is in need of a moral criterion and a justification of such a criterion. If A asks B not to do X because X is morally wrong, B can ask for a justificatory reason for that demand. Those moral demands are generally seen as *overriding*. In case of competition between eudaimonistic interests and religious convictions with moral demands, the moral point of view is of prior importance. Moral demands are categorically binding. The moral significance of human rights, for example, cannot be understood without such an understanding of moral demands. Respecting human rights means taking into account the interests of people even if they have no special relationship to me. And respecting human rights is always of prior importance, even if it has negative impact on my economic interests or is in conflict with religious convictions. By their nature moral demands are necessarily connected to the need for a moral criterion, a criterion that enables us to distinguish morally right from morally wrong actions. And with the introduction of each possible criterion we are immediately confronted with a need for a justification of this principle. So the whole discussion of moral principles and their justification is internally connected with the whole idea of moral obligations and the nature of morality.

Before we have a closer look at those moral principles we have to face some fundamental criticism of the idea of a moral obligation. Very different perspectives ask whether the whole focus of modern moral philosophy on duties, rights and universally binding moral principles is misguided. Elisabeth Anscombe has criticized the entire modern concept of morality (across the very different approaches of Kant, Mill and Hume: Anscombe 1958; O'Hear 2004). The whole idea of a moral obligation and of a moral law, according to Anscombe, must presuppose a lawmaker in order to be meaningful. Without a transcendent instance, the notions of moral law and moral obligation lose content and meaning. In the same vein some communitarians criticize the universalist idea of morality, that results in the concepts of human dignity and universal human rights, for being derived from a hypertrophic concept of rationality with the side effect of destroying the moral traditions of particular communities (MacIntyre 1981).

Not all attempts to rehabilitate virtue ethics, nor all critics of universalist moral approaches, are so fundamentally critical of the modern concept of morality. Some authors propose to combine universalistic *normative ethics* with the much broader approach of an *ethics of a good life* (Habermas 1992; Krämer 1991). Kant had already tried to reconcile the strict idea of moral obligation with the idea of happiness and the highest good. Martha Nussbaum tries to combine a perfectionist ethics, with its notion of universal human rights, with the concept that human rights protect and support those capabilities that are necessary elements of human flourishing for all of us (Nussbaum 1990 and 2000). (Neo-)Hegelians criticize the strong opposition between the rational demands of a (Kantian) morality and the historically realized forms of *Sittlichkeit*. We can discuss the implication of that notion and whether it presupposes some kind of moral sphere *sui generis*.

Some theories of an 'ethics of care' are more radical and propose replacing the whole universalist language of rights and duties, because this language is the result of an atomistic and individualistic concept of society, insofar as it derives from one specific historical concept that cannot provide a sufficient basis for a universalist idea of morality (Gatens 1998 and Held 1993). Some authors see the notion of 'care' as more fundamental than justice and obligation, because the idea of rights and obligations already presupposes an attitude of caring. Others are really arguing for a substitution of traditional terminology in normative ethics. But in general the 'ethics of care' also accepts that moral standards form an evaluative standard of their own.

The reason that ethics introduces the whole idea of moral principles and discusses the justification of those principles is internally related to the idea of a genuine moral standard, morality as a demand with overriding character that places some limitations on our actions. The idea that the moral point of view provides its own evaluative perspective, not reducible to self-interest and the perspective of one's own happiness, has to face the need to formulate a moral criterion for differentiating between morally good or bad actions. And the introduction of such a criterion necessarily confronts us with the need to justify it. Therefore the critics of principle-based approaches have in the first place to answer the question of whether or not they accept the fundamental presupposition of morality as a genuine evaluative standard. It is irrelevant whether we think that virtues are important as well, or that moral demands are not the only important aspects in life. If we deny the whole idea of moral demands, then we do not have to talk about the need for moral principles, about top down or bottom up. We first have to discuss our concept of morality.

But if we accept the idea that there are moral obligations that are categorically binding, then the critics of moral principles have to show how we can recognize these obligations in the absence of the idea of moral principles and their justification. A moral theory that accepts the idea of morality but denies the need for moral principles is a particularist theory (Dancy 2004). Moral obligations from Dancy's point of view are necessarily related to the specific contexts of action. We have to identify our moral obligations in concrete circumstances and abstract principles cannot help us to do that. The value of Dancy's book is of course that he has tried to elaborate systematically the perspective of a theory that defends a morality in terms of duties without reference to moral principles. But it remains unclear how we could identify concrete moral obligations without reference to more general rules and principles. It seems that the criteriological problem is transferred from the general level of principles to the more concrete level of our faculty of judgment. How can it be possible to formulate moral judgments that can be acceptable to all of us, if our faculty of judgment can only come into action in concrete circumstances without guidance from general orientations? Is that only possible in the context of moral traditions that can guide our moral judgments? But such a reference to moral traditions would ignore the situation of moral pluralism.

3.   MORAL PRINCIPLES

We have seen that the reference to moral principles is not simply an accidental characteristic of modern ethical theories but a necessary implication of the idea of a genuine moral point of view. However I did not show in detail what the moral point of view contains. I only assume that the moral point of view demands the limitation of our possible scope of actions, overrides competing evaluative aspects, and can be in competition with our self-interest. We do not naturally act in accordance with moral demands and therefore a kind of criterion and its justification is needed. Such a criterion is especially necessary in the context of bioethics where we are faced with moral and not only eudaimonistic questions, where the discussion takes place under conditions of existing moral pluralism, and where moral traditions do not provide us with adequate answers to this challenging question.

Against the background of these considerations it is not surprising that the discussion of moral criteria or principles has played an important role in the history of bioethics. But a closer look at the kind of principles used in bioethical debate shows a broad diversity of meanings. What are called 'principles' do not have the same meaning, the same discursive function or the same justificatory status. In fact the different kind of principles do not have very much in common. To begin with, I only want to mention in passing some *sort of principles* that are only related to very *specific contexts* or to the action of *specific professions*. I do not want to discuss these in detail here because this kind of principle is theoretically not very interesting. Perhaps in this context we should not talk about principles but about rules, but in the discussion this distinction is not made very clearly. If we are dealing for example with specific principles or rules to guide physicians, technicians or researchers, these principles are only context- or profession-specific applications of more general codes of conduct or moral considerations. The rule that physicians in emergency situations should act "in dubio pro vita", for instance, is a formulation of pragmatic guidance in a specific kind of situation where physicians do not have much time to weigh different alternative actions carefully. Such a rule by itself has no justificatory power concerning the moral quality of the corresponding actions. This rule is only a concretization of a more general moral principle. Similar kinds of rules can be found in discussions of technology or ecology, like the precautionary principle, the principle of biodiversity or sustainability. These principles are of course morally relevant but their moral validity depends on more fundamental moral considerations.

Philosophically more interesting is the difference in the meaning of moral principle in the *principlist approach* of Beauchamp and Childress and in a *one principle approach* such as the Kantian. The status of these principles, their meaning, the justificatory impact and the form of application are very different. But we will also have to look at other forms of one principle approach by authors such as Engelhardt or Singer.

To begin with, the four principle approach refers to *several, non-hierarchically* organized principles that have to be *weighed* against each other. A single principle does not override all other points of view but is only of relative moral importance. In comparison to one principle approaches these principles are *relatively concrete*. If we use principles like 'autonomy' or 'nonmaleficence' we are referring to different kinds of actions or situations. Of course those principles have to be interpreted in order for them to be used in concrete situations, but we have an idea of the circumstances or situations in which 'nonmaleficence' or 'not harming' can be applied. The validity of these principles is not guaranteed by a philosophical justification but by the fact that they are generally held to be morally relevant. Although we differ on all kinds of convictions, according to Beauchamp and Childress we all agree that we have to show some respect for the autonomous decisions of agents, that we have to avoid harming others, and so on. We only differ on the relative weight of these principles in concrete circumstances and on some non-moral aspects of the concrete situation of action. There is possible disagreement over whether or not in concrete cases the autonomy of the patient is more important than our duty of beneficence. Furthermore we can have morally relevant disagreements concerning our knowledge of facts in the world or concerning metaphysical assumptions, for example doubt whether or not newborns are autonomous persons and so on. In moral judgment we have to discuss all that, but our normative judgment will be made – according to this approach – with reference to this kind of moral principle. Since they have to be weighed against each other they are only prima facie valid as central elements of a generally shared common morality. Ethics does not have to justify them but to provide us with a coherent interpretation and application of these principles.

A philosophical framework for such a principlist approach can be found in a Rossian theory of prima facie duties. Ross holds the position that the only points of reference for moral debates are some prima facie duties, such as the obligation not to lie, not to steal, not to kill etc. Ross assumes that we cannot reconstruct the interrelation between these duties systematically; they are not an application of a higher principle, therefore not hierarchically organized. Since there is a plurality of prima facie duties, a process of weighing them against each other is necessary in order to decide on our actual duties. Referring to the intuitionism of G.E. Moore, Ross is of the opinion that we cannot justify these prima facie duties but presupposes that it is intuitively plausible that these principles are morally obligatory and therefore they are not in need of further justification. Thus, his theory makes some cognitivist presuppositions since he assumes that we have some moral knowledge, and that prima facie duties are guiding us to this knowledge. But this basic moral knowledge is only intuitively accessible.

Intuitionist approaches have been criticized for very different reasons but they seem to be undergoing a kind of renaissance nowadays (cf. Audi 2004; Roeser 2002; Stratton-Lake 2002). Critics point to the following problems. First of all, the function of these prima facie duties as moral criteria is doubtful if there are no criteria for the weighing these (non-hierarchically organized) prima facie duties. For concrete moral judgment we would need a nonarbitrary way of weighing these duties. But even if we do not take the problem of application into account, there are some problems with the claim that these intuitions can legitimately be seen as basic assumptions for a moral evaluation. Since it is claimed that those duties are *intuitively* plausible it would be necessary to have some evidence on a pre-theoretical level. And there are at least two different problems with that claim: the problem of the *identification* of those moral intuitions, and the problem with the claim that they are intuitively *evident*. First of all the intuitionist must be able to identify these principles or duties without a kind of constructivist act. But do we not already need a criterion in order to *identify* some principles as *moral* principles? Why do we think some convictions articulate moral obligations and others do not? Why are we morally obliged to respect the private property of other people? Why is that an issue for morality? Perhaps it is only a matter of some social conventions that people have property. Why is it morally important that we protect private property? Do we not need a criterion for distinguishing between moral and other convictions?

But even if we could find an answer to that problem, why would it be evident that these duties are really moral demands? Perhaps they are only the results of prejudice and bad education. And we have to be aware that they must be evident on a pre-theoretical level, since the intuitionist does not want to refer to a criterion that would need its own justification. How can such an intuitionist approach deal with moral demands that are not widely shared? Are we obliged to avoid harming animals? In reality, we do not agree on this. Is the fact of disagreement already an argument against the legitimacy of such a moral claim? How can such an intuitionist approach be open to moral progress? And we have to mention here that this problem of unshared moral demands is especially urgent in bioethics where we are dealing with moral claims in areas where no old traditions and experiences exist.

So the question is whether it is legitimate to refer to a pre-theoretical level of moral intuition as justificatory reasons in moral argumentation. Modern intuitionists often try to harmonize a kind of intuitionist basis with some sort of coherentist methodology. Robert Audi for example tries to avoid a solely dogmatic status of basic intuitions. According to Audi an intuitionist methodology has to be a process in which intuitions and their theoretical systematization are necessarily interrelated elements of moral judgment. The prima facie duties of Ross are a kind of first instance for the moral evaluation. But the process of interpretation and reflection is also a necessary element for the intuitionist. We need normative theories in the process of interpretation, systematisation and application of basic intuitions. And therefore the intuitions are open for further development. Audi even proposes using the categorical imperative as a guiding criterion if there is a conflict between prima facie duties. But the relation between the basic intuitions and normative ethical theories has to be interpreted according to 'reflective equilibrium'. Audi's approach has the advantage of making the relation between intuitions and normative theories more dynamic. But even this approach presupposes some sort of cognitive status of basic moral intuitions on a pre-theoretical level. The basic intuitions have to be seen as valuable, must have some sort of legitimatory power that is in no need of further justification. But why is the conviction that each individual has some basic individual rights a valid moral claim while the conviction that the individual may be sacrificed for religious purposes is not? Why is that only a result of cultural prejudice?

To summarize the problem again: if intuitions are to have a legitimatory task for normative theories we must be able to *identify* these intuitions as moral demands and to give *evidence* of their legitimacy without referring to a normative ethical theory. Otherwise the validity of this theory has to been shown beforehand.

What I have discussed for Ross is mutatis mutandis the problem of the mid-level principles of Beauchamp and Childress as well. Here we have not only the problem of how to know that these principles are really morally demanding. We also have the problem of identifying the content of such principles without a normative theory. If we have a principle like Singer's equal respect for the interests of everybody, Kant's categorical imperative or Mills' greatest happiness principle, the theoretical framework gives us some indication of the interpretative context of this principle. For Kant the internal connection between the categorically binding character of moral demands and the autonomous person as the basis of this self-legislation of morality, determines the possible applications of this normative concept. But the notions of 'autonomy', 'beneficence', 'maleficence' and 'justice' are totally underdetermined in terms of the content. Since they are not used against the background of one normative theory we have no indication of how to use them. Is 'harming' only a concept we can apply to human beings or to all sentient beings? There is no normative theory to help us decide on the semantic content of (morally wrong) harm but only pre-theoretical evidence. Is killing a fly (morally wrong) harming or not? If we do not have a consensus about it before our moral dispute starts, the principlist approach will not help us answer this question. The situation is similar for 'autonomy'. What is the content of such a normative principle? What is protected through this notion, what kinds of actions are forbidden? Of course we cannot expect to get a list of actions that are forbidden via the principle of autonomy. But if such a notion is to function as a criterion in moral deliberations, we must have some possibility of determining its content. In that respect there is a fundamental difference between the mid-level principle and a principle like that of Kant and Mill. "Autonomy" in the line of Beauchamp and Childress is a notion that is undetermined in terms of its possible applications. The content of this notion is known as long as we use it in contexts where the established forms of moral or legal regulation, or at least a kind of general moral consensus, exist. For example in clinical contexts we are used to seeing respect for the 'informed consent' of the patient as the application of the content. And the more the context of a moral question is already morally regulated, the more it will be obvious what "autonomy" means. Is each kind of intervention in the private sphere of individuals already an infringement of their autonomy? Are only persons autonomous beings? How far does the use of the principle carry some anthropological presuppositions with regard to the capacities of those entities that can be prescribed as autonomous beings?

Even more problematic is the application of *beneficence*. Why, to what extent and against whom we could have some duties of beneficence? Concerning autonomy we could at least try to formulate some necessary conditions of application. We could ask, who is an autonomous being? Even if finding these conditions does not answer the question of their moral significance, we at least have some indication of the circumstances in which we can use the notion at all. The content of beneficence, however, is even more difficult to determine. The possible *object* of beneficence is unclear: Animals can benefit from our behaviour but so can persons, religious entities and perhaps landscapes. The *kind of action* is not clear since the possibilities of acting to the benefit of others are unlimited. So why should we assume that we have a prima facie duty to act for the wellbeing of others? To avoid a misunderstanding, I want to be explicit that I do not think it is impossible to give content to such a notion. And I do not share the opinion that we have no obligations to support people in need. But I do not think that the four principle approach can be helpful in any way, because the theory does not provide us with unambiguous guidance for the interpretation of the principle. If the principlist approach is our only guide, each interpretation is equally legitimate. In fact only the dominant values of a given political and cultural context will determine the limits of the beneficiary acts that we expect from each other. But we will have no criterion for deciding on moral legitimacy if we do not agree on those limits. But are we not looking for normative argumentation because we want proof of the validity of moral judgments and the legitimacy of moral norms?

The ambiguous or arbitrary content of such mid-level principles has the result that the process of weighing between the principles becomes an unclear game of 'labels'. The theory does not provide us with a criterion to order the principles hierarchically, and it does not give us a non-arbitrary guideline to determine the content of the principles. And furthermore, the only reason why we should assume that these principles are normatively valid principles is that they are part of a 'common morality'. Since this common morality is only a factual consensus we do not know why it should have a normative force. And since – furthermore – this common morality is pre-theoretical, we do not have a criterion for knowing which convictions are elements of such a common morality. In situations beyond established moral traditions the mid-level principles are no longer helpful, especially if we are dealing with normative questions in a cross-cultural perspective.

4.   ONE PRINCIPLE APPROACHES

The term 'principle' means something very different in the context of one principle approaches. Some examples are given here. In his book *Foundations of Bioethics* H. Tristram Engelhardt discusses the problem of finding a moral consensus between people not sharing a common value system. 'Moral friends' have a common reference point in shared values that provide them with a basis to solve moral disagreement. Between 'moral strangers' this shared basis of assumptions does not exist. The only principle that is valid between moral strangers is the prohibition of 'unconsented-to force against the innocent' (Engelhardt 1996: 72ff.; cf. Steigleder 2003). This principle is the starting point of all moral considerations as far as moral questions of public importance are concerned. This principle is not given moral validity by a philosophical foundation. This minimal notion of morality is the result of the very fundamental consideration that all kinds of moral considerations aim to give moral authority to our action. The content of our moral principles or considerations may be different, but violence is incompatible with the whole concept of moral authority – hence the exclusion of "unconsented-to force against the innocent" from morally legitimate actions. Engelhardt gives a second principle of beneficence that we can ignore here because it is plainly of less importance.

According to Engelhardt his first principle has normative priority compared to values that are only defensible within the context of a special value system. The way he introduces the principle determines its moral importance and gives some indication of its application in concrete contexts. Innocent persons (conscious agents) are protected against unconsented infringements.

Peter Singer begins his moral considerations with a reference to the "principle of equal consideration of interests" (Singer 1993: 21), a kind of impartiality requirement. According to Singer it is an implication of the 'moral point of view' that we have to give equal weight to the interests of everybody. For Singer no further foundation is necessary since this idea of impartiality is a necessary element of the whole idea of a moral point of view. The task of his book *Practical Ethics* is to elaborate the consequences of this principle. The formulation of this basic principle determines the possibilities of application: all beings capable of developing interests count, and all their interests are of the same importance.

Both concepts can be criticised on the grounds that their highest principle can be seen as reductive and the justification of the principle can be criticized; and I would support such criticism. But at least they take the task of ethics seriously. If we formulate moral demands, we must have a criterion for deciding in a non-arbitrary way which are legitimate demands. And such a criterion needs to be justified.

Kant's 'categorical imperative' is another example of a highest moral principle. The principle determines the requirements for maxims that are morally acceptable. With the 'end in itself' principle Kant offers a basic content for these principles. Since it is the highest principle it is open to more specific application, as he shows in the *Metaphysics of Morals* where he offers some specification of the principle in the legal sphere and in more private life. In comparison to the principlist approach Kant's principle is at the same time more open and more determined concerning its content. The categorical imperative is more open because the idea is more open, and the idea of an autonomous person as an 'end in itself' is a principle not only for some contexts of actions but for all kind of human action. As Beauchamp and Childress see autonomy as one principle and non-maleficence as another, the notion of autonomy is smaller. Therefore autonomy in the principlist approach is more or less equivalent to informed consent. For Kant autonomy is the basis of the moral law and is therefore a notion that is important for all moral considerations. At the same time the content of his principle is more determined, because it has to be seen in the context of his justification. With the foundation of his principle Kant demonstrates the philosophical horizon that determines how the autonomous being is an end in itself. His principle and the notion of autonomy are therefore not empty labels open for any kind of content, but there is a theoretical framework as a reference point for discussion of a possible content of moral duties. Since the autonomous person is the basis of all our moral obligations (the lawmaker), this person is the non-arbitrary starting point of moral considerations. At the same time Kant refers not only to the factual consensus concerning the moral validity of his principle, even though he thinks that our moral consciousness in everyday life comes to no different conclusions than the philosopher does. Instead of referring to factual consensus he shows why and in which respect his principle is morally valid, taking seriously the idea that with a moral obligation a claim is made that cannot be taken for granted.

One remark has to be made concerning a criticism that is often made in bioethics against such an approach. It is criticized for being a top-down approach that deduces moral judgments from a single highest principle. That is a clear misunderstanding. Kant never defended the idea of a 'deduction' from the categorical imperative to a concrete moral judgment. The categorical imperative is a principle that we have to use for the examination of our maxims. But the maxims themselves are not deduced from the categorical imperative. There must be some maxims that can be examined in order to come to a concrete moral judgement. Since these maxims are not in a deductive or inductive relation to the moral principle, there is no deductive relation between the categorical imperative and the moral judgement.

In this context we could also refer to the moral principles offered by Karl-Otto Apel or Alan Gewirth (cf. Illies 2003). Apel's theory in the first place offers a procedural approach to the legitimization of moral claims by referring to the idea of a communication situation in which every participant has equal opportunities. It is unclear in Apel's writing whether this argument only results in the justification procedure or whether that implies some basic moral content as well (Apel 1980 and 1990). In his book *Reason and Morality* Alan Gewirth has developed a justification of a moral principle that he calls the *Principle of Generic Consistency*. According to this, it is logically necessary for all agents to accept that they are obliged to act in accordance with the necessary conditions of agency. It is our moral obligation to respect the rights of all agents to those goods that are necessary for their agency. Gewirth shows in detail the implications this has for different areas of ethics and political philosophy (Gewirth 1978 and 1996; cf. Steigleder 1999). Both Apel and Gewirth use the notion 'principle' in a similar way to Kant.

5.   MORAL PRINCIPLE(S) AND BIOETHICS

It is the aim of this paper to show that the notion of a 'moral principle' means something very different in the principlist approach and in the one principle approach. I wanted to show that the status and use of the idea of a moral principle is to be seen in the whole context of the task of moral philosophy dealing with moral obligation, which implies demands that cannot be taken for granted.

The central argument against a one principle approach is the fact that there is no consensus about such a principle. The validity of such a criticism depends on the idea that bioethics in fact has the task of creating a consensus or helping to find one. First of all one can doubt whether an academic discipline can have the task of creating a consensus. For me this is extremely counterintuitive. But in any case principlism can only explicate a consensus that factually exists. If we have no consensus on whether or not we have an obligation to provide people in the Third World with a medication, the principle of beneficence will not help us find such a consensus since its interpretation is so problematic.

A one principle approach cannot hope to create a consensus directly since in any application of such a principle fundamental discussions about the justification of the principle will take place. But such an approach will contribute to the examination of the validity of arguments for the moral demands with which we address each other. The task of modern societies and political institutions in finding agreement about the appropriate way to deal with all difficult issues in bioethics will not in this way become easier. But we can hope to have a more transparent and argumentative way of dealing with factual moral disagreement. Such an approach takes seriously the fact that in a modern world we no longer know what we owe to each other morally. If we do not know it, we have to think about ways in which we *could* know it. We might even have doubts about whether we can know it at all. But still, if we are taking that doubt seriously we are already discussing the legitimacy of our mutual expectations. Bioethics is not only a way of dealing with some side-effects of the life sciences. Bioethics is a discipline that takes seriously the idea that dealing with questions of life and death can confront us with fundamental disagreements about our moral obligations. What questions could be more worth investigating?

References

Annas, Julia. *The Morality of Happiness*, Oxford 1993.

Anscombe, Elisabeth. "Modern Moral Philosophy." *Philosophy* 33, 1–19, 1958.

Apel, Karl-Otto. "The *A Priori* of the Communication Community and the Foundation of Ethics. The Problem of a Rational Foundation of Ethics in the Scientific Age." In: *Towards a Transformation of Philosophy*. G. Adey and D. Frisby (eds.). London, 225–308, 1980..

Apel, Karl-Otto. *Diskurs und Verantwortung. Das Problem des Übergangs zur postkonventionellen Moral*. Frankfurt am Main, 1990.

Audi, Robert. *The Good in the Right. A Theory of Intuition and Intrinsic Value*. Princeton, 2004.

Beauchamp, Tom L. and Childress, James F. *Principles of Biomedical Ethics*. New York: 1979, 5th edition 2001.

Brink, Bert van den. *The Tragedy of Liberalism. An Alternative Defense of a Political Tradition*. New York, 2000.

Burg, Wibren van der and Willigenburg, Theo van (eds.). *Reflective Equilibrium*, Dordrecht, 1998.

Clouser, K. Danner and Gert, Bernard. "A Critique of Principlism." *Journal of Medicine and Philosophy* 15, 219–236, 1990.

Dancy, Jonathan. *Ethics Without Principles*. Oxford, 2004.

Daniels, Norman. *Justice and Justification. Reflective Equilibrium in Theory and Practice*. Cambridge 1996.

Düwell, Marcus and Steigleder, Klaus (eds.). *Bioethik. Eine Einführung*. Frankfurt, 2003.

Engelhardt, Tristram H. Jr. *The Foundations of Bioethics*. New York, 2nd edition 1996.

Gatens, Moira (ed.). *Feminist Ethics*. Dartmouth,1998.

Gert, Bernard. *Morality: Its Nature and Justification*. New York 1998.

Gert, Bernard. *Common Morality. Deciding what to do*. Oxford 2004.

Gert, Bernard; Culver, Charles M. and Clouser, K. Danner: *Bioethics: A Return to Fundamentals*. New York, 1997.

Gewirth, Alan. *Reason and Morality*. Chicago, 1978.

Gewirth, Alan. *The Community of Rights*. Chicago, 1996.

Gordjin, Bert "Das Klonen von Menschen. Eine alte Debatte – aber immer noch in den Kinderschuhen." *Ethik in der Medizin* 11, 12–34, 1999.

Habermas, Jürgen. „Vom pragmatischen, ethischen und moralischen Gebrauch der praktischen Vernunft." In: *Erläuterungen zur Diskursethik*. Jürgen Habermas (ed.). Frankfurt am Main 1991, 100–118.

Held, Virginia. *Feminist Morality. Transforming Culture, Society, and Politics*. Chicago/London, 1993.

Hurka, Thomas. *Perfectionism*. New York/Oxford, 1993.

Illies, Christian. *The Grounds of Ethical Judgement. New Transcendental Arguments in Moral Philosophy*. Oxford, 2003.

Jonsen, Albert R. and Toulmin, Stephen. *The Abuse of Casuistry. A History of Moral Reasoning*. Berkeley, 1988.

Jonsen, Albert R. *The Birth of Bioethics*. New York/Oxford, 1998.

Krämer, Hans. *Integrative Ethik*. Frankfurt am Main, 1992.

MacIntyre, Alasdair. *After Virtue. A Study in Moral Theory*. Notre Dame, 1981.

Nida-Rümelin, Julian. "Theoretische und angewandte Ethik: Paradigmen, Begründungen, Bereiche." In: *Angewandte Ethik. Ein Handbuch*. Julian Nida-Rümelin (ed.). Stuttgart 2–85, 1996.

Nussbaum, Martha. "Aristotelian Social Democracy." In: *Liberalism and the Good.* : Bruce Douglas and Gerald Mara (eds.). New York 203–252, 1990.

Nussbaum, Martha. *Women and Human Development. The Capabilities Approach*. Cambridge, 2000.

O'Hear, Anthony (ed.). *Modern Moral Philosophy*. Cambridge, 2004.

Roeser, Sabine. *Ethical Intuitions and Emotions: A Philosophical Study*. Amsterdam 2002.

Ross, David. *The Right and the Good*. Oxford, 1930.

Singer, Peter. *Practical Ethics*. Cambridge, 2nd edition 1993.

Steigleder, Klaus. *Grundlegung der normativen Ethik. Der Ansatz von Gewirth*. Freiburg/München, 1999.

Steigleder, Klaus. "Bioethik als Singular und als Plural. Die Theorie von H. Tristram Engelhardt. Jr." In. *Bioethik. Eine Einführung*. Marcus Düwell and Klaus Steigleder (eds.). Frankfurt am Main, 72–87, 2003.

Stratton-Lake, Philip. *Ethical Intuitionism. Re-evaluations*. Oxford, 2002.

Timmons, Mark. *Morality Without Foundations. A Defense of Ethical Contextualism*. Oxford, 1999.

Chapter 7

DANGER AND MERITS OF PRINCIPLISM
Meta-theoretical Reflections on the Beauchamp/Childress-Approach to Biomedical Ethics

BETTINA SCHÖNE-SEIFERT
Münster, Germany

1. INTRODUCTION

Undeniably, the most influential book in modern bioethics so far has been *Principles of Biomedical Ethics* by Tom Beauchamp and James Childress (¹1979-⁵2001). This may even prove true beyond Anglo-American discourse. Although to my knowledge only one translation of what has already become a classic has yet been made (a Spanish translation of the 4th edition appeared in 1999), the book has been the basis of much debate and research within European countries. Since its publication in 1979, this remarkable book has covered most of the problems and controversies in the field of normative bioethics, providing sensitive analysis, telling cases, and careful reflections thereon.

Throughout its four subsequent editions, *Principles of Biomedical Ethics* has been updated, keeping up with changes both in biomedical reality (for example the AIDS topic was newly included in the third edition of 1989) and in bioethics literature. While the authors have continuously attempted to extend and improve their arguments (see the various prefaces), their substantive approach and the book's architecture have by and large remained constant. The unchanged architecture does not come as a surprise to anyone familiar with the work, since Beauchamp and Childress have organized their whole approach with reference to four mid-level principles, which in their view govern biomedical ethics. These four principles – (1) respect for autonomy, (2) nonmaleficence, (3) beneficence, and (4) justice – are considered to be *prima facie* binding, as they lack lexical or hierarchical

C. Rehmann-Sutter et al. (eds.), Bioethics in Cultural Contexts, 109–119.
© 2006 *Springer. Printed in the Netherlands.*

order and are in need of much context-sensitive analysis and balancing. With each of the four mid-level principles providing the title for one extensive chapter, the ground is laid for the crucial business of careful subsequent interpretation. These reasoned attempts to develop more specific normative rules and to solve conflicts of principles in the light of specific situations were the book's major achievement throughout its first three editions. The issues analysed range from medical paternalism to the appropriate rules for organ allocation.

The book's reception within academic bioethics has been enthusiastic and intense. As critics attest, since its first edition it has "pervaded biomedical ethics" as "the dominant theory" (Gert et. alt. 1997: 71). Its framework has been adopted not only for various textbooks, case collections, and anthologies, but also for the well-known *Intensive Biomedical Ethics Courses* that have been held for 25 years at the Kennedy Institute of Ethics at Georgetown University (Beauchamp's current and Childress' former home institution). To the thousands of former participants from throughout the United States and from abroad, the four principles are known as the "Georgetown Mantra".

Taking as it does a rather liberal general standpoint and focusing on individual patients' self-determination ("respect for autonomy"), the book has, of course, also invited much criticism with regard to specific normative positions. These issues, however, are beyond the scope of this paper. Rather, I shall concentrate here on the other – connected – level of critical reception, namely that concerning the authors' meta-theory of their normative ethics.

In the late 1980s, Anglo-American bioethics in general became more self-critical of its methods and meta-presumptions. Since then, the work's principle-framework has been criticized from several different perspectives. Among the most serious critics have been Danner Clouser and Bernard Gert, who accused the Beauchamp/Childress principle approach of lacking theoretical unity and being unable to provide substantive normative guidance. It was Clouser and Gert who, in a pioneering critical paper of 1995, coined the term "principlism" for the target of their meta-theoretical attacks (1995: 219ff). This denotation seems to me to retain some of its authors' critical connotations. However, Beauchamp and Childress themselves have since then used the term "principlism" for their methodological account, obviously considering it as neutral a term as consequentialism or the other "-isms" of ethical frameworks. I will use it here in this neutral sense.

Borrowing a definition from David DeGracia (2003: 227), I understand "principlism" as a family of ethical approaches that refer to two or more basic normative principles, at least in part without fixed ranking. Thus understood, there are other principlist frameworks in both general ethics and bioethics, for instance those elaborated by W. D. Ross or Robert M. Veatch (1995). "Principlism" in this sense is quite a loose denominator. It conjoins rather different variants, each of which might deserve different treatment and raise different questions. Some objections, however, are directed at principlism as such, and at the Beauchamp/Childress framework in particular, as the most prominent version of bioethical principlism.

Certain critiques deny altogether that ethical principles have a major role in moral reasoning and justification. Such claims have been made by proponents of virtue ethics, casuistry, narrative ethics, and feminist perspectives (see for example DuBose et. alt. 1994), and cannot here be treated in any detail. It must be sufficient to emphasize that Beauchamp and Childress have made efforts at various points to show that some allegedly rival approaches are – from their perspective – supplements and enrichments rather than competition (^{5}2001: Chapt. 2 and 9; Beauchamp 1995).

In this paper, I shall concentrate on those critics who share with Beauchamp and Childress (and with me) an explicit rationalist presumption for ethical judgments, in the sense of firmly believing in the usefulness of systematic theoretical underpinning for moral reasoning. Clouser and Gert, together with co-author Charles Culver (1997), and Beauchamp's younger colleague David DeGrazia (2003) have provided several interesting critical arguments from this perspective. In the next section, I will characterize Beauchamp and Childress' understanding of their methodology as developed throughout the various editions. In particular, I will explain the coherentist nature of their approach and its key concept of *specification*. Thereafter (Section 3), some of the rationalist objections, that have been recently been made, will be reported and tentatively evaluated.

2. THE BEAUCHAMP/CHILDRESS (EVOLVING) META-THEORY OF PRINCIPLISM

2.1 The earlier accounts

The first three editions (^{1}1979, ^{2}1983, ^{3}1989) of *Principles of Biomedical Ethics* do not elaborate in detail on questions of methodology. Neither choice nor status of the four principles, nor the appropriate ways of filling "the gap between abstract principles and concrete judgements" (Beauchamp 1995: 183) are treated as major problems or even significant obstacles. Rather, the book sets forth basic ethical theory in two introductory chapters. Rule utilitarianism and rule deontology are presented as the two most plausible rival theories – each favoured by one of the two co-authors. Continuing the endless abstract battles over the flaws, difficulties and relative merits (all briefly discussed) of these two competitors was not seen as a fruitful way of approaching the nagging normative problems that had to be dealt with in actual or hypothetical cases, in policy making, or in guiding medical institutions. Beauchamp and Childress stressed that the four principles offered promising middle ground for solving these problems, capturing basic strains of commonly shared morality and allowing for some convergence of seemingly incompatible theories.

Rather than spending too much time on second order questions, Beauchamp and Childress started to prove their common ground hypothesis by *performing* the tasks of context-sensitive interpretation of those mid-level principles. Their success in reaching *common* rules and judgments, despite starting as proponents of *different* general theories, seemed an – albeit indirect – proof of their meta-theoretical starting point.[1] Moreover, the positive reception and adoption of the book's framework seemed to provide additional support. Mid-level principles served as a fruitful starting point for numerous debates and convincing judgments and policies in the field of bioethics (for example the policies requiring patients' informed consent for all medical interventions).

2.2   The developed accounts

I have argued above that throughout the three earlier editions, the major achievement of *Principles of Biomedical Ethics* was its sensitive contribution to or even ground-setting in substantive debates on numerous problems in bioethics. In the two subsequent editions, Beauchamp and Childress made a second major contribution to bioethics by deepening and extending the meta-theory of their principlist approach. They provided interesting views on moral epistemology, in particular on the issue of how to "apply" principles to concrete normative problems. They also reflected on some of the questions of a coherentist theory of moral judgment, and on the possibility, status, and nature of a *common morality*, allegedly shared at all times and in all cultures by all persons who are committed to morality (compare Beauchamp 2003). In this short paper I cannot possibly discuss these issues with the depth and diligence they deserve. However, it seems important to stress the considerable development and enlargement that Beauchamp and Childress have meanwhile provided to their theory of appropriate reasoning and justification in the field of (bio)ethics. In this regard, though not in most of its substantive approaches, *Principles of Biomedical Ethics* has indeed changed a lot. In the first three editions the two theory-introducing chapters made up about 15 percent of the whole. In the fourth edition these chapters were basically kept, but doubled in length. In the fifth edition, the substantive part of the book has been framed by two initial and two final chapters on concepts, theories, and methods of justification. Together they now make up 30 percent of the book, which is still 450 pages long.

It is contested whether, between the third and fifth editions, Beauchamp and Childress have "changed methodology" or rather the way in which "methodology is presented" (DeGrazia 2003: 219). Gert, Culver and Clouser attested after the publication of the fourth edition that: "Beauchamp and Childress have changed their theoretical account considerably. Their fourth edition (in 1994) has accommodated so well to the criticisms of principlism that Ezekiel Emanuel entitled his review of the book 'The Beginning of the End of Principlism'" (Gert et. alt. 1997: 72). I tend to disagree with those who diagnose radical changes or even a different theory in the book's two recent editions. Rather, I understand Beauchamp's/Childress's growing theoretical underpinnings as a thoughtful attempt to *reconstruct* their way of reaching constructive bioethical justifications and judgments. Since their principlist approach to substantive ethical issues has basically remained the same throughout the book's history and has been fruitfully adopted by many in the field, this reconstruction must be taken very seriously, more so than many free floating abstract (meta-) theoretical accounts.

There are certainly ambiguous passages in the various editions, which invite misunderstandings and suggest incompatibility rather than continuity. The question of major theoretical breaks in the various editions would have to be answered by careful comparative analysis of the texts – a task beyond the scope of this paper. Altogether, I would describe the theoretical achievements of the later editions as a continuously developing attempt to reflect and make explicit their long-practiced method. The book's critics as well as recent literature have contributed to this process, and the work is thus becoming increasingly rich and interesting. The key concepts in Beauchamp and Childress' matured meta-theory are *coherentism* and *specification* which are the subject of the following subsection.

2.3 Coherentism and specification

It is the fourth edition (^{4}1994) that introduces two key terms of ethical epistemology and methodology: "coherentism" and "specification".

The latter is a Hegelian term reintroduced by Henry Richardson, 1990, to whom Beauchamp and Childress in turn refer. To specify a principle is understood as making it determinate in content. It leads to more concrete rules or judgments, while maintaining "a transparent connection [...] to the initial norm that gives moral authority to the resulting string of norms" (Beauchamp 2003: 267). As Beauchamp emphasizes, "[t]here is always the possibility of developing more than one line of specification when confronting practical problems and moral disagreements. Different persons and groups will offer conflicting specifications" (ibid. 268). Keeping faith with the relevant governing norms while attending to the factual aspects of the concrete situation is, according to Beauchamp and Childress, not enough to justify a specification: it also has to maximize the *coherence* of the overall set of relevant beliefs.

Beauchamp and Childress now favour more clearly than in the early editions a coherentist model of filling the gap between principles and cases, of developing sets of moral rules. This view, currently favoured by many authors including myself, is the middle-ground alternative to top-down (deductivist) or bottom-up (casuist) models of justification, both widely recognized as largely implausible (see ⁵2001: 385ff.). A coherence model is also known as "reflective equilibrium" – a label introduced by John Rawls. In these matters, Beauchamp and Childress have again been able to draw on recent literature, in particular the work of Norman Daniels (1996). The core idea behind his favoured version of a "wide reflective equilibrium" is a process of bidirectional adaptations of judgments and cases ("bottom") to rules or principles ("top"). Moreover, the aspired equilibrium has to be *broad*, in the sense of including judgments on a range of various types and levels of relevant beliefs, for example on virtues, moral commitments, motivation, the function of ethics, background moral concepts, and so on. The required process of specification, interpretation and revision is assumed to be the open-ended business of moral justification, which aims to increase the mutual consistency and justificatory power of the various normative elements.

2.4   Reference to "common morality"

The earlier editions already mention common morality as the source of the four principles (later "clusters of principles", ⁴1994: 37). The fifth edition pays somewhat more attention to this concept. Basically, it seems to be an empirical presumption that some clusters of rather abstract norms – principles – are universally shared, and thus provide firm ground for beginning the justificatory search for new norms covering new problems. However, even according to the authors' own understanding, these passages leave "unsettled problems that we would have to address in a more complete account of this theory" (⁵2001: 406). For instance, as DeGracia (2003) has pointed out, one would need a clearer account of the common morality's nature and status, in particular of its relation to social consensus. Is common morality considered authoritative for pragmatic reasons? Or is the consensuality of common morality – unconvincingly – taken to guarantee the correctness of its normative content?

Tom Beauchamp, this time speaking only for himself, has recently provided some interesting answers to these critical questions. He takes common morality to be a minimalist list of action norms and virtues that are in fact accepted by all persons "committed to the objectives of morality" (2003: 263). This comes close to the claim that common morality[2] is universally accepted – only amoralists, immoral persons, or people heavily distorted by ideologies are said to make an exception. As in the *Principles*, common morality is to serve as a basis for the development of more concrete moral rules and judgments, policies and institutions. These more specific results in turn belong to the content of *particular* morality. Particular moralities, held by different people in different cultures and at different times, vary considerably in content.

The above question about what justifies the norms of the common morality is here given an answer with reference to what Beauchamp calls his "pragmatic" approach to moral justification: "What justifies the norms of the common morality is that they are the norms best suited to achieve the objectives of morality" (ibid. 266). I will come back to this point at the end.

3.   SOME PAST AND PRESENT OBJECTIONS

As stated above, I can only address a very limited range of objections. In favour of a systematic and reasoned way of justifying morality and giving a prominent place to the norms for action, I will concentrate on critique that presupposes these background assumptions. Even so, I can by no means do justice to all the critical concerns that have been published.

Elaborate protest against the (early) Beauchamp/Childress approach has repeatedly been voiced by Clouser and Gert with particular reference to "… the widespread popularization of principlism throughout the biomedical world, where it is not dealt with as carefully as it is in the hands of Beauchamp and Childress" (1997: 72). They argue that:

> "…the principles of principlism primarily function as checklists, naming issues worth remembering when one is considering a biomedical moral issue. 'Consider this … consider that … remember to look for …' is what they tell the agent; they do not embody an articulated, established, and unified moral system capable of providing useful guidance" (ibid. 75).

The upshot of this critique is threefold. Clouser and Gert see the four principles firstly as too abstract and empty to provide any help, secondly as lacking justificatory ground through systematic connectedness, and thirdly as prone to replacing serious and rich bioethical reasoning with a simple-minded tool with which to "do" bioethics.

As Beauchamp and Childress ([1]1979–[5]2001, 1995) and others (see Buchanan 2000) have repeatedly argued, the first point clearly misunderstands the methodology of the *Principles*. Even in the early editions, when coherentism as the endorsed back theory of ethical reasoning was in its infancy, the reasoning itself as performed by Beauchamp and Childress over hundreds of pages did at least *practise* it. Of course, a mechanical top-down application of principles to cases does not work. But such deductivism was never in the minds of Beauchamp and Childress. Instead, what they described briefly and realized at length, was the context-orientated *specification* of the principles. Hence, as Beauchamp formulated in 1995 (183): "… the progressive filling in and development of principles and rules, shedding their indeterminateness and thereby providing action guiding content".

The second point, regarding a lack of systematic unity between the principles is of more serious concern. It not only results from an intellectual or aesthetic desire for a more integrated ethical theory, but it refers to the standard problem of normative pluralism. How can one justifiably balance principles in the frequent cases of conflict, when there is no hierarchy or meta-principle for appropriate balancing (comp. Veatch 1995)? Does this alleged deficit not invite intuitive and subjective rather than justified normative results? So far, I have not touched on the question of how to solve conflicts of principles at all, but it is obvious that many of the most heated contemporary debates in bioethics, for example the debate on the moral acceptability of voluntary euthanasia, can be formulated as conflicts of principles (respect for autonomy versus nonmaleficence).

Again, Beauchamp and Childress would refer to coherentism, arguing that we simply do not have better ingredients in moral life than *prima facie* norms, and we have no better tools for solving conflicts than a reasoned search for broad coherence between considered judgments, accepted norms and endorsed values on different levels and in various contexts. However, convinced deontologists who insist on the absolute authority of at least some ethical norms cannot agree upon this defence, or at least not its first part. Others would even leave *this* dispute between deontologists and non-deontologists up for coherentist analysis (Daniels 1996: 338). These concerns obviously point to problems in coherentism. Can we really do without any fixed point of orientation when establishing the reflective equilibrium to which we aspire?

Beauchamp and Childress certainly do not provide a fully developed theory of coherentism (see also critical: DeGrazia 2003). Probably no one does as yet, and these are issues for ongoing research.

In this context, I want to come back to an interesting argument recently forwarded by Tom Beauchamp. As mentioned in the last section, Beauchamp holds, that "...what justifies the norms of the common morality is that they are the norms best suited to achieve the objectives of morality" (2003: 266). This is his attempt to justify the authority of commonly shared ground by other means than the mere fact of social consensus. But what is meant by "objectives of morality"? Beauchamp gives a clear and to my mind plausible answer: "The objectives of morality, I will argue, are those of promoting human flourishing by counteracting conditions that cause the quality of people's lives to worsen" (ibid. 260). Morality, he thus holds, stands in the service of human flourishing, nothing more and nothing less. What this effectively means depends on the underlying concept of "human flourishing". When human flourishing is considered to be empirical well-being, the functionality view of morality is a non-deontological premise. Some coherentists would certainly be ready to include this concept in the mass to be equilibrated through a process of specification and revision. However, in the position recently elaborated by Beauchamp, it rather seems to have a grounding status. But these are matters for another discussion.

The third point made by Clouser, Gert and Culver is the perpetuation of what they call "the anthology syndrome" (1997: 75). They claim a dangerous abuse of the Beauchamp/Childress *Principles* by interdisciplinary players in the field of bioethics, who simplistically subsume problems, cases, and conflicts under the four principles and take this to be "theory". Certainly many of us witness such behaviour in some contexts. Whether this is a real danger to bioethics, and whether some shortcuts to sound arguments would not be used anyway by certain people, seem to me to be open empirical questions.

4.   CONCLUSIONS

In my eyes, *Principles of Biomedical Ethics* has contributed greatly to the new field of biomedical ethics. I share Tom Beauchamp's retrospective view that:

> "In the early history of modern bioethics, principles were invoked to provide frameworks of general guidelines that condensed morality to its central elements and gave people from diverse fields an easily grasped set of moral standards. Principles [...] contributed a sense that the field rests on something firmer than disciplinary bias or subjective judgment" (1995: 181).

Moreover within philosophical ethics the mid-level principle approach has provided a way to circumvent the unending abstract debates between proponents of rival ethical theories. As we all know, these battles, fertile as they can be on a sophisticated level, regularly turn into deadlocks when they are fought between "confessors" who accuse each other of having the wrong theory.

If there has ever been a real danger that bioethics might be impoverished by a public misuse of the principlist approach, by the turning of mid-level principles from starting points of coherentist reasoning into simplistic substitutes for theoretical reasoning, this danger has long been averted by the loud critical warnings against it. In addition, the more recent meta-turn of bioethics has heightened the attention given to questions of methods and premises of normative justifications in the field. Rather than objecting to this trend, the authors of the *Principles* have, in the book's later editions, made considerable use of this body of thought and have made contributions to it, even if there are open questions and contested answers – in the field as well as in the book.

In summary, when approaching and describing the field of biomedical ethics – as *Principles of Biomedical Ethics* does – it makes sense to start by articulating general and almost universally acknowledged principles. It is debatable and not of great importance whether one should start with the famous four principles, or with three instead (for example fusing, as Frankena did, the principles of nonmaleficence and beneficence), or with a greater number of action-governing norms. In any case, as Buchanan et al. have recently commented: "the real work of ethical reasoning is quite complex and arduous and […] principles have to be argued for, refined, mutually adjusted to one another, and embedded in a coherent ethical theory that is sensitive to cultural, economic, and political contexts" (Buchanan 2003: 375). Beauchamp and Childress, I am sure, wholeheartedly share this view. Moreover, over the years and editions they have made significant contributions to the development of this still underdeveloped methodological understanding – both directly and indirectly by provoking fertile criticism and by reacting to it. Hopefully there are further editions to come.

References

Beauchamp, Tom L. and Childress, James F. *Principles of Biomedical Ethics*. New York, Oxford: Oxford University Press, [1]1979, [2]1983, [3]1989, [4]1994, and [5]2001.

Beauchamp, Tom L. "Principlism and its Alleged Competitors." *Kennedy Institute of Ethics Journal* 1995, 5, 181–98.

Beauchamp, Tom L. "A Defense of the Common Morality." *Kennedy Institute of Ethics Journal* 2003, 13, 259–74.

Buchanan, Allen; Brock, Dan W.; Daniels, Norman and Wikler, Daniel. *From Chance to Choice: Genetics and Justice*. New York: Cambridge University Press, 2000.

Clouser, K. Danner and Gert, Bernard. "A Critique of Principlism." *Journal of Medicine and Philosophy* 1990, 15, 219–36.

Daniels, Norman. *Justice and Justification: Reflective Equilibrium in Theory and Practice.* New York: Cambridge University Press, 1996.

DeGrazia, David. "Common Morality, Coherence, and the Principles of Biomedical Ethics." *Kennedy Institute of Ethics Journal* 2003, 13, 219–30.

DuBose, Edwin R.; Hamel, Ron and O'Connell, Laurence J. (eds.). *A Matter of Principles? Ferment in U.S. Bioethics.* Pennsylvania: Trinity Press International, 1994.

Frankena, William K. *Ethics.* Englewood Cliffs: Prentice-Hall, 2nd ed., 1973.

Gert, Bernard; Culver, Charles M. and Clouser, Danner. *Bioethics: A Return to Fundamentals.* Oxford: Oxford University Press, 1997.

Jonsen, Albert R. "Casuistry: An Alternative or Complement to Casuistry?" *Kennedy Institute of Ethics Journal* 1995, 5, 237–52.

Pellegrino, Edmund D. "Towards a Virtue-Based Normative Ethics for the Health Professions." *Kennedy Institute of Ethics Journal* 1995, 5, 253–78.

Quante, Michael and Vieth, Andreas. "Welche Prinzipien braucht die Medizinethik? Zum Ansatz von Beauchamp and Childress". In *Bioethik: Eine Einführung.* Marcus Düwell and Klaus Steigleder (eds.). Frankfurt am Main: Suhrkamp Verlag, 2003, 136–151.

Richardson, Henry. "Specifying Norms as a Way to Resolve Concrete Ethical Problems." *Philosophy and Public Affairs* 1990, 19, 279–310.

Veatch, Robert M. "Resolving Conflicts Among Principles: Ranking, Balancing, and Specifying." *Kennedy Institute of Ethics Journal* 1995, 5, 199–218.

---

[1] A limited pioneer experiment of this kind had already proved successful earlier, namely in the work of the first US National Commission for the treatment of certain bioethics issues. This Commission, for which Beauchamp served as staff and Childress as consultant, had to develop normative guidance for biomedical research on human subjects. In its *Belmont Report* (published in 1978) the Commission identified and interpreted three general principles (respect for persons, beneficence, and justice) as relevant for its task.

[2] To the content of common morality belong such norms as the prohibitions to kill, lie, or steal; obligations to rescue the endangered, to keep promises, or to treat persons with equal moral consideration; and the virtues of nonmaleficence, lovingness, or gratitude. Obviously, the content of this common morality does not consist in the four (clusters) of principles. However, there are obvious connections (Beauchamp 2003: 260f.).

Chapter 8

THE JOURNEY FROM ETHICS TO LAW
The Case of Euthanasia

BRIGITTE FEUILLET–LE MINTIER
Rennes, France

Over the last hundred years, biomedecine has undergone a genuine revolution. New powers have been vested in doctors. Medicine now not only offers treatment but also the creation, or even alteration, of humankind. From medically assisted conception producing *in vitro* embryos, to gene therapies which transform one's genetic makeup, procedures have developed enormously.

However, in parallel with this development, cultural concerns have come into play. The philosophy of human rights permeates attitudes at the heart of our Western society. Individuals claim respect of their rights not only to guard against their infringement but also to attain happiness.[1] Thus medicine is now used to fulfil numerous desires (for children, bodily changes,[2] a certain quality of life, even the control of one's death).

This situation led first doctors, and then society as a whole, to a greater awareness of the questions inherent in this evolution. First, can all biomedical practices enabling the treatment of individuals be legitimized at any price, particularly those involving transformation of human beings, such as germ line gene therapy?[3] Second, can science and its tools be used to satisfy all human desires?Answers are not easy. They are primarily ethical because of the need to reconcile paramount interests such as treatment or the respect of individual liberties, and other equally important values such as respect of basic personal rights or those of the human race as a whole. It is this questioning that gave birth to bioethics.[4]

However, ethical reflection has occasionally demonstrated its limitations. Fears expressed during ethical debates have sometimes been such that the need for social regulation has arisen. Bearing in mind the stakes, the need to prohibit, or at least to regulate some practices and procedures has become urgent. The transition from ethics to law has become inevitable. France has started the ball rolling.[5]

C. Rehmann-Sutter et al. (eds.), Bioethics in Cultural Contexts, 121–128.
© 2006 *Springer. Printed in the Netherlands.*

The transition is in principle protective, since it responds to a demand for regulation. Isn't the purpose of law to ensure social order and protect people? But even that truism is somewhat simplistic as transition may nonetheless necessitate "strict supervision". This means that in certain conditions where ethical reflection brings evidence for a modification of existing law that, in its current form, bans a practice, we need to be aware of the importance of this decision, and that those who aim to take the decision should be aware of the implications of this path. In this paper I will look at such a transition in a broad sense to demonstrate its complexity (I), and then pursue that demonstration by considering a particularly thorny ethical question: euthanasia (II).

1.   THE INTRICACIES OF THE TRANSITION FROM ETHICS TO LAW: THE BASIS OF 'STRICT SUPERVISION"

Many biomedical techniques made possible through scientific and biotechnological advances bring their own risks. Conscious of this, society is seeking a set of legal rules.

However, the ethics–law transition requires time for reflection as the need for a legal framework must always be weighed.Such reflection cannot therefore be carried out under the emotive glare of the media or under pressure from those rallying to a cause, however good it may be. The admission of certain biomedical practices shapes the future of humankind and society. A certain detachment must therefore come into play in the face of needs or fears highlighted by ethical consideration. This distancing, dictated by caution, requires the *complete* examination of each practice. However, the current tendency to respond speedily to social demands has frequently led to a focus on only certain aspects of these practices, and to the neglect of others. It is important therefore to reintroduce the idea of an exhaustive study of biomedical practices. Two requirements must be fulfilled.

In the first place, a full examination requires clearly defining the intended procedures on humans in order to avoid incorporating practices whose bases are not the same. For example, considering gene therapy boils down to tackling the question of allowing genetic engineering for the sole aim of care and treatment. The precise definition of this practice makes it possible to exclude from the argument all genetic engineering which might have other aims such as modifying only the physical features of an individual (size, weight, etc). Similarly, as regards euthanasia, it is essential to define clearly the hypotheses envisaged insofar as the wish to die may be formulated in extremely varied situations. Euthanasia may arise at the end or during the course of life,[6] to avoid physical suffering or in response to a refusal to live.

But this precise definition is not sufficient and must be accompanied by examination of *all* aspects of biomedical practices. Many medical procedures fulfil the basic requirement of care and treatment. Even if this argument, granted extremely important, deserves to be put forward, it cannot by itself block out all other aspects, principally the risks inherent in a practice. Within this framework, the economic consequences of allowing numerous practices must be systematically brought home to those who conduct the ethical debate. Indeed, ignorance of such considerations may lead to reducing the vigilance essential to objective examination of a situation. Genetic testing today can tell whether someone is carrying some diseases or is predisposed to developing others. It is not difficult therefore to imagine the financial returns which the marketing of such tests could represent and, incidentally, the pressure which the companies concerned could exert. It is as well to remember that the biotechnology sector is now stronger than the Internet-based new economy.

As for euthanasia, the economic facts turn on the cost of accompanying the dying. Awareness of this is essential. In the context of controlling health expenditure, the risks of giving precedence, without admitting it, to economic motivation are not illusory.

The complete examination of a biomedical practice, essential to any transition from ethics to law, also rests on a second requirement. It seems appropriate to distinguish objectives which may be achieved by legal standards. For instance, if a community authorizes euthanasia in a text of law it can be in order to authorize a person at the end of her life to ask for help to die. But it can also have the goal of permitting a young person who is completely paralyzed after a traffic accident to ask for euthanasia. If we decide to adopt a legal rule we need to see exactly which objectives we aim to reach. Do we want to authorise euthanasia only for end-of-life situations or in other circumstances as well? If we don't attend carefully to this multitude of potential objectives for a general authorisation of euthanasia, the rule may also be utilised in situations nobody had previously imagined. This provides the motivation for a careful surveillance before adopting such new rules. We need to determine the potential conditions (hypotheses) and we need to assure a "strict supervision" of the transition.

The law can prohibit, regulate or authorise a practice formerly prohibited. *A priori*, whatever the goal, the transition from ethics to law seems reassuring. Indeed, the need to draft specific legal rules which will prohibit or supervise biomedical practices implies that the protection ensured by (national or international) laws on humans rights is insufficient and that additional guarantees are necessary. However, such reasoning deserves to be qualified. It is valid when a legal rule prohibits or regulates a practice. On the other hand, it is no longer appropriate where the law's objective is to authorise a practice formerly prohibited. In such situations, the law approves a procedure, which has been condemned as an attack on basic values, and thus represents the evolution of our society as concerns the values which it expects to defend.

For example, French law has ended up legalising abortion,[7] or medical research for non-therapeutic reasons on volunteers,[8] both of these human procedures formerly prohibited as an assault on the embryo or a breach of the interests of those subjected to experimentation. The euthanasia debate comes into this category. There are some calls for legislation authorising, in certain circumstances, euthanasia; a practice currently illegal in France as being contrary to the basic principle of not killing others.

The aim of this discussion, particularly in the case of these examples (abortion, biomedical research without therapeutic goals, or euthanasia) is not of course to judge these procedures. It is simply to bring to the fore the idea that authorisation of hitherto illegal practices must be subjected to increased vigilance, insofar as they overturn the values to which we are attached."Strict supervision" of the transition from ethics to law, when it concerns authorising previously prohibited practices, is all the more important in the face of the current social tendency to yield meekly to pro-research arguments justified by the benefit to the wellbeing of humanity, respect of individual choice or economic exigencies.[9] If the importance of the goal of care and treatment is obvious, if respect of personal freedom is one of the foundations of any democratic society, if the economic situation is not to be underestimated, these objectives cannot be pursued *unless* they do not, in parallel, lead to totally unacceptable attacks on humankind or basic human rights. What, indeed, would be the benefit to a society in which a large number of diseases are treated at the price of discrimination leading to social exclusion of many people, or of the remodelling of the human species? The conflict between the benefits and risks of procedures must therefore be carefully studied.

The idea of "strict supervision" of draft laws to authorise hitherto illegal acts deserves to be defended when the French legislature clearly expresses its willingness to give science the go-ahead for such procedures, with the proviso of subsequent repeal of authorisation if the practices reveal drawbacks. Even if the will of the members of French Parliament relates only to "reversible decisions",[10] it testifies to the prevalence of policies concerned with satisfying immediate needs.

A perfect illustration of the necessity for "strict supervision" of the ethics to law transition in biomedical procedures, particularly in the case of legalising formerly illegal acts, is to be found in debates surrounding euthanasia.

2.   EUTHANASIA: AN EXAMPLE OF THE COMPLEXITY OF THE ETHICS TO LAW TRANSITION

Is it necessary to say that euthanasia poses important ethical problems? This question covers extremely varied concrete situations. To demonstrate the complexity of legal transition, only the case of those at the end of their lives and, more precisely, that of terminal patients, will be tackled here.[11]

This set of problems simultaneously affects the right to life (protecting life, not ending it), human dignity (reducing suffering) and respect of personal freedom (controlling one's life).

Ethical consideration of the question of euthanasia at the end of one's life, in France particularly, has led some to demand legislation.[12] It is an issue where adoption of legal standards would have the aim of authorising an act which is currently forbidden and is, therefore, a procedure requiring "strict supervision". French substantive law is based on the criminal condemnation of euthanasia.[13] The law forbids the killing of another and the consent of the person concerned changes nothing. In French law, the consent of the victim is no defence.

To apply the process suggested above, namely to effect a complete examination of the situation, requires analysing the request by terminal patients to be allowed to die. Such an analysis should not be limited to tackling the question of conflict between respect for the person's autonomy and respect for life, but in practice such debate tend to focus on this question. Willingness to undertake complete examination of the situation must result in considering, prior to questions of conflict of interests, the foundation of such a request.

In the great majority of cases involving the end of life, the request is based on suffering (physical, mental or both). It is thus interesting to seek possible solutions at the stage of formulation of the request to die.

If it concerns physical suffering, doctors should do everything possible to stop or relieve suffering.Specialists in pain relief know that today most pain can be alleviated or controlled. Assuming responsibility for physical suffering therefore is primarily a problem of competence and, incidentally, medical training. In a concrete way, the solution can be sought through the better training of doctors. But this direction can only be pursued if society gives the means to the medical community not only for training but also for creating structures to accommodate people in these situations (e.g. creating palliative care units).

As to the assumption of responsibility for mental suffering, which can also be behind a request to die, it too needs skill on the part of medical personnel and financial investment by society. Such skill results from giving doctors or carers training in psychology or in developing medical multi–disciplines. he presence of psychologists is becoming increasingly important. Nonetheless, such specialised training and creation of posts have a social cost which cannot be ignored in the euthanasia debate.

It would seem therefore that the means to combat the problem before any request to be allowed to die, does exist, but at some cost to the community. State investment in such an undertaking is essential. But this vision is too simple, because the State is comprised of citizens who make fundamental social choices. The euthanasia question leads us to wonder individually and then collectively whether we are ready to defend such projects and to accept the financial sacrifices that such choices will involve.

These explanations aim to show why the requirement for a precise definition of the situation under scrutiny (here, euthanasia at the end of life) and examination of all the issues is so important. Indeed, as regards euthanasia, the danger is of launching directly into the question of a conflict of values, i.e., as to whether the patient's freedom can override respect of life, whereas this euthanasia issue can be considered first through studying means to avoid the request to die itself.

In the absence of such profound examination, the real danger is in neglecting this aspect of the discussion and, incidentally, measures which could be taken beforehand (dealing with pain) especially when these solutions have a significant cost. By ignoring such a vital part of the debate and focussing on the conflict of values, the route to regulation allowing euthanasia is obvious. However, the difficulty inherent in this route is that the essential condition laid down, namely the patient's consent, is not easily attained in practice. Indeed, in the case of end of life euthanasia, either the person is unable to give such consent, or gives it in circumstances such that (because of the effects of suffering) the wish may not be expressed freely. An investigation carried out in the Netherlands (where euthanasia is now de–penalised under certain conditions) revealed that in a quarter of cases, people had been "euthanised" *without* their explicit consent.[14]

Moreover, faced with the cost of relieving physical and psychological pain, it is important to ask oneself whether the conditions which the law might lay down to regulate euthanasia are likely to be loosely interpreted, and if so, who would complain. We must not forget that euthanasia is a final and irreversible act. Those subjected to it cannot thereafter demonstrate the possible error of such a step! The transition from ethics to law as regards euthanasia must, therefore, be placed under "strict supervision".

The transition to legal regulation consequently requires caution. But this "modern virtue"[15] does not imply conservatism or opposition to progress. In truth, "responsibility" must involve refusing practices when the risks outweigh the benefits (e.g. in the case of reproductive human cloning), but also allowing them when they do not. If a previous ban is to be suppressed by introducing exceptions, or removed (as currently in the case of euthanasia), it is important to be very attentive. This does not mean that we should never proceed in this direction but that we should be aware of the weight, the importance and the potential consequences of these options before taking them.

The importance of ethical reflection is therefore obvious. Its improvement, by observing the need for the most rigorous study of biomedical practices, simultaneously requires that such reflection increasingly be led by ordinary people. In the knowledge that uncertainty and doubt are the cornerstones of ethical debate, consensus must come from within society itself. Currently however, most ethics committees leading the debate are composed largely of doctors or scientists. Although their presence is essential in that they are most qualified to present medical procedures, it is unacceptable that scientists be simultaneously "judge and jury" of issues which relate to them first and foremost. The nature of the questions with which we are finally confronted, namely human questions, makes them the exclusive domain of those who must consider their responses. In consequence it is our duty, as members of the human community, to participate in this important debate.

---

[1]   A "right to happiness" is even postulated by J. Carbonnier. "Droit et passion du droit sous la Vème République", [*"Rights and the passion for rights under the 5th Republic"*]. Flammarion: Forum, 1996, p.121.

[2]   This is the aim of cosmetic surgery or of surgery demanded by transsexuals for transforming their anatomy.

[3]   Genetic modification applied to germ cells is carried on to subsequent generations.

[4]   "Naissance de la bioéthique aux Etats-Unis", [*"Birth of bioethics in the USA"*] of F. Isambert. Prospective et Santé, November 1984, p. 63. – "L'éthique et la vie" [*"Ethics and life"*]. F. Quere, published by Odile Jacob, 1991.

[5]   Notably with the so-called "bioethical" laws of 29 July 1994, which are currently subject to revision by the French Parliament. Numerous states are nevertheless content with ethical reconsideration without recourse to legislation.

[6]   Even the young may demand the right to die on the grounds of disability or lack of the desire to live.

[7]   Law n° 75-17 of 17 January 1975 relating to voluntary abortion.

[8] Law of 20 December 1988 relating to the protection of persons who offer themselves for biomedical research.

[9] Economic backwardness is a frequently invoked argument.

[10] Parliamentary debates on the admissibility of embryo and stem cell research. Euthanasia would not therefore be included.

[11] See above on the necessity, in terms of ethical reconsideration, of clearly distinguishing all cases. Besides terminal illness, the question of euthanasia arises in other circumstances: a demand to be allowed to die made by a severely disabled person who can no longer endure their life or, more generally, a demand from someone who, irrespective of age or physical condition, no longer wishes to live. The scope of the debate on euthanasia is therefore potentially very wide.

[12] Report of the National Ethical Consultative Committee of 27 January 2000 on dying, death and euthanasia.

[13] Euthanasia may be punished as voluntary homicide, poisoning or failing to assist a person in danger.

[14] Enquiry carried out in 1995 although, since 1994, the Netherlands no longer prosecutes doctors who practise euthanasia under certain conditions. "Dictionnaire Permanaent de Bioéthique et des Biotechnologies, 3 Euthanasie, acharnement thérapeutique et soins palliatifs", ["*Permanent Dictionary of Bioethics and Biotechnologies, 3 Euthanasia, the medical prolongation of life and the relief of pain and suffering*"], n° 34.

[15] "Petit Traité des grands vertus", ["*Little Treatise on the great virtues*"]. A. Comte-Sponville, PUF, Perspectives Critiques, 1995, p.41.

Chapter 9

RECOGNITION AND RESPECT FOR PERSONS
A Personalistic Interpretation of Kant's Categorical Imperative

ROBERTO MORDACCI
Milan, Italy

> La fin de l'inquiétude,
> c'est la fin de la moralité,
> et de la vie personnelle.
> E. *Mounier*, Le personnalisme, 1950

The approach sketched in this paper develops a personalistic interpretation of Kant's Categorical Imperative. The main difference from the purely Kantian perspective is the attempt to introduce a somewhat richer notion of the human person than the one used by Kant himself, one that is suggested by the phenomenological analysis of the experience of being a person. Such a notion highlights three highly relevant dimensions of the fundamental principle of morality, rather hidden in Kant's wording of it. The first of these is the moral significance of the mutual *recognition* of each other as persons (i.e. as a subject and an individual). The second is an account of the idea of *respect* as a disposition of the will rather than as a merely emotive response to an awareness of the moral law. Thirdly there is the *intrinsically intersubjective* value of the moral reasons for action, a feature that helps to overcome the rather "monological" appearance of Kant's imperative. I will first of all try to formulate the fundamental principle of morality in a way that emphasizes the centrality of the idea of "respect for persons", and outline an interpretation of it within the boundaries of a Kantian approach. Then, I will present an argument concerning the issue of euthanasia based on that principle, partially reinterpreting one of Kant's arguments against suicide.

C. Rehmann-Sutter et al. (eds.), Bioethics in Cultural Contexts, 129–143.
© 2006 *Springer. Printed in the Netherlands.*

1.   PERSONALISM ON A KANTIAN BASIS

The notion of person is explicitly used by Kant in the second formulation of the imperative, which runs as follows: "Act so that you treat humanity, whether in your own person or in that of another, always as an end and never as a means only" (Kant 1785).

For reasons that will soon become clear, I will first of all restate the Kantian imperative as follows: *always treat every person as an end in him or herself and never as a means only*. This version bypasses the reference to humanity, since Kant seems to imply that humanity is to be treated as an end *in every real person*. The reference to humanity as a general concept gives the impression that it is an abstract, impersonal value. But if humanity is really given in persons only, then the imperative commands us to treat persons as ends, as real instances of living individuals of a rational nature (where rational nature is the same as humanity). Reference to "humanity" runs the risk of being confusing.

We can now introduce a second modification. Assuming that to treat a person as an end and not as a means only is, in a strong and deep sense (which I will try to clarify later), to *respect him or her*, we might reformulate the imperative as follows: *always act so that you respect every person, yourself or any other* (or in short: *respect every person, yourself or any other*). Kant does not use the verb "to respect" (*achten*) in this sense, because he thinks that respect (*Achtung*) is a feeling (*Gefühl*) and therefore a merely passive affection, albeit one with an intellectual basis (Kant 1785 II; 1788 I, I, III). A different interpretation of respect, more consistent with a phenomenological notion of person justifies, as I will contend, this use of the verb in the imperative.

We may call this the *principle of respect for persons*, and consider it as the fundamental principle of morality. In a Kantian framework, such a principle is un-derived and so does not need a deductive *justification*, but it can be shown reflectively that all the works of reason in guiding action (i.e. practical reason) spring from it. Kant suggested that any rational choice implies conformity to certain necessary requirements that can be expressed only by the idea of a Categorical Imperative. These are: firstly conformity to the formal requirement of *universality*; secondly the material requirement of a *necessary end*, which for morality must be an end in itself, and only rational natures are ends in themselves since only they can act for freely chosen purposes; and thirdly the requirement of a *complete determination* of the will in the light of a mutual accordance between rational beings as ends in themselves (the "kingdom of ends").

In this sense, Kant uses the notion of *person* to indicate every individual instance of a rational nature in its dignity of an end in itself.

Thus it seems wrong to argue that Kant's notion of person is simply that of a "mere logical subject of acts of reason" as Max Scheler contended in chapter VI of *Der Formalismus in der Ethik und die materiale Wertethik* (Scheler 1913-1916). Individuality is indeed excluded from humanity as a general concept, but not from persons as such, whom Kant always considered as individual instances whose common nature implies both universal and individual traits.

It is extremely important to note that the fundamental principle is a requirement of practical rationality *per se* (*"ein Faktum der Vernunft"*), and that ethics, in this perspective, is not *derived* from any anthropology or metaphysics. Normativity springs from practical reason alone, not from a theoretical account of the person. This is the basic difference between the view presented here and neo-scholastic versions of personalism (such as that of Maritain), which derive their moral principles from the notion of human nature. In this perspective, the foundations of ethics lie in anthropology and metaphysics, and it is necessary to demonstrate the teleological structure of Man and his orientation towards the ultimate good in order to justify the fundamental principle of morality. In a Kantian perspective, this is not needed. While it is true that in order to understand what the fundamental principle implies we need an adequate notion of the human person as the scope of our actions, the foundation of the principle – its normative force – is clarified by rigorous critical reflection on the common use of practical rationality. Furthermore, this is a way of avoiding the naturalistic fallacy: normative principles derive from the functioning of practical reason alone, not from theoretical concepts or the natural properties of objects (even persons). Initially we have the Categorical Imperative, and we see that human persons are the necessary objects of its command; but we then have to unravel the complex notion of a person in order to understand all the practical implications of the imperative.

As I have already said, there are two major differences between our principle of "respect for persons" and the Kantian wording of the Categorical Imperative (in its second formulation): the use of the verb "to respect", and the central, rather than incidental, position occupied by the noun "person". I think that these changes can be justified in a perspective that is still typically Kantian, although they introduce views that Kant did not explicitly hold.

Let us start with "respect". Kant's notion of respect refers to the feeling aroused by the awareness of the moral law in rational creatures: it is a special kind of feeling, caused not by sensibility but by practical reason alone. Respect is the emotional response to the awareness of the moral law within ourselves, and it becomes the *motive* of practical reason, generated by reason itself. Kant does not seem to think that respect could be more than that. Yet there is no reason to confine ourselves to this rather reductive understanding of respect. We may also consider respect as a *practical disposition*, not simply a feeling but an attitude of the will that consists in considering every person as an end in him or herself. In fact, we may transform our emotional response into an attitude of the will, precisely through reflection. As Christine Korsgaard suggests (Korsgaard 1996), it is by way of a "reflective endorsement" that we adopt the maxims that guide our actions. In this case, we reflectively adopt the maxim that to treat every person as an end *is* to respect them. Here, respect is not just a feeling; it is also an attitude belonging to the rational dimension of the person, a deliberate direction of the will towards treating the person as an end. To respect persons is not to use them simply as means[1], and it results from a deliberation about how to act rationally, from an interpretation of what is entailed in the first principle of morality. This might not be a dramatic shift in meaning, but it is important since it makes clear that "respect", in this sense, is not merely a passive attitude but rather an active power that is translated on a practical level to the recognition of the other person as an end in him or herself.

But there is more than this: respect implies that the agent *recognizes* the object of his action as a kind of reality *deserving* that disposition. I must realize that I am a person and that the other is a person, in order to respect myself or any other. But here we have another step in the direction of intersubjectivity: recognition in this sense implies a *dialectic*, which Hegel developed in the *Phänomenologie des Geistes* (Hegel 1807), and which has been further explored by phenomenological and existentialist thought. The need for recognition is the source of any intra- and interpersonal relationship, and its core is the permanent polarity of self and other, of identity and difference, inscribed in each person (myself as another), and in any form of communication (the other as a self).

The dynamics of recognition also recall the issue of *personal identity*. The recognition of my nature as a person and of my personal identity as "myself" (as distinct from any other) both depend on the encounter with other persons (or selves), and on the interplay between my choices, actions and feelings and those of others. At the same time, without the recognition of myself for *who* I am I cannot truly be respected by others. To be treated *impersonally* appears to us as a form of disrespect to which we react in ways not so different from those we adopt to protect our physical integrity.

The dialectic of recognition, implied in the notion and practice of respect, opens a relational, dialogical dimension within the apparently strictly monological[2] Kantian principle. If this holds, we might even restate the first formula of the Categorical Imperative, and say that the test for my maxim is:

"Could this maxim be willed by any other person as an instance of respect for every person?"

To summarize, respect implies recognition, and recognition requires *reciprocity*. I cannot respect any other if I do not recognize the other as a self (a person), nor can I respect myself if I consider myself a mere thing. In this sense, recognition of the other as a person is a necessary step once the principle of respect is brought to consciousness through reflection. Furthermore, I cannot pretend to consider myself a person, and therefore the object of respect, if I am not willing to respect other persons as well. Reciprocity in this sense seems to be implied in my claim for respect. In the words of Thomas Nagel, considering oneself as "a person among other persons" appears to be a prerequisite of morality (Nagel 1970). If, as I contend, treating humanity as an end in itself means respecting every person, then the explicit mention of the idea of respect seems to give a clarification to the Categorical Imperative in a way that emphasises intersubjectivity and relationality. We might now reformulate the fundamental principle (in a way which moves slightly away from Kant) as follows: *respect yourself as another, and any other as yourself.* The similarity of this formula to the Golden Rule is illuminating, since that rule can be considered the most traditional and revered expression of the moral principle of reciprocity, even if we admit, with Kant, that the Categorical Imperative must be kept clearly distinct from the Golden Rule. For this reason, it seems, Paul Ricoeur has suggested that the Golden Rule plays a mediating role between the abstract *unity* of the logical form (universalizability) of the first formulation of the Categorical Imperative and the *plurality* introduced by the mention of persons in the second formulation (Ricoeur 1990).

This brings us to the need for further development of the notion of person, along lines that take us beyond the Kantian texts. For instance, what does it mean to *recognize* a person as such? When is such recognition required from us, before which "kind" of individual being, by virtue of which particular characteristics? And what does recognition imply on a practical level? Is any recognition between persons possible before or beyond *dialogue*? And if it is not, does this imply that respect, in requiring recognition, also requires a dialogical dimension, which Kant seems to leave rather in the shadows?

Usually, Kant applies the word "person" to rational beings considered as ends in themselves. "Personality" is the quality that makes individuals of a rational nature worthy of respect: it is the mark of their "dignity". Kant simply states that this dignity is the fact that "the rational nature is an end in itself", while no other reality in the world is. This is why we call every rational nature a person and not a thing. It seems then that "person" indicates the idea of a rational nature specifically *seen from a moral point of view*, as "that reality which is intrinsically worthy of respect". The concept of person is primarily a moral one, and it seems that Kant would admit something like an *intuition* or an *intuitive knowledge or recognition* of persons, at least in the sense that in the presence of fully developed persons we immediately recognize their value as ends in themselves, even though we may not yet be able to apply the notion of person in marginal cases.

In the second formulation of the Categorical Imperative, the word "person" does not indicate the rational nature in an abstract sense, but every single concrete individual *in which* we find humanity. Humanity must be considered as an end in itself, in every person, yourself or any other. The incidental phrase pointing to persons makes it clear that the ends (Kant says "the matter") of our actions are not abstract notions as humanity or rational nature, but real individuals. Yet, Kant's concept of person is unquestionably rather abstract, meaning in a strict sense: "any occurrence of a rational nature as an end in itself". Although this notion is very rich, and echoes the classical definition of the person given by Boethius (*rationalis naturae individua substantia*[3]), it does not make clear the role assigned to *individual identity* in the moral norms based on the fundamental principle. The classical definition explicitly mentions individuality (*individua substantia*) as a principle of distinction identifying the single instances of the idea of a rational nature.

2.   PERSONS

The idea of person is certainly a rather complex one. It has a number of different historical backgrounds, which make its boundaries rather fuzzy. To mention only two of these backgrounds, "person" has both a theological background in the debates concerning the Holy Trinity and the identity of Christ, and a juridical one in the sense of the subject of rights in Roman law. The history of philosophical thought hosts a number of different (and not always compatible) notions of person. While persons used to be viewed as substances in the Medieval tradition, Locke contended that they were streams of consciousness, and Hume denied that persons are anything real at all.

Yet the concept of person is not an arbitrary one. We immediately recognize the difference between things and persons, and we need a theory (indeed a rather bizarre one) in order to deny it. I do not need to *infer* from perceptual data that I am a person and that I meet other persons in the world, apart from things. I know this immediately from my experience. What I may not be able to do is to express in an exhaustive definition all the variety and depth of the characteristics that fully developed persons deploy. It is for this reason that I may wonder in some marginal cases (which do not appear immediately as persons) whether what I am considering (an embryo, an irreversibly comatose patient) falls within my definition of person[4]. The idea of a person is self-evident from the experience of meeting other persons, but not every instance of a person is immediately evident.

The analysis of the personal experience and knowledge developed by phenomenologists in the first decades of twentieth century (notably by Edmund Husserl, Max Scheler and Edith Stein[5]) suggests that two features of persons are immediately apparent: *subjectivity* and *individuality*. Every person appears as a subjective point of view on the universe, and at the same time as an individual reality whose presence as such is not just that of a mere thing. The problem of "personal identity" can be seen as the attempt to mirror the intrinsic unity of these two features, one pointing to a potential for knowledge and action which is *common* (subjectivity), the other pointing to the irreducible *particularity* of each instance of a personal reality (individuality). *What* and *who* the person is are two questions that are inseparably connected. Phenomenologists contend that the issue of recognition and the dimension of reciprocity are part of the answer to those questions, and we can take account of this: recognizing myself and other persons as persons is part of my being a *subject* and an *individual*.

There are other features which are commonly recognized as typical of the person and which recur in debates concerning the notion of person: a *human body*, an *unrestricted desire, reason, and free will*[6].

Each of these realistic dimensions of the person expresses the features of subjectivity and individuality. The human lived body (the *Leib* of the German phenomenologists) is a centre of subjective knowledge and action, and at the same time it is a principle of individuation (*principium individuationis*) of each person.[7] Desires are subjectively lived as motives (something animals do not experience) and they are experienced as specifically ours. Furthermore, desire is a subjective opening towards a totality of objects that I as an individual intend as the objects of my desire. Reason is the opening of the subjective point of view on the universe and it is given in a particular time and space, that is, from an individualized point of departure. And free will is what makes me an acting and responsible subject, and it is also what enables me to design, at least partially, my character as an individual.

3.   RESPECT FOR PERSONS

Respecting persons means respecting each of these features as a part of a complex and unitary whole. The various specific duties to oneself and to others systematized by Kant in the *Metaphysik der Sitten* (1797) have a common basis in the fundamental principle of respect for persons *as wholes*. The duty of not taking one's life at will, for example, has normative force if and only if taking one's life at will is contrary to treating oneself as a person, that is, as an end in oneself.

A small but significant number of actions, if described accurately, can be shown to be intrinsically contrary to the principle of respect. Traditionally these include: actions that are contrary to one's *duties towards oneself* (such as suicide and arbitrary self-mutilation); and actions that are contrary to one's *duties towards others* (homicide, violence, lying and breaking promises relative to permissible actions) (see Donagan 1977). Duties towards oneself and others are what Kant called *perfect duties. Imperfect duties* do not forbid certain sorts of actions, but prescribe general ends – *self-perfection and beneficence* – for which the agent is free to find any kind of permissible action leading to the indicated goal.

It is important to note that the justification of the duties concerning these actions *does not depend*, as it does in some recent versions of intuitionism (which call themselves coherentism, see Beauchamp and Childress 2001), on their being in a "reflective equilibrium" with each other and with our considered judgements[8]. In the Kantian framework, each duty excludes a certain kind of action as intrinsically contrary to respect for persons, and therefore contrary to the internal requirements of practical rationality. To act in ways contrary to the fundamental principle is inconsistent *per se*, and not because it alters the balance between our considered judgements. If considered judgements have a rational justification, it is because they are consistent with the fundamental principle of morality, which becomes visible upon reflection. The normative force of this principle is self evident, while considered judgements need to be critically assessed in their relation to the fundamental principle. Similarly, the traditional duties are known in the "rational common knowledge of morality", from which Kant himself starts in the *Grundlegung zur Metaphysik der Sitten*. But the value of the traditional duties as moral guides for action does not come from "common morality" as such, but from their foundation in the principle of respect. This is why the personalist theory I am sketching *does not* endorse a coherentist perspective (as principlism does), or an intuitionistic epistemology like the one endorsed by Sidgwick (1874), Moore (1903) and Ross (1930), but rather a specific – indeed Kantian –form of intuitionism, limited to the intuition of the Categorical Imperative and the recognition of persons (Wood 1999).

The fundamental principle can be seen to entail two forms of respect: *negative* (or *passive*) and *active*. This distinction mirrors the one between perfect and imperfect duties: *negative respect* means not violating any person, abstaining from all forms of disrespect (killing, harming, lying etc.) that imply treating persons as mere means; this rule is prescribed by perfect duties towards oneself and others. *Active respect* means promoting the ends of persons. Since persons are ends in themselves, their ability to set and pursue goals must be favoured with any permissible means, as long as this does not cause disproportionate discomfort to the agent. These goals are those commanded by imperfect duties. So when morality directs us to treat every person as an end in him or herself, it commands an attitude that is at the same time passive and active: both non-violation of personal autonomy in making choices, and assistance with all permissible means in the pursuit of a person's rational ends and therefore the promotion of his or her happiness. To respect therefore means both to recognize the limits that the personal nature of beings (ourselves included) sets to our agency, and to favour the ability of persons to fully develop their potentialities and fulfil their desires and personal objectives. Respect cannot be merely a matter of abstaining from interference, rather it is an endeavour to make room for others so that they can realize themselves as persons. A decisive part in this attitude must be played by all possible forms of dialogue, through which we can generate an understanding of ourselves and of others that permits our development and our choices as persons.

The relational dimension of the notion of personal identity implies that the individual may have personal reasons for action (*agent-related*, not agent-relative reasons)[9], but that he or she is ready to state his or her reasons in such a way that they can be understood and possibly accepted (or refused) by others. In order to act morally, one does not need to act only on impersonal, "agent-neutral" reasons. On the contrary, one must be able to will *one's subjective maxim* to be a universal law, that is, one must consider one's personal (agent-referred) reasons as concerning oneself acting in these circumstances, and be willing to face rational criticism from any other for one's acting on that ground (i.e. for one's reasons). The test of universalizability does not imply abstracting from personal (individual) traits of character and history at all; it requires that I can consistently will that any rational agent may understand and approve of my reasons as convincing ones. In order to do this, I must do my best to make my reasons understandable and acceptable for any other well-informed and sympathetic agent. I cannot *will* to live a thoroughly incomprehensible moral life, one that *could not* be recognized as a good one by rational agents. Without at least possible recognition from others, an agent's reasons fail to qualify as moral (in the sense of morally valid). This also implies a dialogical dimension: the *other* to which I am presenting my reasons is someone who can be a real person, although he or she exemplifies at the same time all the rational beings who could reject my reasons as irrational, and who therefore might object to my acting on them (see Scanlon 1998). This rejection, if reasonable, constitutes *in itself* a valid reason *against* my acting according to my maxim, a reason that should always be taken seriously because of the respect due to any rational being.

4.   EUTHANASIA

This is why "autonomy of the will" is accorded respect as an expression of practical reason, and not as a mere exercise of the power of free will. The biggest misunderstanding inscribed in the libertarian arguments for euthanasia seems to be the neglect of this point. We cannot simply identify free (in the sense of arbitrary) choices with morally good ones; freedom of the will is not the exercise of *liberum arbitrium*. As such, the latter is just a power to act, not a criterion for action. The issue Kant takes as the basic question of his inquiry is: "when is the will really free?" His answer is: "when it is autonomous", and this means: "when you can will consistently", that is, when practical reason is unfettered. For where do I find the internal rule of my will, if not in its ultimately rational structure? And how could I ever hope for others to recognize and accept my *moral* reasons for action if they are an expression of passion rather than practical reason?[10]

Therefore, it is because I will my free choice as a reason for action, to be understood and possibly shared by other rational agents, that I must be able to formulate my maxim as a universal law. This is also why the principle of respect also refers to oneself: the respect I owe myself appears as what I owe myself-as-another, that is, as a person.

From this perspective, the judgement on euthanasia implies first of all a judgement on suicide: is it rationally permissible to kill oneself? Do I respect myself as a person in committing suicide (or in asking for assisted suicide or euthanasia)? And, secondly, can it be an instance of respect for a person to accept a request for assisted suicide or euthanasia?

Now, Kant's argument against suicide is based on the practical contradiction of a will deciding to destroy itself: a rational will should will this maxim ("destroy yourself at will") as a universal law. Kant maintains that this is absurd, because it would imply that anyone *should* commit suicide in similar circumstances. Kant's argument in the *Metaphysik der Sitten* (1797 II: *Tugendlehre* I, b. I, c. I, § 6) is rather mixed, and connects three different lines of thought. Firstly, he rejects the Stoic ideal of the wise man who can bravely choose to leave life when he thinks it has become useless.

This courage, Kant thinks, is rather a reason not to take one's life, since it witnesses my superiority as an agent in comparison with all the tragedies of life. We might say it is more consistent with the Stoic ideal of the wise man bravely to bear any adversity, rather than to escape from life in order to avoid suffering. Secondly, Kant states that it is impermissible for anyone to take his life at will, because he has duties. To evade all one's duties cannot be accorded as a right, for no one could give an authorization freeing anybody from all their duties. And thirdly, deliberately to destroy the subject of morality, which is an end in itself, would directly mean to treat it as a mere means. This line of argument, which restates the one sketched in the *Grundlegung zur Metaphisik der Sitten* (sect. II), can be reworded as follows: killing oneself in order to put an end to suffering would imply the use of one's body as a mere means while one is alive. The unity of my person would be split into an acting individual (my will) deciding to kill another (my living body). It is this overtly dualistic outcome that cannot be willed as universal law. This line of argument is the one I find more convincing: the idea of killing oneself introduces a complete division between the will and the life of a single person, as if I could detach myself from my life and yet affirm myself as a *living* will. On the contrary, respecting myself implies, in this line of argument, preserving my unity as a person or, as we might say, my unity as an embodied rational will. This amounts to a strictly deontological rejection of the moral permissibility of suicide, assisted suicide and euthanasia.

This argument appears strictly monological. If we try to frame it in relational terms, we have something like this: taking my life at will cannot be a way of respecting myself as *another whole person*; and taking anyone's life cannot be a way of respecting any other as myself. The reason is that it would mean destroying the *unity* of persons. My reasons for asking for euthanasia, when presented to any other rational being, imply a request to recognize the possibility of a definitive split between the will of an autonomous agent (my decision to die) and his or her individual existence as an *embodied* rational will. The paradox is that the will, in order to be the operative principle of action, asks for the suppression *of itself* as a concrete individual bodily existence. Kant says that this paradox is an overt contradiction, since it implies the separation of body and will, as if I were composed of two separate entities. In relational terms, presenting my reasons for euthanasia implies a request to give a possible and reasonable meaning to this opposition of body and will *within* the person.

Now, the point is that relational terms introduce contextual and individual features in a much stronger sense than just the idea of a rational nature. My personal identity is made up of relational, historical and cultural features that can give specific meaning to my choices. My reasons for action have complex backgrounds, in which terminal or degenerative illness may well play a prominent role. The conditions of suffering a terminal illness may really be undignified, as they threaten the normal functioning of the body and other human faculties (rationality, desire and freedom). Suffering can be sadly increased, and the prospect of losing all control over my mental abilities and my bodily functions may seem a progressive destruction of my dignity, so I may be willing my death not really as an escape from life, but as the end of an enormous burden for those I love. It is understandable that enduring these conditions may be unbearable for some persons; they may reasonably consider that to be kept alive in these conditions amounts to disrespect for themselves and for their relatives. In some cases, there seem to be good reasons to desire one's own death, as Kant himself admits. Yet he contended that this does not mean that *killing* (oneself or another), even in these conditions, may be *willed* as a universal law. What we may wonder, having given more weight to contextual and personal features, is whether the effective conditions of suffering a terminal illness never really make a difference.

We should note, however, that this is not the way in which the principle of respect initially directs us to address the question of terminal illness. The immediate requirement concerning terminal illness, deriving from the principle of respect, is that we must do our best to improve the conditions of life in these situations. We have a duty (and a very strong one) to do our best to preserve the dignity of persons in any circumstance and with all the means we can deploy. This amounts to a strong refusal of so-called therapeutic obstinacy, which appears as a typical expression of disrespect. Contrary to what happens in a natural law approach, in this perspective we do not even need a principle of double effect to justify terminal sedation or the withdrawal of disproportionate treatment (see Donagan 1991). Withdrawing all futile treatments and making extensive use of palliative care and support is *required* by the principle of respect, even if it shortens life.

From this point of view, euthanasia looks like a defeat of respect, a declaration that relationship with the other as a living person is no longer possible, not even simple proximity until the end. Direct killing (of oneself or of another) appears from a general point of view to be a misunderstanding of one's autonomy: it is not in willing my death (or that of another) that I can show my autonomy. Yet terminal illness has the potential to reduce my personality to almost nothing: when my body is collapsing, my faculty of desire is possessed by pain, my reason is sinking, my freedom is annulled, and all this has continued for days and will not cease until I die, being alive is a curse. I should not have been *kept* alive to this point. Would it then be permissible to kill me? That situations like these may occur is a defeat of medicine and of human relationships. So at such a point, killing and letting die may not be so different. In cases like this, however, it does make a difference whether I treat the suffering person with terminal sedation (letting die) or I directly and intentionally kill a patient as a way to terminate suffering. The latter, it seems, cannot be willed as a universal law, since it splits the unity of persons.

To admit euthanasia as a reasonable and normal way (although regulated through a law) of dealing with the suffering and pain of terminal illness would be to infringe the rule of respect. Apart from considerations concerning slippery slopes (especially in the area between voluntary and non-voluntary euthanasia), which do have some basis, respect for the dying cannot be expressed primarily by the practice of killing. On the contrary, we should exclude such a practice precisely in order to make dying more dignified, less tragic and more humane. The life of persons comprises death as its end, and the main reason that euthanasia has become such a hotly debated issue is that medicine seems to have lost respect for the limits of human existence, and we as a culture seem to have lost the resources for finding a meaning for death. Our finitude is something we are less and less likely to bear peacefully. But this issue extends well beyond the very narrow borders of bioethics, and maybe, even beyond the wider borders of philosophy.

References

Beauchamp, T. L. and Childress, J. F. *Principles of Biomedical Ethics*. Oxford: Oxford University Press, Fifth Edition, 2001.

De Monticelli, R. *L'avenir de la phénoménologie – Méditation sur la connaissance personnelle*. Aubier: Paris, 2000.

Donagan, A. *The Theory of Morality*. Chicago: Chicago University Press, 1977.

Donagan, A. Moral absolutism and the double effect exception: reflections on Joseph Boyle's 'Who is entitled to double effect?' *Journal of Medicine and Philosophy* 1991, 16, 495–509.

Hegel, G. W. F. (1807) *Phänomenologie des Geistes*. Ed. by W. Bonsiepen and R. Heede [*Gesammelte Werke*, hrsg. im Auftrag der Deutschen Forschungsgemeinschaft, IX]. Hamburg: Meiner, 1980.

Husserl, E. (1921-1935) *Phänomenologie der Intersubjektivität: Texte aus dem Nachlass*. Ed. by I. Kern [Husserliana XIII] Den Haag: Nijhoff, 1973.

Kant, I. (1785) "Grundlegung zur Metaphysik der Sitten." In *Gesammelte Schriften*, Berlin: Reimer, 1902–1942, vol. 4.

Kant, I. (1788) "Kritik der Praktischen Vernunft." In *Gesammelte Schriften*, Berlin: Reimer, 1902–1942, vol. 5.

Korsgaard, C. M. *The Sources of Normativity*. Cambridge: Cambridge University Press, 1996.

Moore, G.E. *Principia Ethica*. Cambridge: Cambridge University Press, 1903.

Nagel, T. *The Possibility of Altruism*. Oxford: Clarendon Press, 1970.

Nagel, T. *The View from Nowhere*. Oxford: Oxford University Press, 1986.

Parfit, D. *Reasons and Persons*. New York: Oxford University Press, 1984.

Rawls, J. *A Theory of Justice*. Cambridge: Harvard University Press, 1971.

Ricoeur, P. *Soi-même comme un autre*. Paris: Seuil, 1990.

Ross, W.D. *The Right and the Good*. Oxford: Clarendon Press, 1930.

Scheler, M. (1913-1916) "Der Formalismus in der Ethik und die materiale Wertethik."
In *Gesammelte Werke*. Bern: Franke, 2nd ed., 1954.

Sidgwick, H. (1874) *The Methods of Ethics*, 7th ed. London: Macmillan, 1907.

Williams, B. (1956-57) "Personal identity and individuation." [Proceedings of the
Aristotelian Society 57] Reprinted in *Problems of the Self*. Cambridge: Cambridge
University Press, 1973.

Williams, B. "How free does the will need to be?" in *Making Sense of Humanity and
Other Essays*. Cambridge: Cambridge University Press, 1995, 3–21.

Wood, A. *Kant's Ethical Thought*. Cambridge: Cambridge University Press, 1999.

---

[1] A rewording of the Kantian principle in this sense can be found, for example, in Donagan 1977, 65–66.

[2] The Categorical Imperative looks like a monologic principle when it is interpreted as the dictum of an impersonal and universal "Reason", issuing its commands over and beyond any relationship between real individuals. I maintain that Kant himself does not endorse such an interpretation of the principle, but it is true that the Categorical Imperative has been frequently interpreted in a monologic way.

[3] Boethius, De persona et duabus naturis in Christo, III, in Patrologia Latina, 64, 1345. Thomas Aquinas accepted the correction introduced by Richard of St. Victor, who substitutes existentia for substantia, Cf. Thomas Aquinas, Summa Theologiae, I, q. 29, a. 3, ad 4.

[4] This is true even if I accept the argument from potential: according to this, for example, an embryo is indeed a person in the early stages of its development, for it is an entity whose nature is to develop all of the potentialities of persons. What I am arguing here is that the idea of a person appears in the experience of meeting real persons, but the full dimensions of the concept appear only upon rigorous reflection.

[5] See Husserl 1921-1935; Scheler 1913-1916, ch. VI; see also De Monticelli 2000.

[6] The last feature, freedom, is denied by determinists, but we may not need to argue with them if compatibilism is true, that is, if morality is not at odds with the absence of a free will (a thesis Kant would refuse, and I with him). The important thing is that, as Bernard Williams says, "if we are considering merely our freedom as agents, and not the more important question of our political or social freedom, we have quite enough of it to lead a significant ethical life in truthful understanding of what that life involves" (Williams 1995: p. 19). The relevant psychological concepts for ethics are compatible with a naturalistic explanation of the life of the mind.

[7] In the discussion concerning personal identity, there are strong arguments in favour of the necessity for the reference to a single body for individuation. See e.g. Williams 1956-57 and Nagel 1986.

[8] The notion of reflective equilibrium is obviously taken by these authors from Rawls 1971. They blend it with the typically intuitionistic framework of prima facie duties first drawn by W.D. Ross 1930. Needless to say, there is nothing like "reflective equilibrium" in Kant.

[9] The notion of agent-relative reasons, in the sense introduced by Derek Parfit (Parfit 1984), is misleading here: in order to be a pertinent reason for me to act in these circumstances a reason need not be totally agent relative, i.e. valid only for me here and now. A reason can be "agent-related or referred" in the sense that it takes into due account all the relevant features of the agent and the situation in order to specify the description of the act. An agent's moral reasons for action must be referred to the agent, but must be understandable and approvable by any other rational agent as such. Therefore, they cannot be totally agent-relative.

[10] The fundamental issue here is that of the nature of the will. For Kant, the will is practical reason in action; for this reason it can be autonomous, that is, independent of inclinations. For Hume, and for others, it is a passion, an "internal impression", and therefore it is always dependent on sensations and desires. That the latter conception is implausible (as it annihilates morality) cannot be shown here. The question is: would I really be free if every choice were an act of passion? Would I ever experience anything like a categorical imperative, a sense of duty, or what we call "moral reasons for action"?

Chapter 10

RATIONALITY IN BIOETHICS
Reasonable Adjudication in a Life and Death Case of the Separation of Conjoined Twins[1]

DERYCK BEYLEVELD
Sheffield, UK

1.   INTRODUCTION

In the case of *Re A (Children)* the UK Court of Appeal decided that it was permissible to separate two conjoined twins (both of whom would die within a matter of months if they were not separated) because this would give one of them a good chance of survival (whilst involving the almost certain death of the other) despite the objection of the parents.

This paper considers the extent to which decisions in a case like this are capable of being made rationally. To a considerable extent, different views on this depend on different views about the powers of reason in philosophy generally, and more specifically on different views of the powers of reason in dealing with ontological and ethical propositions.

Both Immanuel Kant and Alan Gewirth claim that there is a supreme principle of morality that every being capable of acting for reasons, i.e., every agent, must accept on pain of contradicting that it is an agent. Furthermore, Kant and Gewirth claim that every agent must accept that the necessary and sufficient basis for full moral status is being an agent.

However, I argue that, even on the basis of such a robust position, the decision in *Re A* cannot be determined fully as a matter of *substantive* rationality (where the aim is to reach the objectively true or rationally necessary answer), but must be approached, beyond a certain point, as a matter of *procedural* rationality (which only attempts to apply rationally necessary procedures for reaching answers).

C. Rehmann-Sutter et al. (eds.), Bioethics in Cultural Contexts, 145–162.
© 2006 *Springer. Printed in the Netherlands.*

Using models from Habermas and Ronald Dworkin, the limits of a procedural approach are considered and I argue that focusing on what Gewirth calls "the generic conditions of agency" provides the best basis for a consensus bioethics even if his claims regarding the strict rational necessity of granting rights to these are not accepted.

2.   THE CASE OF RE A (CHILDREN)

In the case of *Re A (children) (conjoined twins: surgical separation*, All ER 2000), the English courts were asked to rule on the lawfulness of separating the Siamese twins, Jodie and Mary.[2] The essential facts are set out in Ward LJ's judgment as follows:

> Jodie and Mary are conjoined twins. They each have their own brain, heart and lungs and other vital organs and they each have arms and legs. They are joined at the lower abdomen. Whilst not underplaying the surgical complexities, they can be successfully separated. But the operation will kill the weaker twin, Mary. That is because her lungs and heart are too deficient to oxygenate and pump blood through her body. Had she born a singleton, she would not have been viable and resuscitation would have been abandoned. She would have died shortly after her birth. She is alive only because a common artery enables her sister, who is stronger, to circulate life sustaining oxygenated blood for both of them. Separation would require the clamping and then the severing of that common artery. Within minutes of doing so Mary will die. Yet if the operation does not take place, both will die within three to six months, or perhaps a little longer, because Jodie's heart will eventually fail. The parents cannot bring themselves to consent to the operation. The twins are equal in their eyes and they cannot agree to kill one even to save the other. As devout Roman Catholics they sincerely believe that it is God's will that their children are afflicted as they are and must be left in God's hands. The doctors are convinced they can carry out the operation so as to give Jodie a life which will be worthwhile (ibid. 969).

At first instance, in the High Court, Johnson J. ruled that separation would be lawful. Unanimously, although without always agreeing with Johnson J.'s reasoning, the Court of Appeal agreed that the separation could proceed lawfully. The parents elected not to pursue the appeal any further. The operation took place, resulting, as anticipated, in Mary's death and Jodie's survival.

The Court of Appeal did not regard the case as an easy one, and it was headline news that the judges were agonising over their decision. The Court made no secret of its difficulties. According to Ward L.J.:

In this case the right answer is not at all ... easy to find. I freely confess to having found it exceptionally difficult to decide – difficult because of the scale of the tragedy for the parents and the twins, difficult for the seemingly irreconcilable conflicts of moral and ethical values and difficult because the search for settled legal principle has been especially arduous and conducted under real pressure of time (ibid. 968–969).

Although the Court declared that it was one of law not morals, this was easier said than done. While the Court could decline to arbitrate between the many competing moral views of the case and could focus instead on the recognised principles of positive law in order to frame the issues, the moral problematic of taking Mary's life in order to save that of Jodie could not be silenced or suppressed.

Broadly speaking, the Court's approach (guided by recognised legal principles) falls into two parts: (i) a consideration of whether, as a matter of family law, separation would be compatible with respecting the best interests of the children; and (ii) if so, whether separation (given that this would involve the foreseen killing of Mary) would be compatible with the criminal law.

At first instance, it was taken as read that both children, crucially Mary, were live and separate persons. Responding to some doubt about this assumption, however, Ward L.J. expressed this view:

Here Mary has been born in the sense that she has an existence quite independent from her mother. The fact that Mary is dependent upon Jodie, or the fact that the twins may be interdependent if they share heart and lungs, should not lead the law to fly in the face of the clinical judgment that each child is alive and that each child is separate both for the purposes of the civil law and the criminal law (ibid. 969).

The baseline for the Court, therefore, was that Mary and Jodie, each as a life in being, enjoyed the same protected status under English law. Accordingly, when endeavouring to apply the (governing) welfare principle, the Court should not open with the interest of one child already prioritised over that of another; the Court starts "with an evenly balanced pair of scales."[3] Having determined that separation would be in Jodie's best interests (giving her the chance of an independent life) but that it would be contrary to Mary's best interests (resulting in the loss of her life), Ward L.J. recognised that "the conflict between the children could not be more acute."(ibid. 1006). To decline to decide the case, Ward L.J. thought, would be an abdication of duty. Yet, how could the balance be struck?

First, the parents' wishes must be taken into account. However, the task of the Court is not simply to endorse the parents' wishes without qualification, nor even to endorse their wishes provided that they are not wholly unreasonable. Whether the parents' wishes are patently irreconcilable with the welfare of the children or a perfectly reasonable view, the case-law requires the judge to make an independent decision as to the child's welfare. Here, although the position taken by the parents of Mary and Jodie was "pre-eminently reasonable",[4] Ward L.J. detected an inconsistency in the way in which they rightly emphasised that Mary's right to life should be respected and yet seemingly minimised the same right on Jodie's part in the light of perceived burdens that would be borne by Jodie and her carers. Thus:

> In their natural repugnance at the idea of killing Mary they fail to recognise their conflicting duty to save Jodie and they seem to exculpate themselves from, or at least fail fully to face up to the consequence of the failure to separate the twins, namely death for Jodie. In my judgment, parents who are placed on the horns of such a terrible dilemma simply have to choose the lesser of their inevitable loss. If a family at the gates of a concentration camp were told they might free one of their children but if no choice were made both would die, compassionate parents with equal love for their twins would elect to save the stronger and see the weak one destined for death pass through the gates (ibid 1009–1110).

Returning to the balance as between the twins, Ward L.J. repeats the pattern of reasoning that he believes the parents should have followed. Whereas Mary is doomed for death, Jodie's actual bodily condition is such that, with separation, she has the chance of life. So viewed, "the scales come down heavily in Jodie's favour." (ibid. 1011). It is again strongly emphasised, however, that what moves the scales in this way is not that Jodie's life is regarded as more valuable than Mary's: – "Mary's life, desperate as it is, still has its own ineliminable value and dignity";[5] nor does the balance turn on the worthwhileness of the lives of these children. Rather, it is the worthwhileness of the separation (the "treatment" as Ward L.J. calls it) that is relevant.[6] So far as Mary is concerned, the separation is worse than futile; but, for Jodie, separation is the key to life itself and to "the dignity of her own free, separate body."[7]

Is this decision, on purely rational grounds, the right decision? A necessary condition for this being so is that the view taken by the court (or the English law for that matter) that Jodie and Mary have full and equal status with respect to a right to life must be rationally required. Within moral theory, however, there are a number of positions about what properties or capacities confer a moral status on beings that makes them rights-holders or otherwise able to benefit from duties to them that others are bound by. One popular view (held, e.g., by Kant 1785) is that beings acquire rights if and only when they become agents (which is to say, when they become beings capable of acting for reasons by being able to guide their conduct by things they value). On such a view, Jodie and Mary can only have full and equal rights status if they are both agents, and there must be doubts both about Jodie and Mary having such a status: indeed, the empirical evidence (insofar as this is relevant) would indicate that neither, at the stage when their plight was being considered by the courts, was an agent. Another view (held, e.g., by Alan Gewirth 1978: 109–110, 122–124, 142–143) is that, while being an agent is necessary for full status, potential to become an agent confers some status and that beings have rights status in proportion to the degree to which they approach being agents. On the other hand, the view pleaded on behalf of the parents of Jodie and Mary was that

> [t]he dignity of human beings [the property by virtue of which they have rights] inheres because of their radical capacities, such as for understanding and rational choice, inherent in their nature. Some human beings, such as infants, may not yet possess the ability to exercise these radical capacities. But radical capacities must not be confused with abilities: one may have, for example, the radical capacity but not the ability to speak Swahili. All human being [sic] possess the capacities inherent in their nature even though, because of infancy, disability or senility, they may not yet, not now, or no longer have the ability to exercise them (Keown 1997: 483).

In other words, beings have full and equal rights status simply by being members of a species some of whose members develop into agents. If such a view seems implausible, then much of this implausibility is removed once it is coupled with acceptance of the religious doctrine of the sanctity of human life, which was well summed by Ward L.J.

> The sanctity of life doctrine holds that human life is created in the image of God and is therefore possessed of an intrinsic dignity which entitled [sic] it to protection from unjust attack (ibid. 999).

Then again, there are views that do not make agency central or even relevant, e.g., such as most versions of utilitarianism for which the capacity to experience pain or pleasure matters, or views that regard being a living organism as sufficient.

Can reason adjudicate between such views or is acceptance or rejection of them merely a matter of subjective commitment? This is not something that I wish to discuss in this paper. Instead, I wish to draw attention to the fact that even within uncompromisingly objectivistic views of ethics (such as Kant's ethics and the Gewirthian view I espouse) there may be problems that limit reason being able to give us right answers to moral questions, at least in application of a moral theory, which I will illustrate on the premise that agency is the qualifying condition for full rights status even when this is coupled with the premise that there is a supreme moral principle (in which agency as the qualifying condition is implicit) that no agent can deny without contradicting that he, she or it is an agent. But first, something brief must be said about Kant and Gewirth on what they claim, respectively, to be the supreme principle of morality.

2.   KANT AND GEWIRTH ON THE SUPREME PRINCIPLE OF MORALITY

Immanuel Kant (1785) and Alan Gewirth (1978), both contend

(a) that morality (whatever other defining features it possesses) sets categorically binding requirements for action;

(b) that there is a supreme categorically binding principle for determining what actions are permissible, required, or prohibited;

(c) that a principle can be justified as categorically binding only if acceptance of (and action according to) the principle can be shown to be rationally necessary in a way that admits of no contingency.

For Kant, the supreme categorically binding principle is the Categorical Imperative – "*Act only on that maxim through which you can at the same time will that it should become a universal law*" (1785: 84)[8], and he seeks to show, by what is generally referred to as a "transcendental deduction", that the Categorical Imperative is a synthetic *a priori* proposition, being "connected (entirely *a priori*) with the concept of the will of a rational being as such" (ibid. 89). According to Gewirth, the supreme categorically binding principle is the Principle of Generic Consistency (*PGC*) – "*Act in accord with the generic rights of your recipients as well as of yourself*" (1978: 135),[9] and he employs a "dialectically necessary method" in an attempt to show that agents and prospective purposive agents[10] contradict that they are agents if they do not accept and act in accordance with the *PGC*.

Gewirth's argument for the *PGC* renders being an agent sufficient for possession of the generic rights (see Gewirth 1978: 109–110). Correlative to Kant's Formula of Universal Law is the Formula of the End in Itself, which requires all persons to treat other persons as ends in themselves and not merely as means (Kant 1785: 91). Since only those capable of being ends in themselves are what Kant calls "rational beings with a will" (equivalent to Gewirthian agents), and being treated as an end in itself is, in effect, to be treated as having fundamental rights, Kant's theory also makes agency sufficient for being a rights holder. However, whereas Kant seems to regard being an agent as necessary to be a rights-holder, non-agents enjoying only indirect protection from the supreme principle of morality (indirect protection being protection as a result of the effects of duties owed by agents to other agents), Gewirth maintains that agents must grant partial agents and potential agents the generic rights in proportion to the degree to which they approach being agents (1978: 122–124, 142–143). Elsewhere (see, e.g., Beyleveld and Pattinson 2000) I have argued, however, that this cannot be correct, because the generic rights are, on account of the way in which Gewirth argues for the *PGC* only capable of being possessed by those who can waive the benefit of the rights, and only agents can do this. Agency is, I believe in both Kant and Gewirth both necessary and sufficient to be a rights-holder at all.

Does it follow from this that neither Jodie nor Mary enjoy the protection of the Categorical Imperative or the *PGC*, except insofar as not granting them protection affects the rights of other agents? I think not, but the reasons why this are so, as I shall now explain, indicate limits on the extent to which the Categorical Imperative or the *PGC* can be used to deduce an answer to the moral dilemmas posed by the plight of Jodie and Mary, even if the epistemological claims made by Kant or Gewirth for them are sound. In the terms that I shall use in this paper, there are limits to substantive rationality in resolving the moral problems involved in this case.

3. INSUFFICIENCY OF SUBSTANTIVE RATIONALITY TO RESOLVE THE PROBLEMS

Suppose, then, that Gewirth's argument for the *PGC* is sound and that the *PGC* is rationally necessary in that no agent can deny the *PGC* without contradicting that he/she/it is an agent. On this premise, what a rational agent may regard as permissible action must not violate the *PGC*. However, there are at least two general reasons why, even if this is so, it must be admitted that answers to ethical questions may be indeterminate, and that, consequently, if a rational answer to an ethical question is defined as one that is logically required by, or logically consistent with, the *PGC*, then rationality will not be able to settle all ethical questions.

First, there might be instances where two incompatible answers are both consistent with (neither logically required, nor logically prohibited, by) the *PGC*; but where the *PGC* nevertheless does not permit the two incompatible answers to be applied indiscriminately. An obvious example concerns whether persons should drive on the right-hand side or the left-hand side of the road. Short of convincing empirical evidence (which, as far as I know is altogether lacking) that the fact that most persons are left-brained and right-handed makes it safer, on average, for them to drive on (say) the right-hand side of the road, I do not see how the matter can be other than morally optional. What is not optional, however (given that the PGC attaches a categorical value to personal safety and bodily well-being), is that persons going in the same direction must be made to drive on the same side of the road within any specific geographical location. In routine circumstances, to allow persons to drive on any side of the road they choose is certainly not morally optional.

Second, it might be the case that even if, in principle, specific answers are required by the *PGC*, the determination of these answers might be attended by enormous complexity or lack of knowledge about non-ethical matters that provide premises in applying the *PGC* to the issues at hand, in consequence of which the most knowledgeable persons acting in utmost good faith, combined with scrupulous attention to the requirements of logic, will nevertheless be capable of disagreeing. In such cases, where lack of knowledge is the cause, the indeterminacy might be only contingent and capable of being remedied at a later date. However, it might be a function of some of the non-ethical matters required to apply the *PGC* being essentially metaphysical (meaning that the propositions concerned can neither be known empirically nor does their assertion involve the upholder in any necessary self-contradiction). Indeed, I believe that a problem of this kind arises in the case of Jodie and Mary.

I have argued elsewhere (see, e.g. Beyleveld and Pattinson 2000) that only an agent ("A") can know directly and with certainty that A is a rational agent. As far as others are concerned, A can know only that they are capable of acting as though they are rational agents. But this does not mean that they are rational agents. They could be automata with no "minds" at all. Conversely, the fact that, e.g., dogs and cats do not display the abilities expected of rational agents does not mean that they are not agents. They could be, even though this appears to be very unlikely. It was further argued that this means (at least where morality is regarded as setting categorically binding requirements on action) that apparent agents must be treated as agents and precaution must be exercised in dealing with apparent non-agents. While apparent non-agents cannot be treated as rational agents when they do not exhibit capacities to display the behaviour of such beings, rationality requires us to have concern for their welfare in proportion to the probability that they might be agents, and this probability is determined by how closely their behaviour approximates to that of rational agents. This is simply because if one assumes that a non-agent is an agent and acts accordingly the error is less serious in *PGC* terms (no agent is denied the generic rights) than if one assumes that what is in fact an agent is a non-agent and acts accordingly (an agent is denied the generic rights) and "ought" implies "can" (correlative to which one has a duty to comply with the PGC to the degree that it is possible to do so and the degree that it is possible to do so is in proportion to how closely the object of the duty approximates to being an agent). On this basis, agents have duties not to harm apparent non-agents that increase, in the case of human beings, from near zero at conception until the newborn displays the behaviour expected of rational agency. I say "from near zero", because within such a frame it is not impossible (just exceedingly poorly indicated) that the human embryo is an agent from the moment of conception, for to establish that it is impossible it would have to be proven categorically that reincarnation, metempsychosis and various other metaphysical doctrines are false (which I do not believe can be done).

Following such reasoning, neither Jodie nor Mary is ostensibly an agent at the time of the proposed separation. It is *possible* that both Jodie and Mary are agents. However, the evidence is that Jodie is a potential ostensible agent (or an ostensible potential agent), whereas Mary is not. From this, it follows that the probability that Jodie is actually an agent is greater than the probability that Mary is actually an agent. Hence, while both Jodie and Mary are owed duties of protection as possible agents under precaution, Jodie must be granted higher status.

It follows that, essentially, the conflict between Jodie's life and Mary's life is on a par with that between the life of a mother and that of her unborn child where both cannot live but the mother can be saved at the expense of the unborn child (when the unborn child cannot be saved at the expense of the mother). There is a difference between the two cases in that the mother could choose to die with her unborn child, whereas Jodie has no (at least ascertainable) capacity to choose to die with Mary, but this difference is not relevant to the issue at hand. The two cases can be made equivalent, for practical purposes, by assuming that the mother has expressed no wish about this situation in the past and is unconscious at the point at which a decision must be made. Where probability of harm and severity of harm are both measured on scales of 0 to 1, these two (analogous) situations can be represented as follows:

Mother	Fetus
moral status = 1	moral status = <1 (i.e. less than 1)
severity of harm = 1	severity of harm = 1
probability of harm = 1	probability of harm =1

Jodie	Mary
moral status = <1 (= z)	moral status = <z
severity of harm = 1	severity of harm = 1
probability of harm = 1	probability of harm =1

So expressed, the conflict between Jodie and Mary must be resolved in favour of the former (as the lesser of two moral evils). However, the moral status (or, as it is now common to say, the "human dignity") of the twins in these terms is not the only consideration; for example, the human dignity of the parents and others can also be affected and this dimension of the puzzle must also be considered.

On the facts of *Re A (Children)*, to separate the twins is to act contrary to the wishes of the parents. It might be contended that this is to violate the generic rights of the parents (and implicitly their dignity as the basis of their generic rights). Furthermore, it might be argued that since the parents are ostensible agents, whereas Jodie (on whose behalf we might argue for the chance to survive) is not, it follows that the twins should not be separated. This, however, is fallacious reasoning. The parents do not have a generic right to do whatever they want. Their generic right to freedom of action gives them rights to do whatever they want provided that they do not violate their duties under the *PGC*. Precautionary reasoning imposes duties on them to protect Jodie as a possible agent (indeed, an ostensible potential agent), which means that only if saving Jodie impacts on their generic rights as such is there any case for giving their wishes preference.[11] Again, putting this in the form of a balance, produces the following:

Jodie	Parents
moral status = <1 (human dignity value under precaution)	1
Severity of harm = 1	<1 (of adverse consequences to parents of Jodie living)[12]
Probability of harm = 1	Probability of harm = <1

Having set up the question in these terms, however, it is not at all obvious how to weigh these competing considerations. For, there are at least two major problems. First, there is an inexactitude about the value to give to any of the variables where it is less than 1. To some extent, ordinal judgments can be made (1 takes precedence over <1) but <1 can't be translated so readily into a cardinal value. Secondly, even if all variables could be given precise values, and assuming that there is commensurability between the same variable on either side of the balance, is there nevertheless commensurability between *different* variables? For example, in the balance between Jodie and her parents, is there commensurability between, say, moral status and severity or probability of harm? Does the 1 value for the parents' moral status equate to the 1 value for Jodie with regard to the severity or probability of harm? In short, where different variables are in play, is there a common tariff allowing for commensurability? Or, if not, is there some other way of conducting the balance?

The difficulty of identifying a common tariff is obvious. Even if one might be able to weigh, say, a low probability of a severe harm against a high probability of a minor harm (and it is not at all clear that one might be able so to translate these options into a common currency), how can one proceed once the probability of agency is factored in? For example, which is the lesser of two evils, a low probability of medium-level harm to a very high probability (ostensible) agent, or a very high probability of medium-level harm to a low probability (non-ostensible) agent? Is it plausible to think that there might be a common tariff allowing this calculation to be made? If not, one might think it more plausible perhaps that the moral status variable should be privileged in the calculation. After all, the precautionary principle is designed to prevent ostensible agents from harming those who, whilst not ostensible, actually are agents – that is to say, the imperative is to prevent the violation of generic rights, whether of those who appear to be or who actually are (appearances notwithstanding) agents. It follows that, if one has to weigh the interests of an ostensible agent against those of a merely possible agent, one must judge that the probability of violating generic rights is higher in relation to the former than the latter. Of course, it is not controversial to say that, other things being equal (or the balance otherwise being in equipoise), one should favour the interests of an ostensible agent. However, one might go beyond this, contending either (i) that, even where one simply does not know whether other things are equal, one should favour the interests of an ostensible agent, or (ii) that, even where other things seem to be marginally inclined towards the merely possible agent, one should still favour the interests of an ostensible agent. On these matters, at least for now, I express no opinion.

The difficulties, moreover, are not yet all in view. Once the balance between Jodie and her parents is resolved, consideration must be given to the interests of any other ostensible agents. To (over) simplify, assume that no other (third party) ostensible agents will be *directly* affected by the decision. Nevertheless, any *indirect* effects (on ostensible agents) of sanctioning the separation of the twins must be taken into account. For example, despite the Court's attempt to disavow making any judgment as to the quality of life of the girls, the decision might have the effect of inviting just such slippery slope thoughts (and encouraging practices that evince a lack of respect for the intrinsic dignity of agents). Such considerations add to the complexity of the case, for "tendency" arguments of this kind are speculative. Moreover, at the end of the chain of indirect effect, we return to the central question of how to compare one kind of (indirect) impact on an ostensible agent with another kind of (direct) impact on a merely possible agent.

If Gewirthian theory does allow for a calculation of this kind to be conducted by way of a *direct* application of the *PGC*, I confess that I do not yet know how this should be done. Specifically, with regard to the balance between Jodie and her parents in *Re A (Children)*, I am unable to offer an answer. This is not because I think that the arguments are evenly balanced (that there is equipoise with the arguments equally weighted on either side); rather, it is simply that it is not known how the balance stands – if this is a case of equipoise, it is equipoise due to complexity and uncertainty.

4.   THE NEED FOR AND LIMITS OF PROCEDURAL RATIONALITY

When, for any reason, supposed master moral principles *inherently* leave what they prescribe in application in doubt, we might say that (within the theory involved) there is no substantively rational solution to the moral problems involved. However, this does not mean that there can be no rational resolution of such problems. The German jurist Hans Kelsen (1967: 195–198) distinguished two ways in which principles can authorise actions/norms. The first way is direct authorisation, or authorisation "according to the static principle", (which occurs when the authorised action/norm can be deduced from a more general principle). For example, "Persons ought not to be burnt alive at the stake" is directly authorised by "Great pain ought not to be caused to people" because burning them alive causes them great pain. The second way is indirect authorisation, or authorisation "according to the dynamic principle" (which occurs when the authorised norm is the result of an authorised procedure). For example, "Whatever norms the British Parliament enacts as laws ought to be applied and obeyed" will authorise whatever norms the British Parliament enacts.

In a moral context, where master principles do not substantively determine answers to moral problems, they might justify procedures for indirect authorisation of answers to moral problems. Thus, they might justify/require the answer to be accepted that is produced randomly (e.g., by the toss of a coin) (which is, arguably, appropriate when the matter is truly optional or in principle indeterminate) or by a body of appropriate persons acting in good faith with proper safeguards against conflicts of interest, having to give reasons, and having their decisions subject to review (which is, arguably, appropriate where the source of the problem is essentially complexity). Such procedural solutions (which I will refer to as "recourse to procedural rationality") seem to be unproblematic when working within an agreed moral theory. In such a context, they are limited only by the proviso that the procedural solution may not validly yield results that directly contradict those direct applications of the master principle that justify recourse to a procedural solution (which, in general will be the most basic rights granted by the master principle). In effect, in

Gewirthian theory, the *PGC* itself and the basic rights granted by the *PGC* are constitutional restrictions on what could be positively legislated in a *PGC* guided jurisdiction (see Beyleveld and Brownsword 1986, Chapter Seven, for further elaboration).

However, I suggest that recourse to procedural rationality is not without its uses even when dealing with disagreements that stem from the espousal of different moral theories/values etc. This is important, because, whether or not moral theories are capable of being true or rationally required, it has to be recognised that that there is no consensus about what is the "right" moral theory. I suggest, therefore, that if cases like that of Jodie and Mary, which centre on moral disputes about which there is no consensus, such as what it is that confers rights-status on beings, are to be "managed" in a fully rational manner, this must at some point be procedural, if it is to be possible at all in practice.

The essence of a procedural solution to a moral dispute is that it requires persons to accept decisions as binding that they disagree with morally. If moral requirements are taken, as they often are (perhaps, characteristically), as categorically binding, or at any rate as overriding all conflicting non-moral requirements, then acceptance of a procedural solution is only coherent if the persons involved have moral reasons to accept decisions that they disagree with morally. In short, what is required is that the consequences of not reaching a decision must be regarded as threatening more important values than the values at stake in the primary dispute. Successful procedural solutions at this level, thus, presuppose a set of core values between the involved parties that is shared between their competing moral theories. The extent to which such a consensus-set exists, limits the efficacy of procedural solutions.

This point can be elaborated by looking briefly at two models of what might be called "procedural ethics". The first of these is to be found in the writings of the German social theorist Jürgen Habermas (seminally 1970). In essence, Habermas presents the view that if answers to moral problems are arrived at in an "ideal speech situation" then these answers are to be accepted as binding (though he even more controversially states that they are to be accepted as true).[13] The ideal speech situation is defined by a number of procedural conditions: in particular, the disputants must be able to present their viewpoints in a way that is free of disparities of power that preclude each of them from having an equal say; the debate must be conducted sincerely, orientated towards the truth without ulterior motives; governed by rules of logical consistency; the disputants must be prepared to yield from their starting position; and answers must be reviewable when new "facts" come to light, etc.

The second model is to be found in the writings of the legal philosopher Ronald Dworkin (1978: 248–252), in his elucidation of what he calls "a discriminatory moral position". Such a position is one

1. for which reasons are given; but which excludes:
   a. prejudices (postures that take into account considerations excluded by "our conventions");
   b. mere emotional reactions (feelings presented in a way that makes them immune support by reasons);
   c. rationalisations (which include: appeals to false or irrational beliefs; considerations that cannot be plausibly be connected to their conclusions; and considerations that the proponent would not be prepared to countenance if proffered by others); and
   d. parroting (repeating what others have said without critical understanding);
2. that is held sincerely;
3. that is consistent with the other beliefs of the proponent;
4. that is not held arbitrarily (i.e., not presented as self-evident when it is generally accepted that reasons for it are required).

I suggest that Habermas' position is overly formalistic (or too purely procedural). The ideal speech situation does not unconditionally provide sufficient reason for persons to accept its conclusions. It only does so if they believe that there are no "right" answers substantively to moral issues, which involves, ultimately, the espousal of relativism of a kind that entails that moral debate is at the substantive level, in the final analysis, not a matter of reason at all, merely of commitment (which renders rational substantive debate pointless as well as disingenuous). On the other hand, while Dworkin's view has a substantive element built in implicitly in the exclusion of prejudice, this merely serves to highlight the fact that the real problem might be that there may be no consensus over "our conventions". This is seriously problematic if "our conventions" refers not only to shared rules of reasoning but also to at least a range of substantive beliefs, *if* the model is intended to provide neutral reasons for accepting answers that *everyone* who is rational must accept. Thus, the existence of "conventions" (equivalent to a degree of consensus) should be viewed merely as setting limits to which a rational solution of ethical disputes is possible in practice.

Viewed in the proper manner, I suggest that the models I attribute to Habermas and Dworkin can be combined to form the formal component of a procedural position that can take us forward. In order to complete the picture, it is necessary, however, to determine whether there are any substantive conditions over which there is sufficient consensus to render such an approach useful and efficacious. An approximation to such a consensus currently exists in the Human Rights Conventions, especially the

European Convention on Human Rights (ECHR), to which all EU Member States are party. While there is nothing explicit in the ECHR about infants, this does not mean that nothing can be inferred about their moral status from the Convention. However, what is central to such an interpretation is the meaning given to "being human". If being human is "being a member of the species *Homo sapiens*" then Jodie and Mary must be accorded human rights under the ECHR. However, there is a problem with this, because many of the rights, which are civil and political rights, are not rights that can be exercised by (thus assertorically attributed to) infants, and even the right that could, most clearly, in principle, be held to apply to them, the Article 2 right to life, has not been extended by the European Court of Human Rights to the unborn. However, even if, as has been argued in Beyleveld and Brownsword 2001 (80–81), that "being human" most plausibly means "being a rational agent" under the ECHR, this does not mean that duties to infants (or even the unborn) cannot be inferred from the Convention rights. Such duties can be derived using the precautionary and proportionality reasoning mentioned above. Indeed, one can go further, for it is also then arguable that the generic rights granted by the *PGC* are implicit in the ECHR.

5. THE GENERIC RIGHTS AS THE BASIS FOR A CONSENSUS BIOETHICS

Apart from Gewirth's argument that the *PGC* is dialectically necessary,[14] a number of other arguments for the *PGC* can be given that are dialectically contingent in that they are relative to premises that persons are not compelled to accept unless the *PGC* is dialectically necessary. For example, it may be argued that anyone who accepts that there are categorically requirements on action must, in consistency with holding this premise, accept the *PGC*. Alternatively, it may be argued that anyone who accepts that anyone who holds that there are categorically binding requirements on action that require equal attention to be given to the interests of all agents must accept the *PGC*. Then, again, it may be argued that assent to the idea that there are human rights under the will-conception of rights must assent to the *PGC*, quite simply because to grant a right to do X requires a right to be granted to the necessary means to X and the generic conditions of action are necessary for the exercise of rights whatever they might be (see Beyleveld and Brownsword 2001: 79–82, 91–94). Parallel to this latter argument, it may be argued that anyone who considers human needs, viewed as needs for action, as the proper justification for fundamental rights must hold that there are fundamental rights to the generic conditions of agency. And, it also follows from the argument from human rights that the

PGC is implicit in the European Convention on Human Rights, from which it follows that the *PGC* is the basis of a consensus bioethics that is based on the Convention, provided that its rights are accepted to be rights under the will-conception.

In making this claim, I do not claim to justify the *PGC* as rationally necessary per se. I merely point out that for those who accept the Convention rights, the rights function as a core of values that rational persons who assent to that core will not be willing to sacrifice just because they disagree over matters that they must, if they are sincere and rational, admit that others who are sincere and rational can disagree with them about.

At this point, it might be objected that ethics and politics are different, and that what I am talking about is politics rather than ethics pure and simple. I agree. In principle, ethics is about what ought to be accepted; politics is about what will be accepted. However, practical ethics cannot ignore features of political reality altogether. Because "ought" implies "can", there is, in general, a difference between what a moral theory requires to be done in ideal circumstances and what it requires when the circumstances place obstacles in the way of its ideal requirements being observed. As one aspect of this, what "ought" to be from an ethical standpoint when all are agreed about the standards that govern moral judgment might not be the same as what ought to be from that ethical standpoint when all are not agreed that this standpoint is the one to employ. In consequence, practical or applied ethics must incorporate a political element in the form of a rational political morality, and I suggest that such a morality (which, in shape, follows the lines of a constitutional democracy) is a procedural ethic with limits set by a core substantive consensus.

It is beyond the scope of this paper to evaluate in detail the reasoning of the Court of Appeal in *Re A (Children)*. However, in brief, it is my view that with the generic rights governing the decision, the Court came to the right decision, though not always for the right reasons. What might be equally important, though, is the manner in which it approached this decision. Provided that the decision was reached in good faith, and there is no reason to doubt that it was, the reasoning that I have used for using the *PGC* as the basis of a consensus bioethics also implies that persons committed to that consensus must accept the decisions reached by Courts operating in good faith unless the decision is contrary to the basic values that underlie the consensus (and, a fortiori, justify recourse to a procedural approach). There is a virtue of civic responsibility that goes alongside a procedural ethics grounded in a substantive core, which is the disposition to accept decisions that command respect even when one believes that they are wrong. Without this, a peaceful pluralistic society cannot exist.

References

All ER. 2000, 4, 961 (UK Court Judgement).

Beyleveld, Deryck. *The Dialectical Necessity of Morality: An Analysis and Defense of Alan Gewirth's Argument to the Principle of Generic Consistency*. Chicago: University of Chicago Press, 1991.

Beyleveld, Deryck and Brownsword, Roger. *Law as a Moral Judgment*. London: Sweet and Maxwell, 1986 [Reprinted by Sheffield Academic Press, Sheffield in 1994].

Beyleveld, Deryck and Brownsword, Roger. *Human Dignity in Bioethics and Biolaw*. Oxford: Oxford University Press, 2001.

Beyleveld, Deryck and Pattinson, Shaun. "Precautionary Reasoning as a Link to Moral Action." In *Medical Ethics*. Michael Boylan (ed.). Upper Saddle River, NJ: Prentice-Hall, 2000, 39–53.

Dworkin, Ronald. *Taking Rights Seriously*. London: Duckworth, 1978.

Gewirth, Alan. *Reason and Morality*. Chicago: University of Chicago Press, 1978.

Habermas, Jürgen. "Toward a Theory of Communicative Competence". In *Recent Sociology No. 2*. H. P. Dreitzel (eds.). New York: Macmillan, 1970, 114–148.

Kant, Immanuel (1785). *Groundwork of the Metaphysics of Morals*. Translated as *The Moral Law* by H. J. Paton. London: Hutchinson, 1948.

Kelsen, Hans. *Pure Theory of Law*. Translated by Max Knight from the 2nd revised Edition. Berkeley: University of California Press, 1967.

Keown, John 1997. "Restoring Moral and Intellectual Shape to the Law After Bland." *Law Quarterly Review* 1997, 113, 481–503.

McCarthy, T. A. "A Theory of Communicative Competence." *Philosophy of the Social Sciences* 1973, 3, 135–156.

---

[1] Some parts of this paper derive from Chapter 11 of Beyleveld and Brownsword 2001.

[2] These were not their real names, but I will continue to refer to them as the courts did throughout.

[3] Per Balcombe L.J. in *Birmingham City Council v H (a minor)* [1994] 1 All ER 12, [1994] 2 AC 212, cited by Ward L.J. at [2000] 4 All ER 961, 1005.

[4] Transcript (www.courtservice.gov.uk), 148.

[5] Transcript (www.courtservice.gov.uk), 147.

[6] This distinction between the worthwhileness of treatment as against the worthwhileness of a person's life draws on Keown 1997. In general, Keown presents a sanctity of life principle as a middle way between vitalism (preserving life at all costs) and a Quality of Life approach (involving [arbitrary] judgments as to the worthwhileness of persons and their lives). The sanctity of life principle, unlike vitalism, does not require treatment to be given, or continued, where it is futile. As Keown is at pains to emphasise though: "the question is always whether the treatment would be worthwhile, not whether the patient's life would be worthwhile" (485).

[7] Transcript (www.courtservice.gov.uk), 147.

[8] These references are to page numbers in the English translation used.

[9] The generic rights are rights to "freedom" and "well-being" – which comprise the capacities required to be able to act at all or with any general chances of success.

[10] An agent is a being who does something voluntarily for a purpose that he/she/it (it) has chosen. A prospective purposive agent is a being with the capacities

required for agency with at least some disposition to exercise them. I, here, use "agent" to cover both agents and prospective purposive agents.

[11] It seems to me that there is no case for saying that moral status as an ostensible agent (with full human dignity) is the only thing that matters. To argue that would be to contradict that any duties are generated by precaution towards those that are not ostensibly agents. Duties to Jodie are generated by the possibility that she may be an agent, not by her being presumed to be an agent.

[12] Compare Ward L.J.'s assessment of the impact on the parents [2000] 4 All ER 961, 1009.

[13] For a discussion of which see McCarthy 1973, 150.

[14] For a comprehensive analysis of which see Beyleveld 1991.

III. CULTURE AND SOCIETY

Chapter 11

THE PUBLIC ROLE OF BIOETHICS AND THE ROLE OF THE PUBLIC

ADELA CORTINA
Valencia, Spain

1. INTRODUCTION: THE PROBLEM OF THE MORALITY CRITERION IN PLURALIST SOCIETIES

The development of moral consciousness in advanced societies has gradually formed two levels of reflection and language within the moral realm: the morality of everyday life, and ethics or moral philosophy[1]. The different moralities active in everyday life attempt to provide direct guidance for action. Ethics also guides conduct, but only indirectly, since its task consists in reflecting on the rational foundations of morality: foundations that are ultimately normative (Cortina 1986: chap.1).

The existence of plural moralities in everyday life (Christian, Islamic, and Jewish moralities, as well as the moralities connected with different versions of Hinduism, Buddhism, Confucianism, etc.) and the existence of a plurality of ethics (eudemonist, utilitarian, dialogic, etc.) poses a serious problem for pluralist societies: *which social authority is legitimised to determine what is right and what is wrong from a moral standpoint?*

Religious associations normally have an institution to determine what is right and what is wrong: churches, doctors of the law, teachers. Political communities also have bodies granted authority to promulgate laws: parliaments. But in the moral field there are neither churches nor parliaments; there is no institution with the legitimacy to determine what is right and what is wrong. If in morally monist societies there are institutions authorised to make moral rules, in morally pluralist societies there are no

C. Rehmann-Sutter et al. (eds.), Bioethics in Cultural Contexts, 165–174.

such institutions and moral republicanism becomes inevitable: the people themselves have to be the protagonists of the moral world (Cortina 2003: chap.8). The problem then consists in determining by what procedure people can jointly set about deciding what they consider to be morally right or wrong, taking into account that the different groups profess different "comprehensive doctrines of good", to use Rawls' term, or as I would prefer it, different "ethics of maxima" (Cortina 2000).

In my view, the only possible method is to set up deliberation processes in the public sphere, that aim to discover gradually together the values and principles of a civic ethics common to the different groups. In morally republican societies the process of deliberation in the public sphere is essential for joint decisions about what is fair and what is unfair, what is right and what is wrong (Cortina 1995; Conill 2003). It is therefore urgent firstly, to analyse what the nature of this process of public deliberation should be; secondly, to clarify in what the civic ethics that should be discovered consists; thirdly, to discern the role that applied ethics, and in particular bioethics, should play in this public process.

2.   PUBLIC OPINION AS A LOCUS FOR MORALITY

With regard to the process of public deliberation, it must take place in the domain of public opinion, this being an essential institution in the civil society of a pluralist political community. Indeed, despite the discussions as to whether civil society identifies itself only with a third sector, distinct from political and economic areas, or whether it is on the contrary essential to include the economic domain within civil society, all agree that the realm of public opinion is an essential part of civil society. Without it no pluralist or enlightened society can exist (Habermas 1962 and 1992; Cohen and Arato 1992; Walzer 1995; Barber 1998). Without it, it proves impossible, among other things, to elucidate those values and ethical principals that a society already shares, which constitute its civil ethics. Extending and reinforcing this domain of public opinion is one of the concerns of contemporary moral and political philosophy.

But it is not easy to determine the traits that public opinion should possess for it to be able to decide in the most appropriate way what is just or unjust. In our own times several models are under discussion, and here I will suggest one that originated in the 18th century liberal republican tradition, specifically in Kant's practical philosophy, which has two essential, but insufficient landmarks today: the model of political liberalism and that of the theory of discourse.

In the republican tradition, at least since the 18th century, the concept of "Public" is linked to the way in which political power is legitimised, because political power is understood to be a *public power, the aims and effects of which are public*, and which thus requires public legitimation (Habermas 1962: chap. 4). In the 21st century we can say that the activities connected with life, of the sort with which Bioethics is concerned, *have public aims and effects and thus require public legitimation*. The fact that a large number of the companies who manage biotechnological resources are financed with private capital does not mean that such companies can act without taking public opinion into account, because the effects of their decisions are public, and thus require public legitimation. The criterion for distinguishing between action requiring public legitimation and action not requiring it is not whether the financing involved is public or private, but rather whether the consequences of such actions are public or private. In this respect the research connected with the human genome and many other fields of research require public legitimation (Cortina 2003b).

In this conception of the public, the Kantian concepts of *Publizität* and "public use of reason" are essential landmarks, enduring today in diverse traditions though with different nuances. The concept of the public is connected with the legitimacy of politics because politics is a public power, and thus its legitimacy can only come from the rule of *rationally* desired laws. A just state cannot be founded on the private and thus arbitrary will of a sovereign or a social group (Kant 1968: 381). The sovereign must promulgate laws taking into account the formula of the social contract, the rational will, which is that of "what everyone might want". And when it comes to determining "that which all might want" public reasoning becomes indispensable.

Indeed, Kant does not leave fulfilment of the social contract in the power of sovereigns without expressly assigning them a "voice of the consciousness" that reminds them of what everyone might want. This is where the *public use of reason* by the *enlightened citizens* is significant, because it is enlightened citizens who must criticise public powers by making public use of their reason. The freedom of the pen is the champion of the people's rights, and "reasoning publicly" is the form of consciousness that mediates between the private and public spheres, between civil society and political power (Kant 1968: 32–42).

From this standpoint the *res publica* is what it is because it has the public good as its business, and also because it advocates as a procedure for attaining public good, the creation of a public space in which the citizens can publicly deliberate on what matters to them. These two facets are absolutely crucial in a conception of the public sphere.

Following this line of thought, Rawls' Political Liberalism explains that public reason is such in a threefold sense. Firstly, because as the reason of equal citizens it is the reason of the public; secondly, because its aim is public good and the fundamental questions of justice; and thirdly because its content is public, and is defined by the principles expressed in the conception of political justice (Rawls 1993: VI). Therefore, in societies with a liberal democracy it is all the citizens who must participate in the process of public deliberation, and not only the enlightened ones. But those citizens, when providing reasons to uphold their positions, must provide only those reasons that can reinforce the overlapping consensus.

It is true that citizens can adopt the inclusive standpoint and put forward reasons which belong to their own comprehensive doctrine of good, but only on condition that those reasons help to reinforce what all can accept, and reinforce the overlapping consensus to such an extent that the model of the public use of reason is the Constitutional Court.

Habermas' theory of discourse is important in that it sets the Kantian formula of the social contract in a dialogue context (Habermas 1992: 432–467). Those who should suggest "what everyone might want" in the public sphere are not the enlightened wise, nor the citizens, but the subjects affected by the political and economic systems, who defend 'universalisable' interests and thus cooperate in the task of forming a common will discursively. Public opinion is made up of citizens who have sensitive antennae for perceiving the consequences of the systems in question, as they are affected by them. The Habermasian conception of the public sphere not only fosters a neutral dialogue, it is a public space created communicatively from dialogue between those who defend universalisable interests, and who are able to reinforce inter-subjectivity.

Participants in the process of public deliberation may adduce reasons of all kinds. They may put forward the inclusive standpoint without the need to attempt to reinforce the pre-existing consensus. And in this respect what Habermas offers is broader than Rawls' proposal. Nevertheless, in my opinion, such a model is not "realistic". The public domain cannot be limited to a civil society consisting only of people guided by universalisable interests. No, civil society is formed by associations of all kinds, and by persons with diverse interests, sometimes universalisable and sometimes private. Not only the enlightened wise (Kant), not only the citizens (Rawls), not only the persons who are guided by universalisable interests (Habermas), but *all those affected* by a decision should take part in a model of public deliberation. In the case of bioethical questions this group will often extend to all human beings, including generations to come, as well as nature, which obviously requires representatives of its own "interests".

However, if admittance to the process of public deliberation cannot be refused to anyone affected by the decisions, how can one carry out this process of public deliberation without falling into a Babel of disconnected moral opinions, or without actually leaving the decisions to those who have the greatest power in the struggle for recognition? The answer is, as I see it, that precisely because the sphere of freedom is greater, the *responsibility* to participate in public deliberation is also greater for those who, having sufficient information in each of the spheres of social life, pursue the satisfaction of universalisable interests therein, aware that ethical norms necessitate that their actions take into account all those affected by them, with no exclusions.

3.   THE PUBLIC ROLE OF APPLIED ETHICS

This is where the role of applied ethics, and especially that of bioethics, is decisive. Applied ethics and public opinion are closely linked, and it is important that they continue to be so in order to allow highly moral answers to moral problems to emerge.

Indeed, in morally pluralist societies there is no single voice authorised to determine what is morally appropriate, which is why people are forced to fashion their own moral judgment by reflection and by following public opinion. That is why in the public sphere it is important to hear the voices of those who, working meticulously and responsibly in their fields of expertise, are concerned that the work being done therein is consistent with the level of civic ethics reached by that society. This is the task that the different forms of applied ethics have performed in the past and which they continue to perform today (Cortina 2003b).

Applied ethics (bioethics, genethics, the ethics of the economy and business, the ethics of the media, ecoethics, professional ethics), among which bioethics is a pioneer, have come about through an attempt to moralise the different spheres of social life[2]. They have gradually been generated in a republican sense, that is, from collaboration between scientific and medical professionals (researchers, doctors and nurses, biologists), ethicists, jurists, and those affected by the decisions that are made in each domain. In each case, the people demand their rights, certain professionals wish to define good practice in their field, and the fear of having to solve all problems in a court of law makes it advisable to create ethical codes, committees and audits: that is, a whole world of *non-corporativist self-regulation* undertaken in an interdisciplinary way.

In order to allow the people and political powers to fashion sufficiently well informed moral opinions concerning essential questions, reflections and information about applied ethics should be introduced into public opinion. This task of enlightenment is not something that should only be performed by the wise, but by all who work in these spheres, who have – or should have – better information, and who are concerned – or should be concerned – with respecting and fostering the *civic ethics* of that society. These are new references that the people can use to form their opinion concerning moral questions.

As I have already stated in other texts (Cortina 1993, 1997, 2003b and 2004), applied ethics, and in this case bioethics, do not adopt the deductive method proper to casuistry 1 or the inductive method of casuistry 2^3, but have instead the circular structural characteristics of a *critical hermeneutics*. They do not start from first principles with intent to apply these, because in pluralist societies there are no principles with common content. Neither do they discover only principles of medium scope from everyday practice, because in any applied ethics there is a certain desire for unconditionality that goes beyond concrete contexts. Instead they hermeneutically detect ethical principles and values in the different spheres of social life, which are modulated differently in each sphere: precisely the ethical principles and values that constitute civic ethics, common to all.

From this standpoint bioethics, like the other applied ethics, has at least a dual structure.

It is, *first of all*, the ethics of a social activity, whether this is biotechnological research or health care. To clarify what this ethics consists of, the neo-Aristotelian concept of "practice", as put forward by MacIntyre (1981: chap.14), is of great use, being defined as a cooperative social activity, which takes its meaning from pursuing certain internal goods, and which demands the development of certain virtues by those who take part in it. In this respect, the internal good of public health care would be expressed in the four goals of medicine proposed by the Hastings Center (Dez. 1996), and that of biotechnologies, in research for a freer and happier humanity.

But secondly, the fact of that activity being carried out in a society that has reached a post-conventional level in the development of its moral consciousness, means that it has to pursue its internal goods respecting a deontological structure: that of the principles and values respected by that social moral consciousness, which are expressed in its civic ethics. It is precisely the fact that this ethics recognises that some types of being or beings have an internal value, which provides a criterion by which to value the consequences of decision. Teleologism and deontologism are not opposed, but instead the special dignity or value of a type of beings (deontological moment) gives meaning to the question about the consequences of decisions *for those beings*, which allows them to be valued.

In spite of Rorty-type contextualist pragmatism, the unconditional is essential for the moral world.

Bioethics constitutes the modulation of civil ethics in the domains of biotechnology and medicine, which forces it to become transnational and to be present in a public sphere that is similarly transnational.

4. TRANSNATIONAL CIVIC BIOETHICS: A WORLDWIDE CIVIL SOCIETY

Indeed, in its application to the question of threatened life, civic ethics would be modulated as a *"civic bioethics"*, entrusted with responding to the great questions and principles, working from values and principles shared by the different groups, or the different "ethics of maxima" within pluralist societies. Discovering these shared values, and from them suggesting responsible answers, is precisely the commitment assumed by the national and international bioethics commissions and associations, and by the ethics commissions of public institutions, finding out as they proceed how a minimum of moral agreement goes beyond frontiers and gradually forges a *transnational civic bioethics.*

In fact, any national commission that attempts to reflect on bioethical problems will take into account the documents prepared by commissions from other countries. Such agreements do indeed explicitly respect national legislations out of respect for the sovereignty of states. But one should remember that pluralism is not something existing between states, but within each one of them, as they are transversal and the minimum ethics is increasingly shared by all.

This is why *Bioethics commissions* increasingly constitute a *"phenomenisation"* of a *"civic"*, and not state *morality*, that joins *citizens (including professionals) from different states.* Discovering these minima makes them bring into play the "public use of reason" through a deliberative process.

5. THE PUBLIC ROLE OF BIOETHICS

As has already been mentioned, bringing out the minima shared by the people and gradually extending the sphere of agreement is essential to the field of bioethics, because it is the people that are affected by the decisions made, and it is immoral to decide on such questions without bearing in mind "what everyone might want."

This can only be done through a continuously dynamic deliberation process, in which solutions are always open to review. This is particulary the case in areas in which scientific and technical progress constantly provides new data, problems and situations that raise doubts about the previous solutions.

Bioethics therefore has a role to play in the public sphere that could be characterised in five ways:

1. First of all it should develop internal interdisciplinary deliberation processes that may be useful as a model for public deliberation.

For example, to evaluate any particular practice ethically, a bioethics commission must take at least the following steps: 1) describe the different aspects in depth from a scientific standpoint; 2) try to bring to light and formulate the ethical values that the different social groups already share in respect to this practice; 3) reveal the ethical principles that guide such values; 4) examine to what extent agreement is already a reality, and where the disagreements start; 5) open up a wide-ranging debate on the points where disagreement arises; 6) attempt to reach the point at which all positions seem morally respectable; 7) offer recommendations for the specific issue from the majority position, while stating any discrepancies. These discrepancies must be of convictions, not of interests, since "moral pluralism" does not consist of a diversity of interests that have to be balanced, but of a plurality of ultimate convictions which nevertheless find points of agreement.

Through this gradual discovery of shared ethical values and principles by which to judge those types of practices that are humanising and those that are not, an increasingly dense civic Bioethics enables the emergence of a pre-existing *ethical intersubjectivity*, as opposed to relativism and subjectivism.

2. It should offer its information and moral convictions to public opinion, through disclosures and statements, in order to foster a reasoning public opinion. This will help to avoid any ideologisation stemming from political or economic interests, and so convert deliberation into a cooperative search for what is most just for human beings.

3. It should gradually construct a transnational civic bioethics. This is already essential, because doing good, respecting autonomy and doing justice – the principles of bioethics – are principles which can only be followed internationally in a world society. The effects of decisions in bioethics often affect not one particular *polis*, but a *cosmo-polis*.

4. It should cooperate in shaping the ethical ethos of societies. Codes, committees, and audits constitute in my opinion a 'phenomenisation' of societies' moral consciousness, the tangible expression that moral consciousness takes form in the different spheres of social life.

5. It should cooperate in shaping an ethos for biotechnological research and public health practice that is fair and wise, thus producing a *"public good"*.

A public good is a commodity whose use is non-competitive. Those who produce it obtain a benefit through it, but at the same time produce a good which is enjoyed by the whole of society with no need to compete for it. In the case of bioethics, a society that enjoys a fair and wise ethos in biotechnological research and public health practice is already cosmopolitan. As Amartya Sen put it, referring to companies, a firm that operates morally is a public good because its morality is contagious, even though it is a company financed with private capital (Sen 1993) As far as we ourselves are concerned, we could say that if technological research and the public health service set out to do good, respect and foster autonomy, and do justice, they constitute the best of public goods.

The public role of bioethics not only consists in participating in public deliberations and fostering the public use of reason, but also in embodying its convictions in everyday life, and so generating a public good.

References

Barber, Benjamin. *A Place for Us.* New York: Hill and Wang, 1998.
Cohen, Jean L. and Arato, Andrew. *Civil Society and Political Theory.* Cambridge, Mass.: MIT Press, 1992.
Conill, Jesús. "El carácter hermenéutico y deliberativo de las éticas aplicadas." In *Razón pública y éticas aplicadas.* Adela Cortina and Domingo García-Marzá (eds.). Madrid: Tenos, 2003, 121–142.
Cortina, Adela. "Civil Ethics and the Validity of Law." *Ethical Theory and Moral Practice,* vol. 3, 2000, 1, 39–55
Cortina, Adela. "Radikale Demokratie und Anwendungsethik." In *Zur Relevanz der Diskursethik.* J.P. Harpes and W. Kuhlmann (eds.). Münster: LIT, 1997, 337–350.
Cortina, Adela. *Ética mínima.* Madrid: Tecnos, 1986.
Cortina, Adela. *Covenant and Contract.* Leuven: Peeters, 2003a.
Cortina, Adela. "El quehacer público de la ética aplicada." In *Razón pública y éticas aplicadas.* Adela Cortina and Domingo García-Marzá (eds.). 2003b, 13–44.
Cortina, Adela (2004): "Virtualitäten einer erneuerten Diskursethik für die Bioethik." In *Neue Tendenzen in der Diskursethik.* Niels Gottschalk-Mazouz (ed.). Königshausen & Neumann, 2004, 65–80.
Habermas, Jürgen. *Strukturwandel der Öffentlichkeit.* Darmstadt/Neuwied: Hermann Luchterhand Verlag, 1962.
Habermas, Jürgen. *Faktizität und Geltung.* Frankfurt: Suhrkamp, 1992.
Kant, Immanuel. "Beantwortung der Frage: Was ist Aufklärung?" *Kants Werke.* Akademie–Textausgabe. Berlin: Walter de Gruyter, VIII, 1968, 33–42.
Kant, Immanuel. "Zum ewigen Frieden." *Kants Werke.* Akademie–Textausgabe. Berlin: Walter de Gruyter, VIII, 1968, 381.
MacIntyre, Alasdair. *After Virtue.* London: Duckworth, 1981.
Rawls, John. *Political Liberalism,* Lecture VI, 1993.

Sen, Amartya. "Does Business Ethics Make Economic Sense?" *Business Ethics Quarterly*, vol. 3, 1993, 1, 45–54.

The Hastings Center Report. *The Goals of Medicine: Setting New Priorities*. 1996, November-December.

Walzer, Michael. "The Civil Society Argument." In *Theorizing Citizenship*. R. Beiner (ed.). NY.: State of New York Press, 1995, 153–174.

[1.] This paper is a part of the research project HUM2004-06633-C02-01/FISO (Ministerio de Educación y Ciencia) and of the research projects of the I+D+I Group 03/179 (Generalidad Valenciana).

[2.] Compare: Beauchamp, Tom L. "On eliminating the distinction between applied ethics and ethical theory." *Monist*, 1984, 67, 514–531. – Apel, Karl-Otto. *Diskurs und Verantwortung*. Frankfurt: Suhrkamp, 1988. – Kurt Bayertz (ed.). *Praktische Philosophie. Grundorientierungen angewandter Ethik*. Hamburg: Rowohlt, 1991. – Lenoir, Frédéric. *Le temps de la responsabilité*. Paris: Fayard, 1991. – Apel, Karl-Otto and Kettner, Matthias (ed.). *Zur Anwendung der Diskursethik in Politik, Recht und Wissenschaft*. Frankfurt: Suhrkamp, 1992. – Lipovetsky, Gilles. *Le crépuscule du devoir*. Paris: Gallimard, 1992. – Cortina, Adela. *Ética aplicada y democracia radical*. Madrid: Tecnos, 1993. – Harpes, J.-P. and Kuhlmann, W. (ed.). *Zur Relevanz der Diskursethik*. Münster: LIT, 1997. – Winkler, E. and Coombs, G. (eds.). *Applied Ethics: A Reader*. Oxford: Blackwell, 1993. – Nida-Rümelin, Julian (ed.). *Angewandte Ethik. Die Bereichsethiken und ihre theoretische Fundierung*. Stuttgart: Kröner, 1996. – Chadwick, Ruth (ed.). *Encyclopedia of Applied Ethics*. San Diego/London: Academic Press, 1998. – Sosoe, Lukas K. *La vie des normes & l'esprit des lois*. L'Harmattan: Paris, 1998. – Kettner, Matthias (ed.). Angewandte Ethik als Politikum. Frankfurt: Suhrkamp, 2000. – Cortina, Adela and García-Marzá, Domingo (eds.). Razón pública y ética aplicadas, Madrid: Tecnos, 2003.

[3.] Compare: Jonsen, Albert R. and Toulmin, Stephen. *The Abuse of Casuistry. A History of Moral Reasoning*. Berkeley: University of California Press, 1988. – Gracia, Diego. *Procedimientos de decisión en ética clínica*. Madrid: EUDEMA, 1991. – Kucewski, Mark. "Casuistry." In: Ruth Chadwik (ed.). *Encyclopedia of Applied Ethics*. I, 1998, 423–432.

Chapter 12

EXPERTS ON BIOETHICS IN BIOPOLITICS

SIGRID GRAUMANN
Berlin, Germany

1.   INTRODUCTION: WHAT IS THE DIFFERENCE BETWEEN BIOETHICS AND BIOPOLITICS?

New biomedical techniques, in particular those such as pre-implantation genetic diagnosis, cloning and germ line interventions, which make the "selection" and "manipulation" of human life possible, are among the most controversial bioethical topics under public discussion. They highlight fundamental differences not only between various interests, but also between basic moral convictions, world-views, and conceptions of the human being. It often seems impossible to reach a consensus among the representatives of divergent positions, and indeed insurmountable disagreements appear to predominate. In other words, a plurality of interests, assessments, convictions, world-views and conceptions of the human being stands in the way of a social consensus on the research and application of these biomedical techniques.

In this context, the establishment of bioethics as an academic discipline can be understood as an answer to the ethical and political controversies caused by progress in the life sciences. It is extremely difficult for political decision-making on legal regulation in the field of biomedicine to find solutions that the general public is willing to accept. One answer to this problem is to delegate the problem solving to experts in bioethics. This means that bioethicists are expected to find "right" answers and, especially if they are members of bioethics commissions, to provide politicians with recommendations for good regulation. Frequent conflicts between representatives of public and academic discourse show, however, that the

C. Rehmann-Sutter et al. (eds.), Bioethics in Cultural Contexts, 175–185.
© 2006 *Springer. Printed in the Netherlands.*

goal of finding solutions for bioethical problems that the public will accept, is often not being attained. Let us now examine these conflicts.

On the one hand, bioethics experts periodically criticise the public discussion of biomedical topics. Their criticism refers specifically to the following issues: discussions are in part carried out with primary reference to strategic interests; "ambiguous terminology" is used (Winnaker 1997); social acceptance is confused with moral legitimacy; arguments are used inconsistently (Mieth 1997); the correct procedure for reaching necessary consensus has not yet been found (Honnefelder 1996); controversial positions are intentionally polarized (Wiesing 1998); some lines of argument are extremely emotionally loaded or linked with intense fears (Winnaker 1997).

On the other hand, the public also criticises ethical discussion. This criticism makes several specific claims: ethical discussion contributes to the de-emotionalising of decisions about human life; it consciously promotes the legitimisation of problematic fields of biomedical research and application; it consists of a proxy discussion between parties that are not directly involved, and so encourages a mode of debate that is in principle undemocratic.

This paper does not intend to resolve the conflict sketched out above between experts and sections of the public concerning the way in which bioethical issues are dealt with in different discourses. My aim is, however, to consider the public debate and the role of experts in this debate – particularly that of bioethicists from disciplines such as Theology, Philosophy, Social Sciences, Law, Medicine etc. As a preliminary move in this direction, I will present some of the results of an analysis of the public discussion in the German media, and then discuss some theoretical aspects of the function and task of the media with respect to such problematic and controversial bioethical issues. I shall conclude with a consideration of the role and function of experts and particularly bioethicists in public debate.

2. THE PUBLIC DEBATE ON IN-VITRO TECHNIQUES IN THE MEDIA

Discussions about controversial biomedical research programmes and areas of application take place in political institutions, in public by means of the mass media and everyday communication, and in the academic world in specialist publications and conferences. My comments here will concern the public discussion in the mass media as it is represented in the printed media.

In order to reconstruct the public debate about the "selection" and "manipulation" of human life, I have carried out research using key words in the archives of twelve national daily newspapers[1] and weekly journals[2] with high distribution rates, which were printed between 1995 and 2002. I have collated and assessed all of the relevant articles on the topic.

First of all I want to present a number of general features of the public debate on the "selection" and "manipulation" of human life. I will then focus more specifically on particular arguments.

The authors of the relevant contributions are for the most part well informed science journalists, familiar with a wide range of issues relating to the subject.[3] In some newspapers, a relatively large number of background articles and interviews have appeared, written by or with the contribution of experts. Four major topics – human cloning, "human breeding", pre-implantation genetic diagnosis, and embryonic stem cell research – were the subject of public discussions between experts, organised by individual newspapers. These expert-debates in the media were: the Dolly debate (1997), the Sloterdijk debate (1999), and the Human-Embryo-Protection debate (2001–2002). They had a clear agenda-setting effect on public discourse.

Further, it is clear that those who, as well as being writers of guest articles, dominate as protagonists and are referred to and cited, are experts in the life sciences, medicine or ethics, and politicians. Professional associations such as the *Deutsche Forschungsgemeinschaft* (German Research Foundation), the *Gesellschaft für Humangenetik* (Human Genetics Society), and the *Bundesärztekammer* (German Medical Association) play an important role. These associations pursue a specific policy in the media in much the same way as politicians. For example, the *Bundesärztekammer* presented a draft of a set of guidelines for the use of pre-implantation genetic diagnosis at a press conference in February 2000. This draft gave the impression that the medical profession held unified position: the application of pre-implantation genetic diagnosis is permissible in individual cases, subject to restrictive self-control on the part of the medical profession. A similar phenomenon occurred with the opinion on embryonic stem cell research published by the *Deutsche Forschungsgemeinschaft* in May 2001.

The voices of other protagonists – representatives of the churches, or of social movements such as the women's health movement, organisations of critical scientists or those representing the interests of people with disabilities – are under-represented in the press in these discussions. One exception is the statement by the Conference of German Bishops on pre-implantation genetic diagnosis, which rejects pre-implantation genetic diagnosis on the basis of the right to protection of human embryos and with reference to eugenic tendencies. The positions of other interested parties are usually only heard in the media when they are argued against – a part of the "bioethics critique" is certainly a result of this. Both the social movements and the Christian communities have their own discourses in their own media. However these have very little influence on public discourse in the mass media.

Contrary to my own expectations, the assumption that the debate about the "selection" and "manipulation" of human life is sensationalised in black and white terms into a conflict between pessimistic conservative and optimistic progressive views turned out to be incorrect. Science journalists report for the most part in a deliberately balanced fashion whereby the opinions of supporters and opponents are usually relatively evenly cited. Pointedly extreme positions are only given exposure in order that they can be countered. Guest articles by bioethics experts on the other hand, usually take up explicit and pointed positions (particularly pronounced in the expert-debates on Dolly, Sloterdijk and human embryo protection).

Also in contrast to guest articles by experts, comparatively few journalistic contributions concern themselves in detail with a single technique or area of application. The "selection" and "manipulation" of human life is usually addressed in a relatively general fashion with reference to individual techniques or areas of application as examples. As far as techniques for the "selection" and "manipulation" of human embryos are concerned, the following tendencies can be observed:

a) The most important criterion for the presentation of a topic in the media is its degree of currency. Mostly this is the publication of commission reports, reports on scientific conferences, and political debates regarding new regulations (for example the new law on reproductive medicine planned in 2000, and the law on embryonic stem cell import in Germany, adopted in 2001). Relatively rare are spectacular events, such as the birth of the cloned sheep Dolly in 1997, or the scandalous speech about human breeding given by the philosopher Sloterdijk in 1999. These usually refer to future techniques such as reproductive cloning and germ line interventions.

b) The different debates overlap in several areas. These include: the discussion on prenatal diagnosis and that on § 218 (the legislation governing abortion in Germany); the debate about the historical experience of eugenics and euthanasia; discussions about in-vitro fertilisation in general; discussions about the economic interests involved in biopatenting. This overlap widens the horizon of the debate and thus leads to the examination of fundamental questions such as the tension between autonomy and social responsibility, the lessons learned from the national-socialist period, and the consequences of an "economic misuse of human life".

c) In contrast to the significant role played by the question of the moral status of human embryos in expert discussions, until the end of 2000 journalists were extremely reserved about addressing this controversial issue in the media; the question was primarily addressed by politicians. In this context, legal-positivistic arguments referring to the Law for the

Protection of Embryos and the constitutionally guaranteed protection of unborn life dominated discussions. Then in December 2000, after the British decision on stem cell research and "therapeutic cloning", the organised expert-debate on human embryo protection began with an article by Chancellor Gerhard Schröder pleading for more liberal regulation in Germany. This was clearly a turning point. Before December 2000, there was for the most part no recourse to the authority of the law, and the issue of the ethical argumentation on which an embryo's right to protection is based, was seldom discussed. Since Schröder's 'chancellor-statement', however, a heated expert-debate about the legitimacy of the German Embryo Protection Act from an ethical and constitutional point of view has taken place directly in the media.

d) Regarding pre-implantation genetic diagnosis, the desire of "genetically 'at risk' couples" to have a healthy child of their own is given high status in the press. One patient was quoted in the *Welt* (28 July 1995) as saying: "All I want is a healthy child". In the expert-debate, but also among those directly affected, the so-called assessment contradiction between prenatal diagnosis and abortion is often raised. "Of course the sick embryos are disposed of when they are only a few days old", argued a future mother in *Focus*, "but isn't that better than aborting them much later as foetuses?" (14 December 1998).

e) In general, paternalistic interventions in individual life choices in the context of family planning are rejected. This position takes on particular significance in its relationship to the "autonomy research" of the women's movement. At the same time, however, the interests of the future parents are called into question. The *FR* writes, for example, that "children come into being as a means of fulfilling the wishes of their parents to a previously unheard of extent – they are 'instrumentalised'" (*FR*, 15 November 1999).

f) The identification of "the potential for misuse" and the danger of a slippery slope to "the manipulation of embryos" (letter to the editor, *FR*, 1 December 1999) plays a significant role in the press. While these kinds of slippery-slope arguments are used in media debate as a matter of course, in the specialist discussion they are extremely controversial. Here visions of the future play an important role. The impression is often given that the "genetic design" of a human being is almost entirely governable. Only occasionally are the limits of the technical possibilities mentioned. The *FAZ*, for example, writes that some ideas about how one technique should be applied are irrelevant because "the interaction of the genes is not amenable to analysis" (*FAZ*, 12 August 1998). Pre-implantation genetic diagnosis and embryonic stem cell research are regularly, and for the most part with critical intention, placed in the context of a vision of a "society without suffering" (e.g. the

Zeit, 2 March 2000); this is answered, for example, by the statement that "the disabled cannot be prevented" (*FR*, 1 December 1999).

g) Until December 2000, the main conflict with respect to in-vitro techniques as presented in the media was between individual freedom with regard to family planning and the development of new therapies for severe diseases on the one hand, and the responsibility of society on the other. On the one side were the unfair burdens and deprivations for the individual couple caused by a disabled child and the postulate of a "moral duty" to avoid suffering in general. The other side emphasised the moral claim of solidarity with sick and disabled persons. So, for example, the *FR* wrote on the subject of pre-implantation genetic diagnosis (15 November 1997): "But if many make individual decisions to be tested, this will change our way of viewing disabled people and our way of treating them ... Here, the social costs of a disabled child are being weighed up against the possibility of preventing his or her existence." In contrast to its relatively minor role in expert discussion, the conflict identified here about the changes taking place in social values and norms was of absolutely central importance in the public debate. Since December 2000, however, the conflict with respect to pre-implantation genetic diagnosis and embryonic stem cell research has been mainly reduced to the postulated right of the embryo to protection on the one hand, and the postulated rights, wishes, and interests of future parents and patients on the other.

In summary, we can make the following claims. In journalistic contributions on in-vitro techniques, a trend could be observed towards formulating the ethical problem as a conflict between a justified desire on the part of a "genetically 'at risk' couple" to have a healthy child of their own, or hopes for new therapies for severely ill persons on the one hand, and dangerous social and cultural changes on the other. The latter was postulated with reference to a society increasingly hostile towards persons with disabilities, the danger of a slippery slope towards a comprehensive instrumentalisation of embryos for research interests, and the vision of a society without suffering and of future children as the objects of parental vanity.

In academic ethical discussion, however, the conflict with respect to pre-implantation genetic diagnosis and embryonic stem cell research was always seen to be between the right of the embryo to protection, and the rights, wishes, and interests of future parents and patients. This corresponds to a limitation of the problem to an individual-ethical level. In public discussion, however, a considerably broader perspective was adopted, and social-ethical questions were addressed. Here, in-vitro techniques were discussed in both a historical (the historical experience of eugenics and visions of the future) and social (discrimination and the danger of a slippery slope) context. The debate about changes in key societal and cultural values and norms was highlighted in the way that conflicts between different communities of interest, systems of belief, world-views and notions of being human were addressed. But with the turning point in public debate in December 2000, the academic individual-ethical approach was also adopted by the media, and this led to a marginalization of psycho-social and socio-cultural arguments.

However, the academic ethical discussion that concentrates on the old pro-life-pro-choice controversy, which began to dominate the public debate at the end of 2000, does not take adequate account of the concerns that were expressed in the public arena. I rather doubt that this can lead to political decisions that the general public is ready to accept.

For a better understanding of this development I will now examine a number of theoretical considerations with reference to the functions and tasks of public debate.

3. FUNCTIONS AND TASKS OF PUBLIC DEBATE

Biomedical research and its areas of application alter social reality. Their legitimisation and, when necessary, regulation is one aspect of political communication. Under the conditions created by a "mass society" (Hannah Arendt), political communication – and political action is essentially communicative action – relies on the medium of the mass media. The media are ultimately the only forum in which political communication, as a public conversation about how common concerns are to be dealt with and ordered, can take place. Alongside other functions, such as entertainment and the use of leisure time, this is a necessary and irreplaceable function of the media, given the conditions created by the mass society.[4] It obliges journalistic activity to take account of public interest, in the sense of a minimal principle that is opposed to the naked power struggle between different interests in the political arena. The ideal outcome is political dialogue, understood as a common search for political solutions (and when necessary compromises) that can be extended across a broad range of issues, and to which all affected parties have good reason to agree (Hügli 1992).[5] Essential moral guidelines for the professional practice of journalism are derived from the principle of public interest – itself already a moral principal – as institution-ethically based structural conditions that make correspondingly responsible journalism possible. These guidelines include: the commitment to truthful and objective reporting, the commitment to accuracy in research, balance in reporting controversial positions, the separation of factual report from opinion, and so on.

In individual cases it may be a matter for debate as to what extent these ethically based demands are fulfilled at an individual or institutional level in public discussion about the "selection" and "manipulation" of human life. I would maintain, however, that my investigations of the public debate demonstrate that in this case, at least in general, no obviously incorrect journalistic conduct is to be found. Recriminations directed at the press on the grounds that no societal consensus on the moral problems associated with the techniques has been reached, would therefore seem to be inappropriate.

As has been shown, the essential difference between the public discussion and the academic ethical debate is that in the former a broader perspective was adopted, and the techniques were addressed in their historical and social context without forgetting the perspective of the individual couple or patient affected. This raises the question of whether such a difference is appropriate from a media-ethical perspective or whether it serves "the public interest". This question is of particular relevance, because of the turning point in the public debate driven by the Human Embryo Protection expert-debate mentioned above. The suggestions made by professional ethicists, that the discussion should be held in a dispassionate and pragmatic manner, are of course all arguments in favour of a narrowing of the perspective (Birnbacher 1999).

Social reality is the "matrix" within which individuals accumulate their knowledge about the real world, construct their systems of belief, make decisions, and act (Hacking 1999). Nevertheless, social reality is only accessible to direct personal experience to a very limited extent. It is increasingly mediated through the media. There is a certain sense in which the media "construct" social reality (Debatin 1997). Social values and norms emerge, are changed, or are newly constructed in public discussion. They thus influence the possibilities for future development of research and application; these possibilities simultaneously reflect back on the development of social norms. This means, however, that the debate as it takes place in the public arena is not only about the social acceptability of specific practices, but also – and always – about the social values and norms, world-views and perspectives on the human being that shape our decision-making and our actions. And since experts, and particularly ethicists, play such a central role in public discussion about biomedical issues, they also have a part in the development of the system of values and norms that a society adopts, and must therefore act with appropriate responsibility. Here the particular authority ascribed to the positions of experts in public debate must be taken into account.

One could argue that ethicists should stay out of the public discussion: they are academics, and their job is bioethics rather than biopolitics. But at the same time it is clear that there is a great need in society at large for ethical reflection, especially in the area of biomedicine. The increasing institutionalisation of bioethics is, of course, not least a response to social controversies about biomedical research. However, this means that bioethics is essentially and unavoidably also biopolitics. One result of this fact is that it is necessary for bioethics to reflect on itself, a process that in my opinion must take place in the tension between the ideals that govern the scientific search for truth and those which govern political communication.

According to Hans Lenk, the genuinely scientific task of bioethics should orient itself according to the internal responsibility for the best possible objective search for truth and its confirmation on the one hand, and the external responsibility to society on the other (Lenk 1991).

The *internal responsibility* of bioethical experts is juxtaposed with the problem of an interdisciplinary approach. To find the "truth" or rather the best answer to ethical questions implied by biomedicine usually requires specialist knowledge from different disciplines. It is not enough that a person participates in a multidisciplinary debate. A philosopher, for example, cannot call himself an expert in the ethical problems of genetic diagnosis without learning genetics, nor can a physician without philosophical expertise. For experts in bioethics this points firstly to the need for interdisciplinary training and secondly to the need to be honest about the lack of knowledge in certain fields.

Another requirement of good (bioethical) research is that it should be critical in the sense of openness about the results. If this is not the case, bioethical research is nothing more than the promotion of the legitimisation of certain scientific goals. Related to this, good (bioethical) research should be as far as possible free of interests. Obviously, this is not completely possible in bioethics. At least some of the participants in the interdisciplinary bioethical discourse clearly have personal interests, for example scientists who want to work in a controversial field such as embryonic stem cell research. Consequently, bioethical research should be aware of the influence of personal interests and refer to a critical reflection of scientific knowledge.

To address these aspects, bioethics' internal responsibility requires an academic space where the influence of political interests is minimised. In Germany there is no such space. For example, academic bioethical debate concerning pre-implantation genetic diagnosis and embryonic stem cell research took place directly in the media. Experts from the different fields discussed their theories in a very polarised way in major newspapers without moderation and mediation by journalists. Obviously, this discussion was guided less by the rules of internal responsibility for finding the best answers from an ethical perspective than by the intention of influencing political decision making. This phenomenon shows that the role of experts in public discourse has not been adequately thought through, which brings me to the next point: the external responsibility of experts.

According to Hans Lenk, the *external responsibility* to society that researchers and in particular experts in bioethics have, points to their role in public discourse with regard to political decision making. The political task of bioethics, in my view, should be directed towards the ideal that governs political dialogue: that of the society's joint search for political solutions that can be generalised.

This is the true subject for debate: how far must responsibility be assumed not only for single clearly definable actions but also for general developments within society that have to be seen in interference with the dynamics of bio-scientific research? Here social sciences and social theory should be included in bioethical research.

In this context, I argue for the definition of the task of bioethics in public discussion as contributing, with the "reflective competence" specific to it, to the clarity of the positions adopted and thus to better communication between the various opposing parties. This means refraining from proxy debates, refraining from changing the agenda (as happened in the Human-Embryo-Protection debate), and addressing the concerns expressed in public discussion – in particular the changes in social values and norms – in ways appropriate to academic ethical discourse.

Finally, I would like to plead for a process of self-reflection in bioethics that takes account of its academic character and political function. This should include the development of a methodological basis for the specific preconditions of bioethics as an interdisciplinary and politically relevant academic field.

References

Berger, P.L. and Luckmann, T. *Die gesellschaftliche Konstruktion der Wirklichkeit. Eine Theorie der Wissensoziologie*. Frankfurt a.M.: Fischer, 1999.

Birnbacher, D. "Klonen von Menschen: Auf dem Weg zu einer Versachlichung der Debatte." *Forum, TTN Technik Theologie Naturwissenschaften* 1999, 2 (November), 22–34.

Dörner, A. "Medienkultur und politische Öffentlichkeit." In *Kultur – Medien – Macht*. Andreas Hepp and Rainer Winter (eds.). Opladen: Westdeutscher Verlag, 1997, 319–335.

Graumann, S. "Die Rolle der Medien in der Debatte um die Biomedizin." In *Kulturelle Aspekte der Biomedizin. Bioethik, Religionen und Alltagsperspektiven*. Silke Schicktanz, Christof Tannert and Peter M. Wiedemann (eds.). Frankfurt a.M.: Campus, 2003, 212–243.

Habermas, J. "Vorbereitende Bemerkungen zu einer Theorie der kommunikativen Kompetenz. Theorie der Gesellschaft oder Sozialtechnologie." In *Habermas*. J.L. Niklas. Frankfurt a.M.: Suhrkamp, 1990, 101–141.

Hacking, I. *The Social Construction of What?* Cambridge, Massachusetts: Harvard University Press, 1999.

Höffe, O. "Strategien politischer Gerechtigkeit: Zur Ethik öffentlicher Entscheidungsfindung." In *Gerechtigkeit. Themen der Sozialethik.* A. Wildermuth and A. Jäger (eds.). Tübingen: J.C.B. Mohr, 1981, 107–140.

Honnefelder, L. "Bioethik im Streit. Zum Problem der Konsensfindung in der biomedizinischen Ethik." *Jahrbuch für Wissenschaft und Ethik* 1996, 1, 73–86.

Hügli, A. "Was haben die Medien mit Ethik zu tun?" In *Medien-Ethik*. Michael Haller and Helmut Holzhey (eds.). Opladen: Westdeutscher Verlag, 1992, 56–74.

Keller, R. "Diskursanalyse." In *Sozialwissenschaftliche Hermeneutik*. R. Hitzler and A. Honer (eds.). Opladen: Leske und Budrich, 1997.

Lenk, H. (ed.). *Wissenschaft und Ethik*. Stuttgart: Reclam, 1991.

Luckmann, T. *Lebenswelt und Gesellschaft*. Paderborn: Schöningh, 1980.

Mieth, D. "Gentechnik im öffentlichen Diskurs: Die Rolle der Ethikzentren und Beratergruppen." In *Gentechnik, Ethik und Gesellschaft*. M. Elstner (ed.). Berlin, Heidelberg, New York: Springer, 1997, 211–220.

Wiesing, U. "Der schnelle Wandel der Reproduktionsmedizin und seine ethischen Aspekte. Von der Prädiktiven zur Präventiven Medizin – Ethische Aspekte der Präimplantationsdiagnostik." *Ethik in der Medizin*. 1999, Band 11, Supplement 1, 99–103.

Winnacker, E.-L. "Wieviel Gentechnik brauchen wir?" In *Gentechnik, Ethik und Gesellschaft*. M. Elster (ed.). Berlin, Heidelberg, New York: Springer, 1997, 43–56.

---

[1] The Frankfurter Allgemeine Zeitung (FAZ), the Frankfurter Rundschau (FR), the Neue Zürcher Zeitung (NZZ), the Stuttgarter Zeitung (STZ), the Süddeutsche Zeitung (SZ), the Tageszeitung (taz), and the Welt.

[2] The Focus, the Spiegel, the Stern, the Woche, and the Zeit.

[3] E.g. Hans Schuh, Volker Stollertz (Zeit), Wolfgang Löhr (taz), Michael Emmrich (FR), Claudia Ehrenstein (Welt), Reiner Flöhl, Barbara Hobom (FAZ).

[4] "Political communication" is, however, not diametrically opposed to "entertainment" in the media: both are a part of "cultural praxis" and both take part in the creation and maintenance of cultural hegemony (Gramsci) as a prerequisite for stable political conditions of governance (Dörner, 1997).

[5] The central idea of a process of shaping public opinion as an appropriate formulation of the morality of public discourse goes back to Habermas, although it is controversial whether Habermas' conception works as a metatheory of ethics (Hügli 1992; Habermas 1990).

Chapter 13

THE CONTRIBUTION OF MEDICAL HISTORY TO MEDICAL ETHICS
The Case of Brain Death

CLAUDIA WIESEMANN
Göttingen, Germany

Medical history always has a presence in contemporary ethical debates. A number of medical ethicists have presented their understanding of the historical origins of medical ethics as a discipline (e.g. Jonson 1998; Carson 1997; Baker 1993–95). This story is important to them because it is intimately linked to their identity as professional ethicists. Most of them believe that during the twentieth century medical ethics awarded patient autonomy its proper place in medicine for the first time in history. Historical evidence also plays an important role in the discussion of particular ethical issues such as euthanasia and genetic diagnosis and/or treatment. In these debates the devastating experience of twentieth century totalitarianism, particularly German National Socialist medicine, still casts its shadow on today's moral debates.

Therefore the question is not: "does medical history contribute to medical ethics?" but rather: "what is its contribution, and how can we describe it?" Whereas philosophical discussions can be very abstract, medical history can show how "things really worked". History is like a large case-book, a textbook full of practical examples and experience, teaching us things that can be applied to today's problems. But what exactly does it teach? Medical history has always been used (and misused) as a particularly persuasive argument in contemporary debates and, moreover, in the legitimisation of contemporary ethical policies: it is a powerful tool in arguments legitimising or de-legitimising ethical claims. Thus, it is evidently important to try to understand the historical dimension of medical ethics rather than to make instinctive, popular or even demagogical use of it.

However, there is no simple way of harvesting the riches of the past. First of all, everything is subject to historical change. Even morality has a

C. Rehmann-Sutter et al. (eds.), Bioethics in Cultural Contexts, 187–196.
© 2006 *Springer. Printed in the Netherlands.*

historical context and changes over time. It is difficult to find timeless and universal ethical principles in medical history when basic notions of disease and illness, suffering and curing are constantly changing. Patients have not always been patients, and doctors have not always been doctors, at least not as we understand the terms today; those who criticise the Hippocratic oath for having an old-fashioned, paternalistic and patronising attitude towards the patient may well be falling prey to an anachronistic fallacy. The ancient physician was not like his modern colleague, and ancient patients were not like their modern counterparts. So the way they met in the curing relationship was consequently also governed by different ethical rules and principles.

But the matter becomes more complex yet. It is not just that ethical theories and principles are historical: historical theories are also ethical. History has a narrative structure; it has to be told in stories that imply meaning. It is usually structured in terms of well-known narrative plots, like drama, tragedy, comedy or the grotesque. These narrative plots convey an ethical meaning; an idea of what is right or wrong, good or bad. In heroic drama, for example, the protagonist defends his idea of what is true and right against the attacks of his enemies. In tragedy, fate is always stronger than human powers, so good is doomed to fail. The grotesque tells us that although we feel that the world should be a better place, it is wise to accept that human struggles are absurd and sometimes even ridiculous.

Medical history is also told in narratives. Most commonly, it takes the *Gestalt* of the heroic drama, in which the hero – for example Robert Koch, Louis Pasteur or Alexander Fleming – struggles for recognition and success. Sometimes it is told as a tragedy, for example when the forces of nature destroy human life as the plague did in the Middle Ages. It is rarely told like a comedy or grotesque – it seems as though the nature of medicine's problems is too serious for such a description. Tragically, the twentieth century has added a new narrative plot to those we already know. The history of Nazi euthanasia, and in particular the Holocaust, seems to transcend every known narrative structure. We cannot tell it as a tragedy – which in any case it was not, given that so many rational human beings were involved in its perfect organisation – nor can we tell it as a grotesque. So from this terrible experience developed a new genre of historical narration: the unique, incomparable and inexplicable historical event (White 1991). Like the other plots, this narrative also conveys an ethical meaning: we will not try to explain something, when explaining it will make the monstrous and dreadful event wrongly appear intelligible and understandable.

These narrative plots have a huge impact on how we perceive the ethical problems of today. Is medical genetics a great endeavour that has shed light on the nature of human beings and that will one day help to overcome the shortcomings of the human body and mind? Or is it proof of man's arrogance and pride, of his vain struggle against the laws of nature (or God)? It is evident that we cannot escape giving history an ethical meaning.

Through historical narratives, medical history is always a force in today's ethical debates. These narratives do not just explain the past; more importantly, they also interpret the future. Thus they structure ethical debates and help legitimise ethical policies. I will illustrate this thesis with an example from the medical history of the twentieth century: the history of the definition of brain death during the last fifty years. I will attempt to identify the narratives that were used to describe the history of brain death and explain their function in contemporary ethical debate, examining what happened during the 1960s and 1970s, as the concept of brain death became increasingly important. I will present the results of my research on the history of the first German definition of brain death, which was in fact published three months before the more famous Harvard Definition of Irreversible Coma in 1968. I will add to my findings some of the research results of medical historians Mita Giacomini and Martin Pernick, who have studied the history of the first brain death definition in America and have reached comparable conclusions (Giacomini 1997; Pernick 1999).

1. NARRATIVES AND THE HISTORY OF BRAIN DEATH

Not surprisingly, the history of the brain death concept is one of competing narratives. The first narrative, that of *certainty and scientific progress*, was championed by advocates of the brain death definition. They told the story of brain death as though it was the story of a lost paradise. Once upon a time everybody knew what death really meant. It was diagnosed when the dying person had stopped breathing and his or her heart had stopped beating. But then progress in intensive care medicine disrupted this idyllic scene, and artificial respiration made it increasingly difficult to diagnose the death of a comatose patient. But science found an answer to this troubling loss of natural evidence. It discovered the true indicator of life or death: the nervous system. And so humanity lost an innocent and natural death but gained a rationally explicable scientific death. And that was even better than before. The medical historian Martin Pernick was the first to judge this popular story of the glorious "discovery" of brain death as a myth. He concluded from his analysis that there had never been a "golden age of heart and lung". People had always feared being falsely diagnosed dead, and buried prematurely. And doctors in the past had also doubted their theories of life and death. But if there had never been a golden age of medical security in the past, why should we believe that medicine is different today?

Critics of the brain death definition, however, told a different story: one of *deception and betrayal*. According to this narrative, brain death was invented by transplant surgeons to suit their needs and demands: the use of brain death as a criterion for death helped to supply organs for transplantation. It was not scientific objectivity but professional interests that governed the implementation of brain death policies.

There is also a third narrative that is particularly prominent in the US, which claims that the Harvard brain death definition was the first medico-ethical document to be issued by medical and non-medical professionals equally, and as such it "struck a blow to medical privilege and autonomy".[1] For the first time the public was invited to take part in medical deliberations on life and death. This, of course, is the narrative of *progress through ethics*, the story of a truly professional medical ethics.

These three narratives show how history and ethics are mutually intertwined. Their advocates take a stance in current ethical debates, and they try to influence the debate by structuring the past. But can we discern which version is true?

2.   BRAIN DEATH AND THE HISTORY OF ARTIFICIAL RESPIRATION AND HEART TRANSPLANTATION

Before we attempt to answer this important question, let me sketch in just a small section of the contested historical period. In Germany during the early 1960s, anaesthetists and forensic pathologists occasionally discussed the need for a new definition of death.[2] At this time, because of the rapid diffusion of artificial respiration during the fifties and sixties, intensive care medicine was being confronted with patients who had incurable diseases and severe brain damage, and yet were kept alive by mechanical ventilation. Moreover, some transplant centres in Berlin and Munich – as in other countries in the West – had begun to transplant kidneys from the newly dead. These organs were procured from patients with cardiac or respiratory arrest who had been "reanimated" for some minutes to prevent the removable organs from lethal destruction.[3] Both practices raised questions about the nature of death. Was it right to call someone who had lost all their brain functions alive, just because their heart would not stop beating regularly? Was it, on the other hand, right to call someone dead whose heart had stopped beating but whose body was "reanimated" and even put on a heart and lung machine to keep their organs alive for explantation?[4]

In 1966, during an International Symposium run by the Ciba Foundation entitled "Ethics in Medical Progress: With Special Reference to Transplantation", the criteria for human death were critically discussed by surgeons, anaesthetists, and lawyers. Interestingly, during this conference and on other occasions, physicians harshly criticised the utilitarian attitude of transplant surgeons towards organ removal from patients with heart attacks. In the spring of 1967, in reaction to this intra-professional critique, the President of the German Society of Surgery set up a "Committee for Reanimation and Organ Transplantation", which was authorised to develop criteria for the determination of "irreversible death" (more detailed in: Wiesemann 2000a, 2000b and 2001).

When news of the first ever heart transplant, performed in South Africa by Christian Barnard, spread around the world in December 1967, the public became aware of the rather high-handed surgical practice of explanting organs from persons who were near death or were presumed to have expired. Barnard himself had taken the heart of a young female victim of a car accident with severe head injuries; no formal criteria for the determination of death had been fixed. In an immediate response to this sensational news, the Harvard Medical School also formed an "Ad Hoc Committee to Study the Problems of the Hopelessly Unconscious Patient". It issued a statement on brain death in June 1968 (Giacomini 1997: 1474). When judging the Committee's activities and its primary intentions, Mita Giacomini, who analysed draft manuscripts and other sources, came to the following conclusion:

"For the Committee members, transplantation was central to the purpose but detrimental to the rhetoric of redefining death. In the final report, the Committee characterized its reasoning process as if it had begun by contemplating the features of death, and ended up conveniently but coincidentally with features consistent with a good vital organ source. [...] In constructing its definition the Committee seems to have begun with the already-familiar characteristics of ideal organ donors" (Giacomini 1997: 1474).

3.   THE FIRST BRAIN DEATH DEFINITION IN GERMANY

In 1968, the German committee also issued a statement on the determination of death, which was published in April 1968, three months before the Harvard definition. This statement revealed even more clearly the dominant role of transplant surgery in the definition of death. It listed criteria for determining brain death (for example, loss of reflexes or a flat EEG), but it also stated that brain death could be "presumed" in cases of fatal respiratory or cardiac arrest. This was the "Barnard solution". Brain death was merely a name for a condition where irreversible and complete brain damage was presumed but not proved. So, German surgeons were encouraged to continue explanting kidneys and livers from non-heart-beating humans.

Now, what is the answer to the question of which narrative of the history of brain death is true? Was it certainty and scientific progress? Or deception and betrayal? Or progress through ethics? At first glance it seems as though the story of deception and betrayal is true. In the 1960s, the utilitarian interests of transplant surgery clearly dominated the discussion about brain death. German surgeons were not really interested in defining exhaustive and comprehensive criteria for the determination of death or brain death, which is why the German statement remained rather vague about which procedure should be performed in order to establish brain death. And what was the public reaction? The first heart transplant had attracted a remarkable amount of attention to the problem of determining death. Had the public noticed that they had been deceived about the true motives of the physicians? How did non-medical experts react?

Surprisingly, German lawyers and theologians by and large did not at all object to heart transplantation or to a redefinition of death. Rather, they declared that these were legitimate tasks of medicine. The famous Protestant theologian and ethicist Helmut Thielecke, for example, declared that he would not object to using the "achievement of modern medicine" to "vitally conserve" bodies for transplantation because this was a question of biology, not of anthropology or theology, or ethics (Thielecke 1970). The redefinition of death was seen as an inevitable side effect of progress in transplant medicine. Viewing this side effect as a serious problem implied opposition to a thoroughly logical and necessary scientific development.

The public perception of transplant medicine and the problem of brain death in the 1960s must be considered against the background of the modernisation of society. German newspapers and magazines generally agreed that scientific progress was intrinsically positive. Even after the untimely death of Barnard's first patient, Louis Washkansky, 14 days after the operation, a leading German newspaper confirmed that the first heart transplant had been "a necessary experiment", and that medicine "owes its progress to men who had not been discouraged by set backs".[5]

As the sociologist Ulrich Beck has pointed out, the instrumental rationality of simple modernisation demanded that side effects were a price that had to be paid for scientific progress, and that set backs should be met with an intensified effort to reach the intended goal. When the surgeon and Nobel Prize winner Werner Forßmann publicly condemned heart transplantation because doctors became slaughterers of their patients,[6] he was harshly attacked by the leftist news magazine *Der Spiegel* for being old-fashioned and irrational. This news magazine proved to be the most radical opinion leader in the process of modernisation. Ethical considerations that disagreed with instrumental rationality were interpreted as atavisms from pre-modern times. According to *Der Spiegel*, water-skiing Barnard, who was seen in discos, danced with Heidi Brühl, and lunched with Gina Lollobrigida,[7] represented a new generation of modern physicians.

My conclusion is that all three narratives prove true. The history of brain death is a story of betrayal and deception about the true motives of transplant surgery. However, it is also a story about faith in scientific progress on the part of physicians and, more importantly, on the part of the public and non-medical experts. Moreover, it is a story about progress through ethics, because those who represented ethical discourse at that time took part in the decision-making process. They agreed to a redefinition of death and were determined to hold a progressive and ultimately modern ethical position.

Yet on the other hand, we could also conclude that none of these narratives proves true. It was not a story about certainty and progress, because scientific progress is always of an ambivalent nature and scientific certainty is rarely achieved. It was not a story of deception and betrayal, because at the time everybody agreed upon the need for a redefinition of death. It was not a story of progress through ethics, because ethicists today do not believe that the definition of death is a purely scientific task.

4. MEDICINE, ETHICS AND REFLEXIVE MODERNITY

These judgements illustrate how our perceptions of modernity and the role of science in modern society have gradually but consistently changed over the last fifty years. Today we live in what Ulrich Beck would call a reflexive modernity, a product of the ongoing process of modernisation, the characteristics of which are completely different from those of simple modernity, but which has evolved from simple modernity "by virtue of its inherent dynamism" (Beck 1994: 2) Simple modernity was the essential basis of industrial society. It was characterised by the uprooting of tradition, the predominance of instrumental rationality, a linear model of progress, and a belief in the rational control of the process of modernisation. Reflexive modernity, however, no longer treats side effects and the risks associated with scientific development as an inevitable attribute of modernisation. Instead they can contradict the supremacy of instrumental rationality. Whereas simple modernity objectified and rationalised scientific problems as an incentive to further scientific research, reflexive modernity manages risks in order to help demystify and de-monopolise science (Beck 1992 and 1993; Schneider 1999).

As scientific hypotheses for risk management become ever more complex they are losing their credibility. The public has to decide between different plausible or probable scientific claims. Reflexive modernity raises questions such as: which risks do we really want to live with? Who is defining which risks are negligible? Who is going to control the side effects of scientific knowledge and technological applications? Risk management becomes an eminently political question. This is the place of medical ethics today. The de-monopolisation of science has made room for medical ethics to influence and sometimes even to control scientific development.

What are considered genuinely ethical problems are a function of historically contingent factors, especially of mentalities. The ethical questions of today derive from how we interpret the goals of the historical process. Medical ethics makes implicit use of historical narratives to identify and explain the problems of today, and to find the solutions of tomorrow. But what we once interpreted as a heroic drama may easily change into a tragedy or even a grotesque.

References

Baker, R. (ed.). *The codification of medical morality; historical and philosophical studies of the formalization of Western medical morality in the eighteenth and nineteenth centuries.* Dordrecht, 1993–1995.

Bauer, K.H. "Über Rechtsfragen bei homologer Organtransplantation aus der Sicht des Klinikers (unter besonderer Berücksichtigung der Krebsübertragung)." *Der Chirurg* 1967, 38, 245–251.

Beck, U. "The Reinvention of Politics: Towards a Theory of Reflexive Modernization." In *Reflexive Modernization. Politics, Tradition and Aesthetics in the Modern Social Order.* U. Beck, A. Giddens and S. Lash (eds.). Cambridge, 1994, 1–55.

Beck, U. *Die Erfindung des Politischen. Zu einer Theorie reflexiver Modernisierung.* Frankfurt/Main, 1993.

Beck, U. *Risikogesellschaft. Auf dem Weg in eine andere Moderne.* Frankfurt/M. 1986; Engl. translation: *Risk Society: Towards a New Modernity.* London, 1992.

Brendel, W. *Anhang to the first German edition of F.D. Moore: Transplantation. Geschichte und Entwicklung bis zur heutigen Zeit.* [Engl. Original Philadelphia, London 1963], Berlin, 1970.

Brosig, W. and Nagel, R. *Nierentransplantation.* Berlin, 1965.

Carson, R.A. (ed.). *Philosophy of medicine and bioethics: a twenty-year retrospective and critical appraisal.* Dordrecht, 1997.

Culmann, H.-M. *Zur geschichtlichen Entwicklung der Todesauffassung des Arztes im europäischen Raum.* Diss. med., Freiburg, 1986.

Engisch, K. "Über Rechtsfragen bei homologer Organtransplantation. Ergänzende Bemerkungen aus der Sicht des Juristen (des Kriminalisten)." *Der Chirurg* 1967, 38, 252–255.

Frowein, R.A.; Euler, K.H. and Karim-Nejad, A. "Grenzen der Wiederbelebung bei schweren Hirntraumen." *Langenbecks Archiv für klinische Chirurgie* 1964, 308, 276–281.

Giacomini, M.A. "A Change of Heart and a Change of Mind? Technology and the Redefinition of Death in 1968." *Social Science and Medicine* 1997, 44, 1465–1482.

Jonson, A.R. *The birth of bioethics.* New York, 1998.

Kaiser, G. "Künstliche Insemination und Transplantation. Juristische und rechtspolitische Probleme." In *Arzt und Recht. Medizinisch-juristische Grenzprobleme unserer Zeit.* H. Göppinger (ed.). München, 1966, 58–95.

Kirchheim, D. and Robertson, R.D. "Biologische Grundlagen und Fortschritte in der Nierentransplantation. II. Patienten- und Spenderwahl, Organkonservierung." *Der Urologe* 1967, 6, 67–74.

Largiadèr, F. "History of Organ Transplantation." In *Organ Transplantation.* F. Largiadèr (ed.). Stuttgart, 1970, 2–12.

Laves, W. "Agonie." *Münchener medizinische Wochenschrift* 1965, 107, 113–118.

Liebhardt, E.W. "Zivilrechtliche Probleme an der Grenze zwischen Leben und Tod." *Deutsche Zeitschrift für die gesamte gerichtliche Medizin* 1966, 57, 31–36.

Mollaret, P. and Goulon, M. "Le coma dépassé (mémoire préliminaire)." *Revue neurologique* 1959, 101/1, 3–15.

Mollaret, P. "Über die äußersten Möglichkeiten der Wiederbelebung. Die Grenzen zwischen Leben und Tod." *Münchener medizinische Wochenschrift* 1962, 104, 1539–1545.

Moore, F.D. *Transplantation. Geschichte und Entwicklung bis zur heutigen Zeit.* Philadelphia, London 1963.

Pernick, M. "Brain Death in a Cultural Context. The Reconstruction of Death, 1967-1981." In *The Definition of Death. Contemporary Controversies.* S.J. Youngner, R.M. Arnold and R. Schapiro (eds.). Baltimore, London, 1999, 3–33.

Pernick, M.S. "Back From the Grave: Recurring Controversies Over Defining and Diagnosing Death in History." In *Death: Beyond Whole-brain Criteria.* R.M. Zaner (ed.). Dordrecht, 1988, 17–74.

Schlich, T. "Medizingeschichte und Ethik der Transplantationsmedizin: Die Erfindung der Organtransplantation." In *Transplantationsmedizin und Ethik. Auf dem Weg zu einem gesellschaftlichen Konsens.* F.W. Albert, W. Land and E. Zwierlein (eds.). Lengerich, Berlin, 1994, 11–32.

Schlich, T. *Die Erfindung der Organtransplantation. Erfolg und Scheitern des chirurgischen Organersatzes (1880-1930).* Frankfurt, New York, 1998.

Schneider, H. "Der Hirntod. Begriffsgeschichte und Pathogenese." *Der Nervenarzt* 1970, 41, 381–387.

Schneider, W. *'So tot wie nötig – so lebendig wie möglich!' Sterben und Tod in der fortgeschrittenen Moderne. Eine Diskursanalyse am Beispiel der öffentlichen Diskussion um den Hirntod in Deutschland.* Münster, Hamburg, London, 1999.

Seiffert, K.E. "Überblick über den gegenwärtigen Stand der Transplantation von Organen und Geweben." *Der Chirurg* 1967, 38, 255–259.

Spann, W. and Liebhardt, E.W. "Reanimation und Feststellung des Todeszeitpunktes." *Münchener medizinische Wochenschrift* 1966, 108, 1410–1414.

Spann, W. "Strafrechtliche Probleme an der Grenze von Leben und Tod." *Deutsche Zeitschrift für die gesamte gerichtliche Medizin* 1966, 57, 26–30.

Spann, W.; Kugler, J. and Liebhardt, E.W. "Tod und elektrische Stille im EEG." *Münchener medizinische Wochenschrift* 1967, 42, 2161–2167.

Thielecke, H. *Wer darf leben? Ethische Probleme der modernen Medizin.* München, 1970.

White, H. *Metahistory.* Frankfurt/Main, 1991.

Wiesemann, C. "Notwendigkeit und Kontingenz. Zur Geschichte der ersten Hirntod-Definition der Deutschen Gesellschaft für Chirurgie von 1968." In *Hirntod. Kulturgeschichte der Todesfeststellung.* T. Schlich and C. Wiesemann (eds.). Frankfurt/M., 2001, 209–231.

Wiesemann, C. "Hirntod und Intensivmedizin. Zur Kulturgeschichte eines medizinischen Konzepts." *Der Anaesthesist* 2000a, 49, 893–900.

Wiesemann, C. "Instrumentalisierte Instrumente: EEG, zerebrale Angiographie und die Etablierung des Hirntod-Konzepts." In *Instrument – Experiment: Historische Studien.* C. Meinel (ed.). Berlin/Diepholz, 2000b, 225–234.

---

[1] Giacomini contests this claim (1997: 1479)

[2] Mollaret and Goulon (1959); Mollaret (1962). Mollaret, who is usually cited as the first to equate brain death with the death of a human being, was in fact rather reluctant to accept a new definition of death. See also: Frowein, Euler and Karim-Nejad (1964); Laves (1965); Spann (1966); Spann and Liebhardt, (1966); Spann, Kugler and Liebhardt (1967). A short description of the early history of the notion of brain death from a medical point of view can be found in Schneider (1970). For a history of attitudes towards the notion of death see Pernick (1988) or Culmann (1986).

[3] Brosig, and Nagel (1965); Bauer (1967); Seiffert (1967); Kirchheim and Robertson (1967). For the history of kidney transplantation see Largiadèr (1970); Moore (1963); Brendel (1970). For the early history of organ transplantation see Schlich (1994 and 1998).

[4] See: Liebhardt (1966); Kaiser (1966) and Engisch (1967).

[5] Die Welt, No. 298, 22. Dec. 1967, 2, my translation.

[6] Frankfurter Allgemeine Zeitung, No. 2, 3. Jan. 1968, 18.

[7] German and Italian singer and actress.

Chapter 14

ALTERNATIVE MEDICINE
A Dispute on Truth, Power or Money?

EVA KRIZOVA
Prague, Czech Republic

> I do not label our official medicine as western medicine,
> but rather as the ruling medicine.
> It has money and power and tries to preserve both.
> *A Czech physician's view*

During the last two decades of the twentieth century, a cultural transformation of modern medicine began to take place in the industrialised countries of the western world. On one hand scientific and technological development accelerated, surpassing its previous limits. On the other, unconventional medicine[1] reappeared after a long post-war period of silence, and its use has become more and more widespread during recent years (Eisenberg 1993 and 1998; Fisher 1994; Thomas 1991). Is it not strange that in an era of technological miracles and heroic medical performances people are again using either traditional or even exotic healing practices? Why do they do so if science advances in leaps and bounds, and applies its new techniques in practice? Is this a symptom of the fall of the rational world-view? What does it indicate? While Erich Fromm could still claim in 1973 that science is considered the highest value in industrial societies, that it is associated with moral correctness (Fromm 1997), at the beginning of the new millennium opinion is much more heterogeneous. In accordance with the sociological description of a postmodern society, different or even contradictory value systems coexist side by side and claim the right to their own existence. On the streets we are confronted with bizarre combinations of traditional and technological elements in human behaviour and appearance. To be individual in one's style has become a new sort of cultural demand. Science is still a respected and prominent social subsystem, but the

C. Rehmann-Sutter et al. (eds.), Bioethics in Cultural Contexts, 197–210.
© 2006 *Springer. Printed in the Netherlands.*

popular adoration of science and its presentation as a symbol of progress in the modern era is neither predominant nor universally shared. Science not only brings great benefits (Porter 1997) and ongoing promise and hope, it is also a powerful social institution with its own interests, resources, morality and culture. Science has become an area of study for sociologists and philosophers, who have analysed the construction processes and revealed the interests and ideologies of science as an institution (e.g. Kuhn 1962). It is not accidental that medicine is used as the best reflection of science. After all, medical findings aim to preserve the highest value – human life – furthermore the confusion between the subject and object is most clearly illustrated in the therapeutic relationship. Lewis noted that man's power over nature implies fundamentally power over another man by means of nature (in: Tuckett 1978). In Michel Foucault's philosophy, the idea that knowledge serves to control and manipulate is central to his elaboration of concepts of bio-power and bio-politics (Foucault 1979 and 1980).

Among the public, sceptical attitudes to modern science are increasing since it does not always provide useful answers to the questions that people view as relevant within the context of everyday life. A mass use of high-tech inventions in practical everyday life is simultaneously accompanied by a new irrationality that may be observed in an unprecedented and spontaneous use of bizarre rituals (e.g. piercing, tattoo). Globalisation and migration only intensify the processes in this huge melting pot. Our perception of reality and its reflection in social behaviour is probably more autonomous and less subordinated to any visible supremacy (not excepting science and medicine) than ever before. The scientific world is viewed as one part of the lifeworld that exists in parallel with the domains of dreams, imagination and everyday life (see Schutz 1962).

It might seem that nothing can stop a diversification within medicine that would result in a similar pluralism to that which has occurred in other social spheres. However, reality differs remarkably at first glance. The current situation is characterised by persistent disagreement between scientific and unconventional medicine, though of different intensity in different countries. Even when the medical community is not united and the situation is conditioned by cultural specifics, a certain degree of conflict or at least of embarrassment over the coexistence can be observed in most western countries (Ernst 2000; Leibovici 1999; Schneirov 2002; Schneiderman 2000; Siahpush 1999a). In the most tolerant ones the concept of integrated medicine veils scientific medicine's ongoing struggle to retain control. Unconventional medicine has also become a matter of political consideration and decisions. The European Commission, the World Health Organization, the British Parliament and the U.S. Congress have all dealt with this issue during last decades.[2] In the transition countries of Central and Eastern

Europe the situation is characterised by an open conflict between the advocates and opponents of unconventional medicine within the medical and scientific community[3]. Representatives of conventional medicine in many countries protest against the diversification of health care services using a variety of means, and try to preserve a monopoly or at least a dominant position in health care provision. According to Hodgkin (1996), medicine is surprisingly unwilling to acknowledge explicitly the postmodernist plurality of values and interests, yet at the same time it is contributing significantly to postmodernist reality by its own practice, particularly technological interventions into human reproduction and death, organ and genetic manipulations, plastic surgery and potentially cloning. But why is the diversification so painful for medicine, and what is hidden under the surface of the arguments?

Unlike other services, health care in Europe is publicly controlled. Organisation of health care financing and provision stems from view that the right to health care is a human right. Unfortunately, the other side of this benefit is the restriction of consumer behaviour, legitimated by public responsibility for user protection and safety. Each health care provider is confronted with the fact that he deals with the most precious value – the life of a human being. Therefore even in the most democratic countries as well as in those with a high-impact private market (such as the USA) many ethical and juridical issues arise about the coexistence of different therapeutic systems. Even if unconventional medicine were to be supervised by corporate associations and a certain level of quality guaranteed, many questions would still emerge as matters for debate. These include: is scientific medicine compatible with any of the unconventional methods or are there some options that should be banned? In case of a combination of different treatment modalities does the complementary medicine intensify the healing effect of an evidence-based therapy or can it paralyse it? In case of a replacement therapy is it ethically acceptable to let clients pay for unconventional treatment while people are made to contribute to the public health care system through taxation or insurance premiums? Is it ethical to treat patients with antibiotics if homeopathic remedies would also help? Is the public health care system responsible for the treatment a patient who was harmed by an unqualified healer or by a treatment delay? What are the limits of autonomy and public/private responsibility? Are unconventional methods eligible as the last or as the first choice? How much are physicians obliged to communicate on this issue?

In order to better understand the professional attitude towards unconventional medicine we need to reflect on some specific aspects of the world of medicine. At the clinical level the sociological background soon becomes evident as being closely related to the historical process of medical professionalisation. Owing to the highly specialised character of medical qualification, the consensus prevailed that physicians were the only experts able to judge which procedures were acceptable and why. As Freidson discovered:

> "the profession is the sole source of competence to recognize deviant performance, and it is also ethical enough to control deviant performance and to regulate itself in general. Its autonomy is justified and tested by its self-regulation" (Freidson 1988: 137).

Medical professionals then, not only possess the erudition, skills and competencies to define what a disease is, but they also have the power to set the rules of practice and control it. Moreover, they determine not only the means of treatment (legitimate healing procedures), but further the license to provide health care services (who is allowed to treat). The medical corporation keeps control of this matter through undergraduate and postgraduate education, certification and licensure. From this perspective, medicine is a special profession that achieved a monopoly of health care service provision to society during the nineteenth and twentieth centuries, and is now able to suppress unwelcome competition. Freidson explains that:

> "when the public is considered too inexpert to be able to evaluate such [i e. medical] work, those dominating society may feel that the public needs protection from unqualified or unscrupulous workers. Having been persuaded that one occupation is most qualified by virtue of its formal training and the moral fiber of its members, the state may exclude all others and give the chosen occupation a legal monopoly that may help bridge the gap between it and laymen, if only by restricting the layman's choice" (Freidson 1988: 74).

Yet, this monopoly was compensated by a special commitment to provide the best possible method of treatment that is in accordance with the progress of human knowledge. It means that society tolerates medical autonomy and its exclusive position as the price of protection from malpractice and unskilled performance. As Goode says:

> "only to the extent that the society believes the profession is regulated by this collectivity orientation will it grant the profession much autonomy and freedom from lay supervision or control" (Goode 1966: 37).[4]

To serve public (common) interests and to provide a high level of altruism is the other side of medical autonomy. Freidson continues that:

> "The profession's service orientation is a public imputation it has successfully won in a process by which its leaders have persuaded society to grant and support its autonomy" (Freidson 1988: 82).

Hence we can better understand why the focus on safety, scientific rigour and objectively measurable effectiveness is so central to the discourse. Ontological and epistemological issues seem to be the alpha and omega of the debate, even when they in fact veil a protection of autonomy and monopoly in health care provision. To respect the scientific evidence becomes one of the ethical commitments of physicians.[5] Moreover, medicine is nowadays called upon to be evidence-based, which means that scientific research and theory are the main basis of its legitimacy. This demand aims not only to protect physicians from being accused of malpractice by unsatisfied patients, but also to protect patients from self-willed experimentation by physicians. Modern medicine postulates a standardisation despite the fact that there is no standardised patient, no standardised diagnosis, and no standardised suffering. Bureaucratic administration of health care financing in the system of pre-paid care, and a continually increasing emphasis on cost effectiveness only intensify this process that is both clinically and economically instrumental. Scientific rigour seems to reduce (not eliminate) the inevitable medical uncertainty (Fox 1957 and 1980). One of the reasons why physicians do not welcome complementary use of unconventional medicine is that simultaneous treatment masks the healing process with hardly identifiable or observable variables that are not under their control. Further, the most serious argument against a replacement therapy is a delay in starting a proper (specifically efficient) therapy.

Nevertheless, although alternative medicine still provokes negative reactions in some groups of physicians, the number of unconventional providers with medical education is growing, as is the number of physicians' referrals to unconventional providers. The medical community is heterogeneous (despite the publicly presented corporate image), and attitudes range from negation of the principle through a passive tolerance to participation in postgraduate education and provision of unconventional medicine. Many empirical surveys have confirmed a significant use of alternative medicine among physicians that is apparently inconsistent with the official statements by medical associations. Among physicians, general practitioners are the closest to their patients and are most tolerant of unconventional practices (Blumberg et al. 1995; Wharton et al. 1986; Perkin et al. 1994; Krízová 2002). Even when standardisation is proclaimed, clinicians admit more or less informally that at the micro-level an individual patient's variations must be taken into account. A discrepancy between the "theoretical" and "clinical" mentality is also evident. According to Freidson (who was probably inspired by Bucher 1962):

> "even within the single profession of medicine, the differences between "client-dependent" and "colleague-dependent" practices are of critical

importance for the way work is performed, and the differences between the practising and the scientific worker's work experience are of critical importance for the way each looks at himself and his work" (Freidson 1988: 75).

But why do some physicians abandon the safe field of scientific evidence-based medicine for unconventional procedures with speculative or unproven effectiveness, and assume a deviant identity? Is it a sign of their personal growth, or does it indicate their professional failure and helplessness? Does it mirror a role-conflict resulting from discordance of ethical commitments to science and patients, or is it their way of using the placebo effect, which was so neglected in recent decades of modern development? Are values and world-views fundamental in changing behaviour, or do physicians simply respond to growing patient demand? Is unconventional medicine a good business, or does it rather complement the public sector with privately provided services? Are physicians obliged to use unconventional medicine when patients wish to do so? If they do not recognise it, are they allowed to criticise the patient's choice? Does a physician's autonomy mean his or her right to withhold a treatment in case of a disagreement about treatment options?

Among the reasons for a physician's inclination to unconventional medicine three main motivations may be assumed: clinical, cultural and economic.[6]

The clinical motivation legitimises unconventional medicine by its positive impact on a patient's condition (even without a theoretical model or testing of specific effectiveness according the scientific rules). Often, a positive personal experience with an unconventional procedure precedes a deeper engagement in education and practice (this is mainly true of acupuncture and homeopathy). Several physicians in the survey cited their own health disorders, or serious illnesses or injuries of family members. A personal confrontation with the limits of orthodox medicine in chronic diseases stimulated their expansion beyond scientific medicine.

"The attitudes towards unconventional methods are related to the position of a physician. Being highly educated and experienced medical experts, who have their commitments, subsistence and sometimes even sense of life within their medical profession they may consider unconventional medicine as a fraud and quackery and an inappropriate competition, as well. As soon as they or their close relatives fall ill and scientific medicine fails, their negative attitude gets weakened or even changes substantially." An authentic opinion from the focus group, translation from Czech by the author, see note 6.

As Ernst (2000) confirmed, legitimacy as viewed by these clinicians, is mainly opinion-based. From their perspective, any relief is useful for the traditional commitment to serve the sick, and this dominates in the hierarchy of obligations. This attitude is close to the pragmatic perspective of the sick. Such physicians try to extend their clinical knowledge and skills in order to improve their performance in the face of the infinite variety of human disorders. They also compare the treatment's side effects. These physicians have coped with their deviant identity by enhanced identification with their clinical responsibility to the sick, and have justified their behaviour by their patients' profit. They believe that they have become better physicians, able to provide more effective help with regard to particular individual patient needs. In this context the association between clinical autonomy and clinical responsibility may arise. Consequently, a wider range of therapeutic means, which can be suited to the particular biographical details of the patient's disorder, make clinical decision-making more autonomous, although less predictable and controllable. At the same time, the physician is aware of his increased accountability for the treatment outcomes, and probably intensifies the attention paid to the entire treatment process.

"A physician who treats by unconventional procedures becomes more emotionally engaged. First, due to the fact that the patient is a private client, second, due to lacking standardisation and verification of the treatment procedure that make the diagnostic and therapeutic process highly individual. This may imply that the physician is less self-assure in the sense of false certainty and may enhance the placebo effect resulting from the personal engagement of the physician. His responsibility increases hand in hand with his higher (i.e. less restricted) clinical autonomy." An authentic opinion from the focus group, translation from Czech by the author.

The second motivation may be subsumed under value changes that are connected with the postmodern cultural and mental transformation. In this respect, spiritual needs and values of both patients and physicians are crucial. Unconventional medicine tries to compensate for gaps in the positivistic approach. A personal religious world-view, a philosophical insight and/or ecological sensitivity (and systematic approach) may influence affiliation with unconventional medicine. From the "green" perspective, unconventional medicine is often identified with soft and natural (non-artificial, non-chemical) processes with a low or non-existent iatrogenic risk.

In our survey, physicians often expressed appreciation of the extended space for collecting personal history that arises from a broader perception of disease and a greater emphasis on subjective reports. In this case, physicians incline to unconventional procedures not because of having observed their efficiency, but rather because the treatment is part of a symbolic and holistic framework that views as clinically relevant those biographical and life-style aspects that are usually neglected by scientific medicine. This means that unconventional medicine offers a therapeutic model that is more congruent with the physicians' perception of concepts of health and disease in general and that offers to overcome the mind-body dualism (May 1998). Despite all its internal diversity unconventional medicine represents a new paradigm of health and disease that lies behind changes in physicians' (and patients') behaviour. The approach consists in a re-definition of health and disease, a new perception of the roles of the sick and the therapist, and the priority of soft and non-invasive treatment (Woodhouse 1997).

The third motivation is associated with economic issues. Although medicine ideally articulates ethical values and goals we cannot conceal the fact that it is also a substantial market or at least an important area for suppliers. Therefore, financial issues also emerge as a legitimate matter of fact. Unconventional medicine is often excluded from public coverage. Hence, access depends on ability and willingness to pay at the moment of use. The more unconventional medicine grows in the private market, the more it becomes a special field of business. Even though economic motivation, as represented by patient demand and additional income, is not crucial in a physician's affiliation with unconventional medicine (because we believe that the primary reasons are cultural), it will most probably become a more important factor in the future, as every health market including this one is threatened by an information imbalance between providers and users and by inequity in access (Ernst et al 2003). Recently, more and more procedures are at least partly covered by health insurance companies, although this is not true in the countries where unconventional medicine is officially suppressed. Health insurance companies in the most progressive countries try to increase the scope of health care benefits for patients, since demand for unconventional medicine is connected with lower total costs and also with better motivation to stay healthy or recover soon. American studies have found that "people who utilize such alternatives tend to be healthier than the general run-of-the mill patient" (Schneiderman 2000: 84). As one Czech general physician found:

> "whenever the patient inquires about unconventional medicine during the consultation it always reliably indicates his/her motivation to recover soon".
> An authentic opinion of a GP in the focus group, translation from Czech by the author.

While the public systems are based on the principle of solidarity, privately provided unconventional services offer the individual the chance to be egocentric and try to restore in a reduced way the meritorious principle. Cost issues relate to the whole system. Modern medicine has become a very expensive machine, and any effort to slow down the cost increase and preserve access to treatment is ethically important.[7] To protect the sustainability of health care in the future is a major ethical responsibility.

CONCLUSIONS

Owing to the service-orientation of physicians who feel an ethical commitment resulting from professional autonomy granted by society, it is not surprising that clinical motivation and altruism prevail in physicians' self-perception. The practice of unconventional medicine has resulted, in the eyes of those who advocate it, in a professional improvement of clinical performance and in becoming "a better doctor". Both opponents and proponents of unconventional medicine use the same justification in their discourse about "a duty to promote good and act in the best interest of the patient and the health of society, (...) and the duty to protect and do no harm to patients" (Ernst et al 2003). This could be termed a heresy, as described by Coser (1956)[8]. Both sides compete in who serves the sick best – goals are identical but the strategies how to reach them differ principally. Future development necessitates a harmonisation of different approaches and will probably lead to a hybridisation (Frank 2004a). This is a critical issue, since as yet integration processes have occurred under the dominance of scientific medicine and have threatened the core identity of unconventional practices (May 1998). The focus on scientific rigour in proving their efficacy is logical but when applied strictly and uncritically it may also be deconstructed as a magical formula against the validation of complementary and alternative medicine (CAM). Ernst emphasizes "without scientific proof CAM is unlikely to survive", but he concludes that owing to the nature of CAM "scientific rigour can only be taken to a certain point" (Ernst et al 2003).

A renaissance of unconventional medicine in the context of mainstream medicine is a symptom of a cultural change that has also had an effect on health care. Medicine as a more rigid and more autonomous social subsystem was less open to such changes, and therefore the process of diversification is significantly more complicated and painful, although still unavoidable. The diversity of opinions and practices also mirrors globalisation, or as Stollberg (2003) suggests, *"glocalisation"* (Glokalisierung) – a synthesis of global and local elements. From this perspective, we can conclude that the coexistence of scientific (evidence-based) and unconventional medicine is the first step to a postmodern health care system. The model of a value-free medicine (as part of a value-free science) is beyond redemption. Some authors expect that treatment styles will be crucially important in patients' choice. (Malone and Luft 2002). Malone and

Luft expect that physicians will self-selectively group together according their shared philosophies of practice, and patients will choose those whose philosophy is most closely matched to their own. Of course, this is possible only in large countries like the U.S. or in big cities.

Medicine has to cope with the "liquid character" of social life (Bauman 2000). This is the challenge it faces at the beginning of the new millennium. Unfortunately, the corporate and institutional interests of medicine and health care may be in conflict with the apparent social demands for change. Considering the growing tension between restricted resources and infinite demand there is a remarkable moral dimension to the whole dilemma – is it morally justifiable to use limited resources for treating a sick person when a similar or even the same goal could be achieved by softer and less expensive techniques? Unconventional medicine creates new tasks for physicians – not directly in the sense of attaining new knowledge and clinical skills, but rather at the level of communication, in their perception of the therapeutic situation, in a redefinition of physicians' and patients' tasks, in the acceptance of the fact that patients' wishes are inherently included not only in health, but also in disease. Physicians are facing new ethical and legal demands associated with the extension of informed consent to the area of complementary and alternative medicine (Sugarman 2003, Weir 2003, Brophy 2003, Sade 2003, Thorne et al 2002, Ernst et al 2001, Monaco et al. 2002). The most painful demand is how to divide and share the power and responsibility between experts and lays. In a certain sense, this would rehabilitate ancient elements of medical care (see Stollberg 2001: 57, 64). Most probably, future development will be neither as smooth as the optimists believe nor as complicated or unattainable as the pessimists think.

References

Bauman, Z. *The Liquid Modernity.* London: Polity Press, 2000.

Blumberg, D.L.; Grant, W.D.; Hendricks, S.R.; et al. "The physician and unconventional medicine." *Alternative Therapies in Health and Medicine* 1995, 1, 3, 31–35.

Brophy, E. "Does a doctor have a duty to provide information and advice about complementary and alternative medicine?" *J Law Med Ethics* 2003, 10, 3, 271–284.

Bucher, R. "Pathology: A study of social movements within a profession." *Social Problems* 1962, 10, 40–51.

Coser, L. *The Functions of Social Conflict.* Glencoe, Illinois: Free Press, 1956.

Eisenberg, D.M.; Kessler, R.C.; Foster C.; et al. "Unconventional medicine in the United States." *The New England Journal of Medicine* 1993, 328, 246–252.

Eisenberg, D.M.; Davis, R.B.; Ettner, S.L.; et al. "Trends in alternative medicine use in the United States, 1990-1997." *Journal of American Medical Association* 1998, 280, 1569–1575.

Ernst, E. "The role of complementary and alternative medicine." *Br Med J* 2000, 321, 1133–35.

Ernst, E.; Cohen, M.H. and Stone, J. "Ethical problems arising in evidence based complementary and alternative medicine." *J Med Ethics* 2003, 30, 156–159.

Ernst, E. and Cohen, M.H. "Informed consent in complementary and alternative medicine." *Arch Intern Med* 2001, 161, 19, 2288–2292.

Fisher, P. and Ward, A. "Complementary medicine in Europe." *Br Med J* 1994, 309, 107–110.

Foucault, M. *The History of Sexuality*. Vol.1. New York: Vintage, 1980.

Foucault, M. *Discipline and Punish: The Birth of the Prison*. New York: Vintage, 1979.

Fox, R. "Training for uncertainty." In *The Student-Physician*. Robert K. Merton, G. Reader and P. Kendall (eds.). Cambridge MA: Harvard University Press, 1957, 207–241.

Fox, R. "The evolution of medical uncertainty." *Milbank Meml Fund Quarterly* 1980, 58, 1–44.

Frank, R. and Stollberg, G. "Conceptualising hybridisation – on the diffusion of Asian medical knowledge to Germany." *International Sociology* 2004a, 19, 71–88.

Frank, R. and Stollberg, G. "Medical acupuncture in Germany: patterns of consumerism among physicians and patients." *Sociology of Health and Illness* 2004b, 26, 3, 351–372.

Freidson, E. *Profession of Medicine*. Chicago and London: The University of Chicago Press, 1988.

Fromm, E. *Anatomie lidské destruktivity [Anatomie der menschlichen Destruktivität]*. Praha: Lidové noviny, 1997.

Goode, W.J. "The Librarian: From Occupation to Profession?" In *Professionalization*. Howard M. Vollmer and Donald L. Mills (eds.). Englewood Cliffs, New Yersey: Prentice Hall, Inc., 1966.

Gürsoy, A. "Beyond the orthodox: Heresy in medicine and the social sciences from a cross-cultural perspective." *Soc Sci Med* 1996, 43, 5, 577–599.

Hodgkin, P. "Medicine, post-modernism and the end of certainty." *Br Med J* 1996, 313, 1568–1569.

Krizova, E. "Nekonvencní medicína z pohledu praktickych lékaru [Unconventional medicine as viewed by general practitioners]." *Prakt. Lék.* 2002, 82, 7, 425–429.

Kuhn, T.S. *Struktura vedeckych revolucí*. Praha: Oikoymenh, 1997 [original version: *The Structure of Scientific Revolutions*, 1962].

Leibovici, L. "Alternative (complementary) medicine: a cuckoo in the nest of empiricist red warblers." *Br Med J* 1999, 319, 1629–32.

Malone, R.E. and Luft, H.S. "Perspectives in accountability: past, present, and future." In *Ehical Dimensions of Health Policy*. Marion Danis, Carolyn Clancy and Larry R. Churchill (eds.). New York: Oxford University Press, 2002, 272–275.

Martin, B. "Dissent and heresy in medicine: models, methods, and strategies." *Soc Sci Med* 2004, 58, 4, 713–725.

May, C. and Sirur, D. "Art, science and placebo: incorporating homeopathy in general practice." *Sociology of Health and Illness* 1998, 20, 2, 168–190.

Monaco, G.P. and Smith, G. "Informed consent in complementary and alternative medicine: current status and future needs." *Semin Oncol* 2002, 29, 6, 601–608.

O´Neale, R.J. "Lords call for regulation of complementary medicine." *Br Med J* 2000, 321, 1365.

Parsons, T. "Social Structure and Dynamic Process: The Case of Modern Medical Practice." In *The Social System*. Talcott Parsons (ed.). The Free Press of Glencoe, 1964.

Perkin, M.R.; Pearcy, R.M. and Fraser, J.S. "A comparison of the attitudes shown by general practitioners, hospital doctors and medical students towards alternative medicine." *J. Royal Society Med* 1994, 87, 523–525

Porter, R. *The Greatest Benefits to Mankind.* London: Sage, 1997.

Sade, R.M. "Complementary and alternative medicine. Foundations, ethics and law." *J Law Med Ethics* 2003, 31, 2, 183–190.

Schneiderman, L.J. "Alternative medicine or Alternatives to Medicine? A Physicians's Perspective." *Cambridge Quarterly of Health Care Ethics* 2000, 9, 1, 83–97.

Schneirov, M. and Geczik, J. D. "Alternative health and the challenges of institutionalization." *Health* 2002, 6, 2, 201–220.

Schutz, Alfred. "On Multiple Realities." In *Collected papers, vol. 1: The Problem of Social Reality.* M. Natanson (ed.). The Hague: Martinus Nijhoff, 1962.

Siahpush, M. "A critical review of the sociology of alternative medicine: research on users, practitioners and the orthodoxy." *Health* 1999a, 4, 2, 159–178.

Siahpush, M. "A sociological critique of alternative medicine: liberation or disempowerment?" *Social Alternatives* 1999b, 18, 4, 57–62.

Stambolovic, V. "Medical heresy – the view of a heretic." *Soc Sci Med* 1996, 43, 5, 601–604.

Stollberg, G. "Heterodoxe Medizin, Weltgesellschaft und Glokalisierung: Asiatische Medizinformen in Westeuropa." In *Kommunikation über Krankheiten.* (Serie Bielefelder Schriften zur Linquistik und Literaturwissenschaft). Gisela Brünner and Elisabeth Gülich (eds.). Bielefeld, 2003.

Stollberg, G. *Medizinsoziologie.* Bielefeld: Transcript Verlag, 2001.

Sociological shifts in the medical profession, Czech Republic Grant Agency, project no. 403-02-0691.

Sugarman, J. "Informed consent, shared decision-making, and complementary and alternative medicine." *J Law Med Ethics* 2003, 31, 2, 247–50.

Thomas, K.J.; Carr, J.; Westlake, L. and Williams, B.T. "Use of non-orthodox and conventional health care in Great Britain." *Br Med J* 1991, 302, 207–210.

Thorne, S.; Best, A.; Balon, J.; Kelner, M. and Rickhi, B. "Ethical dimensions in the borderland between conventional and complementary/alternative medicine." *J Altern Complement Med* 2002, 8, 6, 907–915.

Tuckett, D. and Kaufert, J.M. (eds.). *Basic Readings in Medical Sociology.* London: Tavistock Publ., 1978.

Weir, M. "Obligation to advise of options for treatment – medical doctors and complementary and alternative medicine practitioners." *J Law Med Ethics* 2003, 10, 3, 296–307.

Wharton, R. and Lewith G. "Complementary medicine and the general practitioner." *Br Med J* 1986, 292, 1498–1500.

Woodhouse, M.B. "The concept of disease in alternative medicine." In *What is Disease.* James M. Humber and Robert F. Almeder (eds.). Totowa, New Jersey: Humana Press, 1997.

---

[1] In the title I used the phrase "alternative medicine", since it better illustrates the divergence between mainstream and marginal medicine. However, in the literature the concept of unconventional medicine has recently spread as a more general one. Unconventional medicine is defined as a heterogeneous mix of healing procedures of different historical, cultural and geographic origin. The

concept itself is criticised, e.g. by Stollberg who argues that alternative or complementary medicine is also based on conventions (see Stollberg 2003). A common denominator is the lack of scientific evidence according to rigorous scientific rules, and an unacceptable theoretical explanation of effectiveness. Unconventional methods can occasionally be used as replacement therapy or more frequently as complementary therapy.

[2] The European Commission organised a survey of unconventional medicine and published an expert review (*COST Action B4- Unconventional Medicine: Final Report of the Management Committee*, Brussels: EC, 1998). The Select Committee on Science and Technology of the British House of Lords recently decided upon the regulation of unconventional medicine and discussed an expert study submitted by Exeter University (www.parliament.publications.uk; and: O'Neale 2000). In 2000, the World Health Organization in Geneva published *General guidelines for Methodologies on Research and Evaluation of Traditional Medicine*. In 1993, the U.S. Congress formally established the Office of Alternative Medicine at the National Institutes of Health, now the National Center for Complementary and Alternative Medicine (*Expanding horizons of health care. Five-year strategic plan 2001-2005*. http://nccam.nih.gov).

[3] Before 1989, acupuncture was widespread and provided legally within the state health care system in the Czech Republic, though without the spiritual context of traditional Chinese medicine. Informally, lay healers treated in reduced numbers using mostly energy flow and herbal remedies. After 1989, there was a boom in unconventional medicine as a sign of global democratisation, which was later followed by restriction from scientific and medical representatives in the mid-1990s. Homeopathy was expelled from the national Institute of Postgraduate Medical Education and from medical associations; acupuncture was excluded from the covered benefits and is now provided on a private basis. No private coinsurance exists that would include unconventional procedures.

[4] Goode, quoted by Freidson (1988: 82), uses the wording "collectivity orientation". And: According to Parsons, medical role is collectivity-oriented, not self-oriented (Parsons 1964: 434).

[5] The Ethical Code of the Czech Medical Chamber requires that physicians treat in congruence with scientific evidence. Similarly, one of the American Medical Association's nine principles of medical ethics requires that "a physician shall continue to study, apply, and advance scientific knowledge ... " http://www.ama-assn.org/ama/pub/category/2512.html (last visit June 05).

[6] This part relies on research outcomes collected within the research projects "Use of alternative medicine in the Czech Republic" (supported 1997-1998 by Research Support Scheme), and "Sociological, philosophical and ethical aspects of unconventional medicine" (supported 2001-2002 by the Grant Agency of the Czech Health Ministry). The data are based on quantitative surveys (postal questionnaires mailed to attendants of postgraduate physicians training courses in unconventional medicine and to a random sample of Czech general practitioners), and on qualitative interviews, including a focus group with eight physicians inclining to unconventional medicine.

[7] Siahpush (1999b) mentions the macro-social consequences of unconventional medicine. According to some authors he analyzed in his sociological critique of alternative medicine, one of the reasons for the recent growth of alternative medicine is that the ruling class views it as a viable alternative to scientific

medicine, which has proved to be exceedingly costly. The enormous amount of money that flows into health care is a hindrance to capital accumulation.

[8]   The term heresy is frequently used in the discourse on alternative medicine, see further Martin 2004; Stambolovic 1996; Gürsoy 1996; Frank R. et al.2004b.

Chapter 15

DIMENSIONS OF CULTURAL DIVERSITY OF MEDICAL ETHICS

PAVEL TISHCHENKO
Moscow, Russia

1. INTRODUCTION. ON A BROAD SENSE OF MORAL DIVERSITY IN HEALTH CARE

The title of my article carries a significant theoretical difficulty. In what sense can we speak about cultural diversity and "ethics"? In the work "Modernity – An Incomplete Project" Habermas recalls the idea of Max Weber fundamental to this topic who "characterised cultural modernity as the separation of substantive reason expressed in religion and metaphysics into three autonomous spheres. They are: science, morality, and art" (Habermas 1987: 148). That is why as soon as we start using the word "ethics" which from the time of Aristotle has meant a theoretical (secular) way of discriminating "good" and "evil", we are immediately in danger of losing "cultural diversity" as soon as other forms of moral evaluation are neglected.

For example, the Russian Orthodox Church was never interested in the development of "ethics" because rational principles, rules, imperatives or whatever could have only an instrumental role. The direct way to the idea of "the good" is belief. Meeting moral difficulties in medical practice (as well as in other areas of life) people should first of all recreate their authentic way of "being-in-the-world" (their identity) in prayer, confession and communion. From the Church point of view without establishment of such a basic "pre-understanding" for any ethical understanding and judgement, nothing could be done properly. But for modernity such spiritual pre-understanding is a form of prejudices.

C. Rehmann-Sutter et al. (eds.), Bioethics in Cultural Contexts, 211–227.
© 2006 *Springer. Printed in the Netherlands.*

I think that the same could be said about the "moral identity" of believers in a significant number of other traditional and new religions. Such a position of resistance to the project of modernity in the area of morality is wrongly named "pre-modern" (as Habermas does) because a) their representatives are not going to become "modern"; and b) the Enlightenment idea that history progresses (develops) from a religious to a scientific world view has been falsified by history itself. This idea as such was and still is the way cultural diversity could be easily "coped with" (at this stage I prefer to use the Hegelian term "aufheben") and turned into a hierarchy of stages on a "ladder of development" from underdeveloped to developed forms. At the end of the 19th and beginning of the 20th centuries such arguments (proven by biomedical "objective knowledge") were used in most developed countries to justify of the natural rights of world domination that included the sense of intellectual and moral superiority. Similar ideas, of course in a much less aggressive form, exist in a number of ethical theories. As an example, I could mention Kohlberg`s theory of moral development used by Habermas in his version of discourse ethics. From this perspective: "… there is a universal valid form of rational moral thought process which all persons could articulate, assuming social and cultural conditions suitable to cognitive-moral stage development. We claim that ontogenesis towards this form of rational moral thinking occurs in all cultures in the same stepwise, invariant stage sequence" (Kohlberg 1984: 286). The question is: to what extent is rational thinking is necessary for moral life, and to what extent it is sufficient.

The modern cultural situation puts the notion of the "development" of moral reason in question. The existing healthcare environment is a kind of a "summit" or "bazaar" of a great number of different modern and archaic practices from shamanism via psychoanalysis to gene therapy. As Ulrich Beck eloquently argues, this diversity is not a result of a lack of scientific knowledge, but of its most recent and advanced developments (Beck 2000). The existing deep crisis in the foundations of modern science (e.g. in theoretical physics and biology) has made scientists much more open minded to alternative worldviews than it was the case several decades ago. The same phenomenon is easy to recognise in the healthcare setting.

There is no hope today that education will protect against the influence of "prejudices", but rather makes people more sensitive to them. Daniel Callahan writes: "There is the remarkable popular interest in alternative and complimentary medicine, with evidence that nearly 40% of Americans turn to it for help not provided by a conventional medicine; and the more educated people are, the more they have use of it" (Callahan 2000: 679). Archaic and new alternative and complementary medicine retains its own type of healer-patient relations that could not, in principle, be morally ordered on the grounds of a Western type of ethics. For example, in most kinds of psychotherapy and healing practices "paternalism" (as a specific form of personal dependency) is not an external condition of relations with patients that could and should be morally meliorated. It is the internal structure of healing itself.

Let us summarize our introductory exploration. I am arguing that it is essential to speak in a broader sense about the diversity of moral perspectives in the area of healthcare. The problem of diversity of medical ethics is relevant only to a part of moral community, not to the whole of it. In this segment of the modern cultural environment the idea of "theory" presents the path for moral betterment. Meeting problems people should first try to grasp their nature theoretically and then act in accordance with this theoretical understanding. In a broad sense, the cultural diversity of moral stances incorporates the self-sufficient existence of those moral perspectives that simply do not need by virtue of their nature any theoretical account. This kind of diversity is present not only between countries, but also within all existing countries; it constitutes the specific feature of the cultural situation called by Ulrich Beck "other modernity". And what is most important, those parts that are "external" to ethics could not be situated on a ladder of development as "pre-modern" in relation to the central project of modernity where ethics is the way to distinguish between ideas of good and evil. Relations of different modes of moral life are not temporal, but spatial. They coexist beside each other and in competition with each other in the open space of modern societies.

In this article I am going to discuss first of all a proposal for coping with diversity in medical ethics that was offered by Tristram Engelhardt, Jr. in his "Foundations of Bioethics". Then, I am going to speak about cultural foundations of diversity in a broad sense. Afterwards I point out some examples for political diversity (most essentially in times of war and terrorism). I will finish with an evaluation of the specific role of public in competition of multiplicity of moral perspectives in health care.

2. CONTENT-LESS THEORY AS THE RESPONSE ON THE CHALLENGE OF DIVERSITY

Diversity constitutes both the most specific fact or feature of modern ethical thinking in biomedicine (or bioethics), and a challenge to it. As H. T. Engelhardt, Jr. eloquently wrote: "[m]oral diversity is real. It is real in fact and in principle. Bioethics and healthcare policy have yet to take this diversity seriously" (1996: 3). Taking diversity seriously, Engelhardt diagnosed the danger: "... there is a swarm of alternative ethics ready to give rise to a babble of conflicting bioethics. This circumstance constitutes the foundational moral challenge of all healthcare policy. It brings the very field of bioethics into question" (1996: vii). Refusing both the nihilism of those who presume the impossibility of offering a generally accepted secular ethics, as well as those who dogmatically ignore the fact of diversity and try to create a content-full canonical moral theory for everybody, Engelhardt "… attempts instead to secure a content-less secular ethics. Given the limits of secular moral reasoning, all that is available is a mean (within certain constraints) of giving moral authority to common undertakings without establishing the moral worth on moral desirability of any particular choices" (ibid. viii).

The content-less way of doing ethics is a response to a paradoxical double attitude towards the phenomenon of diversity itself. We could fear it experiencing a kind of fundamental threat and a powerful motivation for any kind of thinking, including that in moral theory. And, we could also evaluate positively the threat to any attempt to find the foundation for morally right cooperation.

For Engelhardt, the first attitude towards the phenomenon of diversity is characteristic of the classical way of thinking and grounded both in the orthodox Judain-Christian tradition and philosophical traditions going back to Plato and Aristotle. To think means to grasp what is common in different things. The second perspective Engelhardt associates mostly with the negative historic experience of attempts to establish unified moral communities by brute force. "The great murderous endeavors of this century from Stalin and Hitler to the Gang of Four and Pol Pot have been born of attempts by force to make states single moral communities. Despite brutal repression, diversity remains" (ibid. 10). The historic account of the value of diversity does not mean for Engelhardt its' metaphysical reevaluation. Just to the contrary, for him the aim of content-less moral theory is to offer a "secular means for coming to terms with the chaos and diversity of postmodernity" (ibid. 10).

Such a theory for Engelhardt is the basic condition for peaceful collaboration between moral strangers in democratic societies. By understanding, his book provides "not simply a political theory, but an account of morality that should guide individuals when they meet as moral strangers to fashion healthcare policy. In the case that, when they so meet, they tend to collaborate in the realm of politics, through *a res publica*, a common thing that moral strangers of diverse moral community can share. Within that perspective nothing can contentfully be shared by all" (ibid. 10).

In this expressive passage we could notice two crucial issues for further exploration. First of all, we recognize the power and authority of the public, or *res publica*. But this recognition is far from being serious or consistent. The power of the public in itself is mere chaos for Engelhardt, or "a bubble of conflicting bioethics". That is why (this would be the second issue) his own mission as a professional in moral philosophy is to offer theoretically grounded "guidelines" that order chaos of public moral attitudes, making fruitful collaboration of strangers possible. But if any experts' idea or theory "should guide individuals" this would mean that individuals themselves are unable to self-organize or cooperate without the *paternalistic* guidance of an expert in philosophy. Individuals taken together are just inert chaotic masses. They need a prophet, a guru or a "father" to become a moral community.

In his introduction to "Meditations on First Philosophy" Descartes wrote that his mission is to convert those who do not believe in Gods existence using secular reasonable arguments. Engelhardt tries virtually the same – using reason to reestablish the unity of people left in chaos and estrangement after the loss of a common faith. Of course the difference is that Engelhardt offered only "empty", "formal" or "procedural" integration. But still this is an attempt to integrate others for their own good, but without any interest in their understanding of their own good, without the participation of those "others". *It is a paradox of the founders of bioethics, that they questioned medical paternalism while staying firmly on the ground of moral paternalism.*

As another example of this paradox, take this passage by Robert Veatch: "What we really have before us is a series of unsystematic, unreflective, ethical stances, or traditions. What we need is some ordering of that chaos we term a tradition, some systematic structuring of medical ethics so that physicians, other health professionals, government health planners, and consumers of medical care – all those who are important medical decision makers – can have some grasp of where they stand and why they may be in conflict with others with whom they interact" (Veatch 1981:5). The philosopher looks like a new Moses, through whose personal theoretical experience the internal order presents itself in "commandments" of principles and rules. The consequences of this inevitable characteristically philosophical thinking in medical ethics is to shift from anti-paternalism in relation to medical doctors to the paternalism of experts in moral theory which I have discussed elsewhere (Tishchenko 1998).

Here, I would like to pay attention to several other points. The position of a theoretical moral observer presented in "The Foundations" is outside the world of the moral community. It is reflected in introductory exploration, but has no role as a participant in the life of the "diverse moral community". To say more, this observer has no need of such participation. His reason is self-sufficient for a theoretical grasp of content-less moral "universals". And, what is important – it needs to be outside, distant in order to reflect on a priori conditions for the peaceful collaboration of "strangers" in the domain of "res publica".

But the public in itself is ignorant, uneducated in moral philosophy and that is why the chaos of public unsystematic stances could not be directly ordered by theoretical philosophical observations. Those results should be transformed into simple "guidelines" and invested in the fabric of communal life in the form of bureaucratic procedures. Just as Kant needed an enlightened monarch as a mediator between reason and society in general, the philosopher in medical ethics needs a bureaucracy for the implementation of constructed (or discovered) norms in the life of modern communities.

Not surprisingly Engelhardt's theory of moral strangers in bioethics was not shared by all in "res publica". Maybe this happened because his theory was not empty enough, and he had smuggled the content of the ideology of American liberalism into his supposedly content-less theory (Pellegrino 1998: 21). Anyway, the development of bioethics after the publishing of "The Foundations" has demonstrated that it is as impossible to discover content-less generally accepted ethics, as content-full ones. We witness a growing diversity of content-less or procedural accounts in medical ethics that leaves no hope for discovery of theory in this area. Does this mean our sliding into a bog of nihilism? Does "anything go" in moral life? Should we fear the fact of diversity and try to order the chaos? Or is it that this diversity is a direct result of the creativity of moral reason and we should wish it to "be fruitful and multiply ..."?

Let us stop at this point and return to the initial situation in order to make clear what is really at stake. First of all, it is a surprise that Engelhardt as a philosopher of the Hegelian brand is so swift to overlook the dialectical reflexive play of "form" and "content" inevitable in this situation. As already noted by a number of scholars the content of American liberalism is easily found in the foundations of Engelhardt's "content-less" theoretical structure of "guidelines" for moral strangers in democratic healthcare.

Second, Engelhardt's distinction between content-full ethics for "friends" and content-less ethics for moral "strangers" is based on their relation to the fact of diversity which could be refuted not only because of the impossibility of separating "form" from "content". Engelhardt also presumes that for a moral community of friends there is, in principle, the possibility of a common content-full theoretical ground. From the historical point of view this presumption is wrong – we cannot find any content-full theory that is more or less shared in any real "moral community of friends". As a matter of basic historic fact, there is no single "moral theory" among Muslims, Jews, Roman Catholics, Russian Orthodox, Marxists or others. Everywhere (of course in different proportions) we can easily find diversity of moral positions and perspectives. But absence of the theory does not mean chaos in moral life. This empirical (historical) argument moves us closer to a couple of essential philosophical questions.

Do we really need a theoretical ground for peaceful collaboration? Is diversity "the threat"? Of course, "every philosophy makes a practical and theoretical claim to totality and that not to make such a twofold claim is to be doing something which does not qualify as philosophy" (Spaemann in Habermas 1995: 16). But the status of such claims could be different. It is reasonable to consider that any theoretical perspective is not a vision from "nowhere". It is always housed in an exact life-world, structured specifically from historical, cultural, ideological and personal perspectives. This makes diversity inevitable, and positions claims to totality in a kind of imaginary utopia. In general any theory presents its own utopian world. For example, the physical theoretical idea of an "ideal gas" marks a state of affairs that could never exist, but in spite of this (to some extent, because of this) is very fruitful for science and the further technological control of existing natural things and processes.

The difference between modern and classical moral philosophy (as well as the difference between modern and classical science) is in their different attitude to the question of the congruence of multiple theoretical (or "utopian") worlds and to the question of the observability of the observer.

After the Second World War in philosophy, and somewhat later in science, we witnessed a swift decline of interest in general theories. That means a decline in the desire to integrate a multiplicity of theoretical perspectives into one homogeneous world from one theoretical perspective. The situation is the same in physics, biology, medicine and philosophy. It is a new much more modest style of theoretical thinking that treats as essentially important claims to totality, but rejects "playing God" – this means, it rejects ambitions to speak for all from the position of an objective observer.

In modern medical ethics this modest style of theoretical exploration without the desire for theory is present in such influential trends as principlism, casuistry, the narrative approach, to some extent discourse ethics and some others. In different ways all of these theoretical perspectives share an understanding of modern medical ethics (bioethics) as predominantly an activity in a public forum. This means that the public cannot not be treated as a kind of chaotic mass that needs an expert to order its unsystematic moral vision. On the contrary, it holds as crucially important such things as moral common sense, morally structured in a narrative life-world, the heritage of communicative practices for the resolution of moral conflicts etc.

When neither the natural world nor the world of moral communities present in a diversity of theoretical perspectives, cannot be theoretically integrated into a single vision (and where there is no desire to do so), the idea of common ground slides out of the "utopian worlds" of theoretical reason into empirical reality of the life world. I would like to agree with A. Johnsen that the beginning of bioethics can be traced to the establishment of the first ethics committees (Jonsen 2001).

Two existential discoveries of the limits to human reason were the creators of the first ethics committees. First of all, it was recognized that medical doctors do not have enough knowledge or experience to speak on behalf of their patients' good. Second, it was also recognized that there could be no expert in moral philosophy whose knowledge would be sufficient to clarify and resolve of moral problems in a healthcare setting. Of course moral experts were invited to participate in the common enterprise of evaluation and decision making, but (just like medical doctors) only on equal terms with lay people. To my mind, ethics committees are a kind of working model of a public forum of bioethics as part of a deliberative democracy. They present the pragmatic foundation, that in principle does not need theoretical guidelines for the common moral life of strangers, as well as friends. This does not mean that there is no need for moral experts' participation. It is always fruitful to give a theoretical discussion of the values that are involved and the procedures that should be used. But acceptance or rejection of any theoretical claim in public forum depends very little on theoretical perfection, but on other factors I discuss later.

Before doing this, I would like to explore some a priori cultural conditions for the integrated life of a modern moral community.

3.   CULTURE OF OTHER MODERNITY

Transformations of practices of biomedicine and discourses of medical ethics act in synergy with basic transformations of modern culture. They pump resources of new cultural impulses and give them back tremendous acceleration. In human beings they provokes new feelings of human power incorporated in biomedical technologies re-creating nature and a new experience of failure in the perceptual, imaginative and conceptual grasp of the nature of this human power. Human beings use this power in order to protect themselves against the threat of disease and other natural "enemies". But are we protected from misuse of this power?

For example, if one day genomics together with other biomedical sciences can fix "genes of death" in our bodies death would not disappear: it would be hidden, preserved and increased by genomics itself as the power of destruction localized in our "cultural genes" of freedom. Freedom of will is our basic cultural value as well as the cause of practices of transgression (including criminals, terrorists etc.). This failure reminds us to some extent of the truth of the Bible – the basic cause of death (as well as cultural progress, we could add) is our "sins" – the moral and ontological condition of human freedom. Humans may control nature, but who (if God is really dead) could control the "controllers"? Rethinking this question in the new cultural situation of biomedicine produces a new *feeling of sublimity* that retains the paradox of human power and ontological as well as existential weakness.

The feeling of sublime, presenting limits to human existence, shapes specifically the cultural *play* of a number of other existential feelings of which fear and hope are the most important. The specific design of this play in modern biomedicine was constituted in the 1960s by *two great existential discoveries leading to the establishment of the ecological movement and of bioethics.*

First of all, we consider the birth of the ecological movement. Somewhere in the seventeenth to eighteenth centuries in European culture the Christian idea of *salvation* was replaced by two initially separate ideas of *health* and *freedom*. This substitution and separation constituted the route to for scientifically oriented medical improvement and political action.

In the realm of biomedicine the basic threat to human identity was recognized in the external world of Nature, and the path to "salvation" was found in the scientifically designed technological control of external natural forces. Nature was blamed as the *enemy* and technology as the *saviour*. That is why the military metaphor was and is so popular in medicine, particularly in the ideology of the treatment of infection diseases and in political rhetoric (such as "war on cancer") and even in commercials ("Our soap kills all known germs!").

The ecological movement turned this linear existential orientation into a circular or paradoxical one. At the core of this existential shift was a fundamental discovery of a new threat to human existence. And what is crucial, it was found exactly where Western thinking was looking for "salvation" – in scientific and technological progress aiming at the conquest of Nature. We began to fear, but we did not give up these technological hopes. The desire of technologic control of natural forces is still the most powerful. But today it is balanced by the opposite desire to preserve Nature, to save it from human beings. The paradoxical play of existential feelings expressively illustrates the text of one advertisement for mineral water I found one day in my mail – "Natural spring water from Russian forests manufactured with the use of the most advanced Japanese technologies".

The diagnosis of the new existential threat was made by science itself; and what is wonderful is that the means of salvation from the new threat are designed by science as well in the growing industry of environmental control and protection as well as the manufacturing of "natural" products. Science has become threat and saviour in one, or as Ulrich Beck says – it has become "self-reflective", "self-doubting" and "self-limiting". The society that is growing on the new existential basis Beck calls the "risk society". The ecological shift has split scientific thinking into conflicting voices of scientific "truths", pro and contra exact technological innovations. Accompanied by the collapse of demarcationist programs in the philosophy of science (programs which attempted to establish a boundary between science and non-science) this shift legalized the multiplicity of truth-oriented scientific discourses. It made science more "open minded" not only in terms of intra- and inter-disciplinary differences inside science itself, but also in relation to non-scientific reasoning (religious, astrological, shamanic etc.). All of these different perspectives are today in permanent "dialogue" (Mikhail Bakhtin) in the evaluation of what is happening and in looking for prescriptions for what should we do in critical situations produced by technologic progress.

The transplantation of ecological thinking into biomedicine presumably happened after the thalidomide disaster (in which children were born without limbs to mothers who used the drug thalidomide during pregnancy), and transformed the pattern of relationships between science and practical medicine. For example, the "development time" for new drugs jumped from several weeks from the moment of synthesis of a new therapeutically active substance at the beginning of the 60s to around ten years by the beginning of the 1980s. "Development costs" escalated 20 times on more. The *Safety*, that is the prevention of the harmful effects of the "saviour" (the drug), had increasingly become the orientation of medical science (Le Fanu 1999: 247).

The second shift could be called "bioethical". It happened because of the intervention of multiple moral discourses into the area of medicine previously under the monopolized control of the scientifically oriented rationality of biomedical science and healthcare practice. The success of this intervention was predetermined by another existential "discovery", in the form of a number of public scandals involving biomedical research. *The human body is not just an "object" of scientific research or medical treatment, but also the "flesh" of a specific person – its owner.* That is why any action touching human flesh has an irreducible moral dimension that is largely invisible from the scientific point of view. This "blind spot" in scientific reason constituted a legitimate place for moral reason within modern biomedicine. At the same time, as its societal application to some extent, political practice has intruded into the world of biomedicine. Abortion, patients' rights, the rights of disabled people, cloning, genetic engineering, stem cell research –

all of these and many other internal issues in biomedical science have become powerful motivations for political movements.

So the conflict of "truths" in the ecological shift was exacerbated by a conflict between scientific and moral reason in a bioethical shift. Not surprisingly, moral reason appeared in this "boxing ring" not as a unified agency but as an internally conflicted crowd of moral points of view, perspectives, values etc. This constellation of moral discourses inside biomedicine is usually called *bioethics* – the word initially created by Van Rensselaer Potter for the ecological obligations to the biosphere as a whole. I think that such a paradoxical existential play of desires, fears and hopes constitutes the basic structure of modern cultural identity – a specific rhythm of *existential repetition*. Modern progress in biotechnology offers "protection" from powerful external natural forces. Meanwhile, the ecological movement and bioethics provide complementary "protection" from the abuses and threats of biotechnology itself.

The cultural diversity of moral stances in healthcare depends deeply on the degree of appropriation by a moral community of the new existential rhythm of a risk society. For example, in Russia, as in other Third World countries, people are more or less informed about the ecological dangers or moral threats of biomedical technologies. But this information does not significantly influence their everyday behaviour because it does not touch the level of existential feelings. These feelings are still those characteristic of classical industrial society, with its predominant motivation of the conquest of Nature through technological control. The same kind of moral diversity may be found within all existing societies.

4.   ETHOS OF WAR AND PEACEFUL MORAL COMMUNITY

The ethos of war constitutes a fundamental limit to the diversity of moral perspectives. Some differences cannot be tolerated. The limit of tolerance justifies and excuses both violence in relation to others and the limitation of democratic rights in relation to public.

Hegel was right to interpret war as an authentic and radical route to human self-understanding and self-development (but in contrast to Hegel, I would argue not necessarily progressive). At any rate warfare is a standard human "solution" to problems of cultural and political diversities. Whatever people we are, war is always with us and in us. In the short periods of peace, it is there potentially, and in most of human history , it is there in actuality. Of course, we differ very much in relation to the question of what kind of supreme value has to be threatened in order to provoke the move from potential to actual war-identity. For some people in some historic periods financial interests are enough (such as the "Opium wars" of China in the 19th century), for others it is national interests (like Russia in Chechnya), racial (Germany in the Second World War), ethnic (the Middle East today), class values (USSR), the values of the "modern liberal market democracy" and "universal values of human rights" (the case of Kosovo), or the fear of global terrorism (the case of Iraq). It makes a difference, but offers no option to which modern societies can be without self-identity of war.

Moral principles for the evaluation of human actions in a state of war could be called an "ethos of war". An ethos of war presumes the basic demarcation among "we" (allies) and "they" (enemies). The moral metaphor of a pirates' ship is relevant to this difference (at least until the middle of the 20th century). The ten commandments are used only by the internal circle of those who are on board of our ship. Others are outside the moral community of humanity. We are "the people" and they are "beasts" to be exterminated if they threaten us, and to be used "as means only" for achieving the great ends of the war (like the thousands of innocent lives in Hiroshima). In the second part of the 20th century this idea was meliorated by a number of international laws and conventions on the moral standards of warfare. But still, the difference in value between "we" and "they" is dramatic. It constitutes the difference between "medical ethics for us" and "medical ethics for them" in the time of war. The Iraq war gives a lot of examples of priority setting in the distribution of scare health care resources. First it was necessary according to the ethos of war for alliance troops to save oil fields (so valuable to "us") and only last hospitals (desperately needed for "them"). Specific moral problems and values of the ethos of war could also be found in the methods of provision of medical help in zones of catastrophes, areas of mass destruction, or after the terrorist use of chemical or biological weapons.

Biomedical professionals participate in war in different ways. Special problems are associated with their activity in the development of chemical, nuclear and bio-weapons. Hidden in secret labs and protected by the prioritisation of "national interests" this area of biomedical science is beyond public control. Evidence of violations of human rights could be found throughout. Justification for all such violations is based on the idea of legitimate sacrifice of human lives and private values on the alter of national interests (or another supreme value) in a state of war.

"One part of the war machine conscripted a soldier, another part conscripted a human subject, and the same principle held for both. In effect, wartime promoted teleological as opposed to deontological ethics; "the greatest good for the greatest number" was the most compelling precept to justify sending some men to be killed so that others might live. The same ethics seemed to justify using institutionalised retarded or mentally ill persons in human research" (Rothman 1995: 2252). The Nuremberg code and the Helsinki Declaration (1964) were attempts to "turn the table" and to make values of peaceful community higher than those of ethos of war. One of the basic principles of "ethos of peaceful community" is just the reversal of the principle of legitimate sacrifice. It goes: "Every biomedical research project involving human subjects should be preceded by careful assessment of predictable risks in comparison with foreseeable benefits to the subject or to others. Concern for the interests of the subject must always prevail over the interests of science and society" (World Medical Association 2000, 1964).

In official self-consciousness of modern societies this move was successful, but the ethos of war survived in the depth of military structures and after "September 11" was resurrected. As an example, we could mention existing international situation concerning the control of development of new methods of biological warfare. In 1972 the Convention on the Prohibition of the Development, Production and Stocking of Bacteriological (Biological) and Toxin Weapons and on their Destruction (BTWC) was signed by a majority of countries. An application to this protocol a "Verification Protocol" that aimed to establish effective regime of inspections was in preparation for a couple of decades after this. But it was "suspended in November 2001 when the United States declared that they would not support nor permit the conclusion of a binding multilateral verification agreement. Among the reasons that US officials cited for the refusal was that the US believes that other countries are cheating and that the US should not be subject to the same standards as the rest of the world, and that the intellectual property of the US biotechnology industry would be put at risk by spying inspectors" (Sunshine Project 2002: 21).

The argument of "spying inspectors" echoes the same argument heard from Soviet military authorities, Saddam Hussein, authorities of Northern Korea and others belonging to the so called "axis of evil". It is based on the ethos of war. It protects development of biomedical research in the area not only from spying enemies, but also from own public control of moral and ecologic standards at use. What ever good ends and justifiable reasons has the unleashed "war against terrorism" – it inevitably expands the area of biomedical research and medical practices that are bound by the ethos of war with all negative consequences for the basic values of international peaceful moral community.

From the perspective of a peaceful moral community war is an extraordinary event. It is a transgression of moral order, its' justifiable or unjustifiable violation. From Hegelian point of view that has significant reasons – it is the way that a new world order is born. Both evaluations have their own justice and power. It is impossible to neglect or prefer either of them.

I consider the ethos of war as an external boundary for the ethos of a peaceful moral community. I now finish with an evaluation of the specific role of public in a peaceful moral community, in competition of multiplicity of moral perspectives in health care.

5.   THEATRE OF BIOETHICS AND SPECIFIC ROLE OF PUBLIC

Presented in a huge number of scientific and public discussions bioethical deliberation looks like a "theatre" that, from my point of view, has its own cultural role. From the perspective of classical philosophy Engelhardt was right to call it "... a babble of conflicting bioethics" (Engelhardt 1996: vi).

I think, keeping in mind Bakhtin's vision of thinking as dialogue, that the conflict of "babbling bioethics" is a precondition for the existence of the bioethics we have today and its most characteristic feature – the dialogical way of reasoning. The word "babble" is instructive, reminding us of some specific points in the history of European culture that help to illuminate the phenomenon of diversity of moral perspectives and the role of the public, putting the "res" (power) of the public in a new light.

According to the Oxford dictionary, – to babble means to speak incoherently, indistinctly or foolishly. I would like to remind you of the time of the Renaissance, when the "fool" had become a popular figure in scientific and philosophical writings. To mention two of the most significant examples: Erasmus Desiderius' (1466–1536) "In Praise of Folly" appealed to the wisdom of "babbling" in order to undermine scholastic teaching and to free space for a new image of Man in the Universe. In Galileo Galilei's (1564–1642) "Dialogue Concerning the Two World Systems", one of the first roles is "a Simple Man" (Symplitchio), – a lay man or "a man from a street" whose everyday wisdom helps Galileo to dispute Aristotles' ontology.

At the very beginning of the New Era we could see "open-minded" science in permanent dialogue with the "public" – the community of ignorant people. For example, Robert Hooke, who developed the basic methodological principles of scientific research for the London Royal Society, insisted that knowledge should be developed to the extent that scientists could demonstrate truth in public. And public demonstrations of scientific experiments were a routine activity of the society at the very beginning of its life.

At the same time we could see the development of similar phenomena in medicine in the "anatomic theatre"; The famous frontispiece of Vesalius' "De Fabrica Human Corporis" shows a scene from an anatomic "performance". Among the spectators we can recognize many leading thinkers and political leaders of the time, including Martin Luther. What was the role of those "fools" – people ignorant of anatomy, crowded around Vesalius? What was his role? He was not their teacher and they were not his students. Both parts participate in the magic of the anatomic theatre. The role of Vesalius was to dissect decomposing disfigured flesh to recreate in the imagination of the spectators the young flourishing athletic body, constructed in accordance with the Pythagorean concept of ideal proportions.

I suppose that the role of the spectators was that of a jury in court hearings. Vesalius like an expert witness attested to the new truth of human nature. His attestation competed with the attestation of a rival theatre that of the Church. And those ignorant in anatomy, the "fools", were to decide which attestation was more persuasive and which less in this historic case of Vesalius contra Galen (the Church). So, the babbling of fools (the public) pronounced a verdict on conflicting expert attestations in ongoing court hearings of history.

I will insist that the same kind of "bubble" constitutes the heart of bioethics as specific intellectual activity. To some extent this issue is acknowledged in the so co called "week principle of publicity" (Henry S. Richardson). In bioethics justification must be offered in terms of reasons that may be publicly stated. The need for "profanation" of professional knowledge of scientific truth or moral goodness is the same as it was in the epoch of Renaissance – the conflict of truths and conflicting ideas of good could be solved only in public "court hearings" (at any rate in civilized manner).

If to change slightly one of Paul Ricoeur sayings – conflict of truths is a goad that sends "experts" to a court of appeal – to make attestations in front of jury composed of lay people (public). In such attestations rival experts (e.g. scientists and moral philosophers) could not appeal to jury knowledge of foundations of the truth or the good they are presenting. They should turn reasoning out of "depth" of knowledge onto "surface" of narratives of public presentations. Due to Paul Ricoeur narration is the "summit" of scientific (description) and moral (prescriptive) reasons. They could meet each other and be in conflict only on the surface of life stories. "The actions figured by narrative fictions are complex ones, rich in anticipation of an ethical nature. Telling a story, we observed, is deploying am imaginary space for thought experiments in which moral judgment operates in a hypothetical mode" (1992: 107).

In this sense attestation of rival moral and scientific "experts" in bioethics in front of public ("jury") is basically a storytelling – an imaginary thought experiments in which expert constructs a version of a "life plan" – how life would change for good or for bad if proposed ideas (new norms or new methods of treatment) would be accepted by the "jury". In order to disapprove this "dramatic" attestation rival experts should create their own imaginary versions of life stories – possible changes (surely negative) in life of lay people. In other words – bioethics looks like a kind of a competition of storytellers or tragic poets in antic theatre. Public (spectators) is the agency that grants gifts (awards) of recognition to those whose "stories" mostly fit their internal moral predispositions. It is the way of permanent co-evolution of public moral identity and ideology of biotechnological progress.

If to put this idea in other words – "bioethical theatre" presents a specific form of cultural selection of competing for survival multiplicity of moral values, stiles of life, and world visions, and on the other hand, – multiplicity of biotechnological projects of solution of human problems. This way humanity achieves contingent, open to reappraisal normative structures that order and stabilize the stream of scientific progress in biomedicine.

6.   CONCLUSION

Moral diversity in health care setting is a symptom of reintegration of three previously separated forms of human reason – science, morality and sense of beauty (working in the theatre of bioethics) into a new structural unity of forming risk society. At the same time this new structural unity opens space for participation of cultural projects independent from the project of Modernity.

References

Beck, U. *Risk Society. Towards An Other Modernity*. [Russian translation] Moscow: Progress-Tradition, 2000.

Callahan, D. "Judging the Future: Whose Falt Will It Be." *The Journal of Medicine and Philosophy* 2000, 25, 3, 679ff.

Engelhardt, H.T. Jr. *The Foundations of Bioethics*. New York, Oxford: Oxford University Press, 1996, 2nd edition.

Habermas, J. "Modernity – An Incomplete Project." In *Interpretive Social Science. A Second Look*. P. Rabinow and W.M. Sullivan (eds.). Univ. of California Press, 1987.

Jonsen, A.R. Paper presented at the Euresco-conference "Bioethics – an interdisciplinary challenge and a cultural project. 8–13. Sept. 2001, Davos, Switzerland.

Kohlberg, L. *The Psychology of Moral Development*. San Francisco, 1984.

Le Fanu, J. *The Rise and Fall of Modern Medicine*. London: Little Brown Company, 1999.

Pellegrino, E.D. and Thomasma, D.C. *For the Patient's Good. Restoration of Beneficence in Health Care*. New York, Oxford: Oxford University Press, 1998.

Ricoeur, P. *Oneself as Another*. Paris, 1992.

Rothman, D.J. "Research, Human: Historical Aspects." In *Encyclopedia of Bioethics, Revised Edition*. W.Th. Reich (ed.). New York: Macmillan Library Reference, Simon and Schuster Macmillan, 1995.

Spaemann is quoted from Habermas, J. *Moral Consciousness and Communicative Action*. Translated by C. Lenhardt and S.W. Nicholsen. Cambridge, Massachusetts: The MIT Press, 1995.

Sunshine Project. *An Introduction to Biological Weapons, their Prohibition, and the Relationship to Biosafety*. Third World Network, 2002.

Tishchenko, P. "The Goals of Moral Reflection." In *Advances in Bioethics*. M. Evans (ed.). JAI Pres INC., 1998, v. 4, 51–64.

Veatch, R.M. *A Theory of Medical Ethics*. New York: Basic Books, Inc. Publishers, 1981.

World Medical Association. *Declaration of Helsinki. Ethical Principles for Medical Research Involving Human Subjects (1964)*, Adopted by 52nd WMA General Assembly, Edinburgh, Scotland, October 2000.

IV. BODY AND IDENTITY

Chapter 16

BODY, PERCEPTION AND IDENTITY

JEAN-PIERRE WILS
Nijmegen, the Netherlands

1.   A SURVEY

The new genetic anthropotechniques turn human life into a biotechnological mega-project. The implications reach much further than those of the older cultural-philosophical insight, that people have always had a 'technical' relationship with their body. The 'body-techniques' of Marcel Mauss determined the non-naturalness, or the *artificial* character of the functions of the human body. Although the body was perceived as part of a never-ending cultural transformation, this transformation encountered its limits in the *positivity* of the biological body, and in spite of their drastic effects on people's behaviour, these cultural techniques remained *external*. The current signature of anthropotechniques makes this biological borderline disappear: the techno-naturalistic symbiosis renders the determination of borderlines redundant. Every limit becomes a *virtual* limit. A radicalised objectification of the body is now the precondition for analytical interventions *in the restricted sense of the word*: the genetic code is viewed as an ensemble of entities that are totally distinct in principle, and which can subsequently be rearranged and recombined in the context of a novel *trans-natural* view. As such the transformation of the anthropological profile in its entirety becomes possible.

From this perspective every identity has a triple index. *Constructivity*, *provisionality* and *preferentiality* characterise the underlying structure of every determination of identity. At this point the grammar of cultural codes becomes decisive while the grammar of natural codes becomes anachronistic. Genetic vocabulary experiences the permanent changes that

231

C. Rehmann-Sutter et al. (eds.), Bioethics in Cultural Contexts, 231–245.
© 2006 *Springer. Printed in the Netherlands.*

cultural grammar originated. A generative grammar – the grammar of culture – alters the texture of the genetic code. These new techniques have an *internal* profile – they render the borderline between culture and nature potentially meaningless. All of this has far-reaching consequences for the character of new identities: they no longer experience resistance because of the partial autonomy and recalcitrance of nature. The non-cultural and non-technological components of nature that are remembered in the old maxim of 'natura naturans' are being transcended; every nature is now a 'natura naturata sive culturata'. It is the perception of the body in particular that changes. The body is mobilised and experiences permanent change. It is approached not in terms of *durability* but of *optimisation*, not with *passivity*, but with *activism* and *interventionism*. Hence identity becomes a *discordant* and a *relativist* concept. From now on identities are *gained* on the basis of a constant *war* against something that refuses to participate in the permanent revolt. Concepts of the self must time and again be relativised and reconstructed according to the preferential order of the moment.

2.   THEOMORPHY, ANTHROPOMORPHY AND TECHNOMORPHY

Whenever ethical conflicts loom on the horizon of our culture, a semantics of creation is mobilised. Most of the time this semantics is mere rhetoric and, moreover, it is hollow rhetoric, used to fill the odd gap. On the one hand, the nature of what is at stake is illustrated, because appealing to creation is tantamount to a self-legitimisation that, in view of the immemorial originality of the instance appealed to, can hardly be bettered. On the other hand, the problem solving potential no longer needs to tackle the issue of legitimisation. Moral views have their foundation *condensed*, as it were, into an abridged version – in a shortcut through the paradoxical concept of the creation before the dawn of mankind. As is so often the case, the popularity of the *word* is in disproportion to the content of the *concept* it embodies: that of creation. The fewer references a word conveys, the greater its power. In general, at the root of the metaphor of creation lies a rather handiwork-like imagery, nurtured by biblical mythology. However, it is all too well known that the road from *Mythos to Logos* is long and winding, even though the modern rehabilitation of the myth has been accomplished. By far the most difficult gap to bridge is that between the myth, which can be reduced to a need for some kind of basic confidence on the part of the creatures, and the moral content that can be distilled from this myth. The bridging that is called for here implies speculation, which makes it impossible to bring the ethical confusion of earthly beings to a satisfactory conclusion. Hence, it might be altogether preferable to refrain from woolly semantics, and to admit that 'creation', at least as a matrix for ethics, has been exhausted. The modern subject has accomplished the expression of its individuality by transforming the theorem of creation into the realm of creativity. Meanwhile the 'creatureliness' of yore has blossomed into a continuous autopoiesis. Even before the Romantic era, self-building was regarded as the perfectibility of non-transmittable individuality. Instead of the modern, abstract appeal for biographical self-realisation, and in place of the once 'allopoietic' passivity of man, literal self-creation by means of the new anthropotechniques has emerged. Both scenarios threaten to overtax us.

In spite of the danger of an over-simplistic classification, one can distinguish three major successive paradigmatic models in European history: Foucaultian archaeologies that have provided a context for every kind of human self-understanding. Each model presents a privileged form principle, a conclusive *'morfè'* that preconditions the content of each determination. I will call the first model *theomorphy*. The key term 'God' determines the morphology of this model. This category was an *attractor*: all predicates used within the bounds of this model could be related to this central notion. We need *not* contemplate the orthodox or heterodox content of the notion of God, but rather the function it fulfils as *attractor* and as *generator* of every meaning *against the background of human self-understanding*. The classical onto-theological model has the following characteristics: originality, lack of alternative, stability and optimality.

In this model, *originality* was linked to the fact that all 'being' is created. 'What is' was perceived as *created* and owing its being to an original *initiative*. 'What is' was in its entirety – literally – allopoietic, and it everywhere demonstrated a dependency on an authority from which it originated, which was, in itself, pre-creational. Because of this *strict* descent, a cosmo-theological entity came into being, which was characterised by *a lack of alternative*. The totality of the cosmos led to the conclusion that the creational instance was omni-potent – God's 'omnipotence' was a 'power over all that is', all-powerful in itself. Omnipotence thus stood for the *range* of the divine power as well as its *intensity*. The cosmos was the exhaustive demonstration of this omnipotence. At this point I should stress, however, that we are still a long way from the late medieval view, prefigured in Augustine's work, that omnipotence actually means that creation could have been entirely different. Here the potential alterity of creation still had to give way to the view of an absence of a potential alternative form for the cosmos. The structure of things, the onto-logic of 'being' represented an 'ordo' that was comprehensive as well as reasonable. This ordo-theology functions as a *guarantee* of the *stability* of the cosmos. It was in fact inconceivable that a fundamental disturbance could disrupt the constellation of this 'ordo'. Every disturbance could be repaired by reshuffling the existing classification, the hierarchical teleology of being. The said classification and teleology concerned relations between species as well as the 'nature intra species', or the nature of the individual being. Because this ordo-structure was *original*, and as such possessed theo-logical foundations of validity, it left a mark of *optimality* on the essence of things. Creation was optimal, and even its tendency towards sub-optimality because of the Fall of Man could not undo this quality. This optimality could be taken as a licence for remarkable optimism: creation is good, what is good must be safeguarded and depicted. This model was thus structural-conservative.

The second model – the anthropomorphic model – initiated drastic changes. On the one hand, there was a total shift in perspective, but because of this very shift the model remained bound closely to the force field of the first, theomorphic model. A quotation from Ludwig Feuerbach makes the totality of the change clear: "Religion is the first self-consciousness of man. Religions are sacred, because they hand down the first consciousness. But what comes first of all in religion – God – is in itself second to truth, because it is only the objectified essence of man, and what comes second in religion – man – must therefore be placed and named as first" (1841; ed. of 1976: 318). It is hard to comprehend the radical consequences of this shift in perspective: in one way religious consciousness was infected from that moment on with an indelible suspicion – that of being a fiction at life's service. The fierce wind of this projection, ultimately considered to be wrong, blew through all notions of transcendence. But it is exactly the inversion that Feuerbach desires – the redefinition of theology as anthropology – which makes the relative dependence of the new model abundantly clear: there is a vivid recollection of the God of the first paradigm, a God that has to be faced and realised within the history of the species.

Nevertheless, the changes cannot be ignored. The characteristic of originality assumed another meaning. After Pico della Mirandola's "De dignitate hominis" man became *auto-creative*. Dignity was no longer to be understood as receptivity to the normative ideal of an essential characteristic, but as the regulative responsibility for the ideal itself. The totality of things may still have been the result of an allopoiesis, of a 'creatio ex deo', but this allopoiesis had one significant exception: humans were paradoxical beings. Humans existed in the possession of a *created self-originality*. The signature of humans 'being created' was already formalised, because every implementation was already the result of a self-conceptualisation. Humans descent, the fact that people are by nature creatures, was still admitted, but this nature was no longer invariable. On the contrary, the lack of alternative now had to yield to the fundamental *variability* of humankind. This view could be articulated as an insight into the *accidentality* of human nature, the essence of which is but one single possibility as compared to the infinite number of potential possibilities in the mind of the Creator.

But yet another paradox arose: the essence of humankind is only *factual*. This was the view of Nicholas of Cusa. The concept of variability in its most radical form was that the essence of humakind consists in humanity's latent essencelessness. Humans could be different time and again, since there was no teleology to restrain them naturally or therefore morally. The Creator was no longer the guarantor of a fundamental inertia, of a perpetuation of the original 'ratio creandi'. Through the dignity of his most important creature God enabled the thwarting of an all-embracing 'ordo-theology'. For where its hierarchy reached its highest point, the stabilisation of the cosmos was guaranteed in the least stable way: human are self-changing creatures, and therefore *essentially unstable*. It became increasingly difficult to defend the optimality of the essential nature of things and of humankind. The *idea of perfectibility* took the place of this optimality. Good could only come into being through the amelioration of the status quo, not by imitation or mimetic identification with an archetype. Humankind was now the starting point of an active world-interpretation and self-interpretation.

As the providential presence of the Creator was driven back step-by-step and as his passivity increased, humankind became the centre of every significance and meaning. There was now room – literally – for this auto-creative activity. From the thirteenth century onwards the Cabbalistic interpretation gained ground in the heterodox ranks of culture, especially Isaac Luria's view that, after his act of creation, the Creator had withdrawn: God 'draws back' or 'contracts' and in doing so has made room for the autonomy of the creation. The Cabbalistic doctrine of the so-called *Zimzum* could be viewed as a form of *implicit* theological legitimisation of the auto-creative competence of man (see Scholem 1976: 285ff). The technical realisation of this competence followed bit by bit. But this legitimisation anticipated the third model, that of *technomorphy*. The "world era of passivity of being comes to an end and an era of activism begins", as Peter Sloterdijk writes (my translation, 2001: 71).

The *technomorphic* model, with which we are confronted in advanced biotechnologies, can be considered as a radical break with the two preceding models. This can be illustrated by looking at the role of technique. In the first, theomorphic model, technical knowledge was seen as an *inadequate approach to the ideal norm of nature*. In the second, anthropomorphic model, technique was seen either as an *imitation* or as an *excelling* of nature. Both models are connected by the idea that technique ultimately belongs to an *essentially different ontological order* from nature itself. There is an *alienation of being* between technique and nature. Compared to nature, technique is 'allotechnical', in a certain sense it is 'contra-natural'. The technomorphic model leaves this double ontological order behind: it causes the line between nature and technique to vanish. "At this moment, for the first time, the threshold is reached beyond which technique starts to be a nature-like technique – homeotechnique instead of allotechnique. [...] It cooperates with life and it penetrates into the self-production of life. [...] Here begins a new form of symbiosis with the old nature, which is as ominous as the first technique. Yet the epi-natures of the second technique will be utterly different from the contra-natures of the first technique" (ibid. 70).

Literally speaking we are dealing with a *sym-biotic* technology. What emerges here is a *techno-naturality*, which is ontologically *indifferent* and in its turn has two variants. On the one hand, there is the *miniaturisation of technology* (all the way to nanotechnologies), which makes it possible to populate the human body to an ever greater extent with organ substitutes and organ optimising equipment that matches the physiological 'background'. On the other hand, there is the technologically induced correction and amelioration of the genetic code, which causes a *comprehensive transformation* of human nature. Advanced technology has an "intra-structural" (Virilio 1994: 108) nature as opposed to a natural substratum. According to Paul Virilio this concerns the "endocolonisation" of the body (ibid. 123). But aside from this evaluation it is instructive that technology becomes 'intra-structurally' active through miniaturisation. It becomes invisible to its 'subject' (or perhaps its 'object'). Because of its *intra-natural potential* this technology is capable of becoming a *secondary subjectivity*: it lies at the basis of the new existence, it is the underlying feature, the 'fundamentum artificialis' of life.

The consequence for the self-image or identity is that the determining identity-creating factor is fully *introjected*: the body is no longer involved in an identity-forming *existential dialogue* through socially and culturally reflected profiles. At this stage cultural icons and social compulsions are – literally – incorporated. From now on the transformed body *is* the identity. The difference between the 'order of the body' and the 'order of culture' is no longer of a *dialectical* nature: it has become indifferent.

The characteristics of the preceding model have been altered: instead of the *created self-originality* there is a *self-created quasi-originality*. But the origin is no longer a 'terminus a quo' in the sense of a 'descent'. A yawning gap now separates 'descent' from 'future'. Indeed it seems that today we are gripped by a descent phobia. 'Descent' is in principle suspected of *inadequacy*. It is merely the material for a future-oriented moulding of the 'natura nove sive technologica'. At the centre of self-perception there is an auto-poietic construction. One can speak of *quasi-originality* here because it has become a 'terminus ad quem': auto-creativity has liberated itself from its natural bonds and carries with it the – once again – paradoxical index of *future naturalness*. *Variability* too has been abandoned since we are no longer dealing with an *essential instability* that is connected to factuality, but with the *conventionality* or *arbitrariness* of the identity profile. Every *facticity* has now become a virtual *potentiality*.

The idea of perfectibility is pushed towards its semantic limit: the origin is not optimised, but rather *exterminated* and abandoned in favour of a *potential identity*. This future identity too will be merely a temporary one. The 'natura nova' will always have emancipated itself from the 'natura prima sive vetus'. What is new becomes a fatality in the wake of progress. In his "On what is new. An essay on cultural economy", Boris Groy writes about the determining power of what is 'new': "There is no way one can break through the rules of what is new, for it is this very breach the rules expect from us. And in this interpretation of the word, the innovation is what we could call the only reality, which is expressed in culture" (in Bachmeier 2001: 42, my translation).

3. PARADIGMS AND THEIR STYLE-CREATING CAPACITY

One can consider these successive models as ideal world-images and self-images. They are so-called 'background hypotheses': basic philosophical and meaning-generating structures that underlie the attitude a culture adopts towards the fundamental questions of existence. These paradigms are also called *symbolic fields*. Other terms that are sometimes used in this context are "cultural condensing space" (Rolf Peter Sieferle, 1989), 'episteme' (Michel Foucault) or 'Paradigma' (Thomas S. Kuhn). In the everyday profile of a culture they are presupposed rather than thematised. They are the semantic membrane through which culture breathes and gives birth to every concrete meaning. In the fundamental archaeological structure of an era they generate a selective repertoire of observation possibilities and observance rules. It is extremely difficult *not* to use these background hypotheses and *not* to give in to their compelling power. Theirs is a *collectivising* authority. This obviously does not mean that they have to be understood as determinants in the restricted sense of the word. While it may be difficult for the individual to escape from their formative influence, he is not entirely at their mercy either. He is able to resist them and his actions can deviate. But within the action range of a paradigm a keynote becomes audible which resounds in all phenomena. This is not inconsistent with the claim that these symbolic fields always have a somewhat fictitious nature. It is exactly their *fictitious* components that make it so hard to escape from their influence. They cannot easily be verified or falsified.

I will try to characterise the three models in another way to make their moral implications even more apparent. The first – the theomorphic – model was *teleological*. Accordingly, every 'being' had a purposive nature and was limited accordingly. The 'final causes' were substantial: they literally embodied all beings and phenomena. Morality amounted to the attempt to represent this teleology adequately and all-inclusively. Even under the influence of Christianity this model did not change fundamentally. Although the theological doctrine of the 'natura lapsa', of nature fallen as a result of the Fall of Man, was a disturbing factor in the harmonic teleology, this irritation was nevertheless soothed by flanking measures. Behind the disturbed order of nature, the 'fides quaerens intellectum' was able to recognise the original purpose. The order could also be considered as a *borrowed* order, warranted only by divine grace. In this way the 'conservatio sui' was in fact a 'conservatio a Deo sive Creatore'. Just as morality was an expression of the natural-ethical order, technique consisted in mimetically reproducing the natural-structural order in the world of 'making', of 'poiesis'. A "controlling principle" (Sieferle 1989: 22), which noted and sanctioned deviations, kept a close watch.

As we have already seen, the second – the anthropomorphic – model was clearly more dynamic and its basic configuration was one of variability, perfectibility and fundamental *autonomy* of the subjects. The understanding that human nature is *plural* developed. Human essentiality consisted in the ability to pluralize essence. And technique could even be seen as a transgression of nature. But it was exactly the ontological alterity of technique that made it impossible to indicate it as a destabilising factor for the order of nature in its entirety. In the background of this model the comforting view of the *oeconomia naturae*, or the *economy* of nature, ensured that technique did not cause a derailment, but rather prevented nature itself from straying (Sieferle 1990). Even when there was a lack in nature, it could be compensated for in a creative way. Humans auto-creativity gave the ability to turn shortage to advantage. It was exactly the fact that humans are 'Mangelwesen', a 'shortcoming being' (Helmuth Plessner) that gives sufficient motivation to do so.

What was still unthinkable can be described as *the* pre-eminent incentive of the third – the technomorphic – model. "The concept of a total anthropogenic crisis of nature, of a self-destructing global process, was still wholly unthinkable in the context of the model of an *oeconomia naturae*"(Sieferle 1989: 33). Charles Darwin threw overboard the comforting basic premise of the second model. After renouncing his initial natural theological position and his belief in the constant character of the species, he too assumed that nature was susceptible to being *disturbed*. This possibility of disturbance was the consequence of the theory of evolution. The new basic belief now proclaimed that every being occupies a *relative* place within the sequence of evolution, and that it must defend its place by means of *mutation* and *selection*. The belief that the phenotypic properties of living beings, their pathologies and their behaviour allow for a mono genetic explanation (the *genetic reductionism* one encounters everywhere today) was the incentive for a eugenic refinement, which has managed to imbed itself in every layer of our culture.

The *ontological* relativism implied in Darwinism, turned into an *ethical* relativism. "One need not be a Kantian to understand that human beings should never be merely means, and certainly not part of a breeding scheme, but that they contain their life's goal within themselves in every situation of life, in every culture and at every moment. At the same time this explains why our culture is on a slippery slope from the moment it starts to think in terms of evolution, naturalistically, futuristically, since it is part and parcel of evolution *per se* to be tempted into relativising an existing generation, confronting it with what the next generation will have achieved" (Sloterdijk 2001: 68).

If the crisis in nature had an *anthropogenic* cause, then technique had to be activated to minimise the crisis – even if it was contrary to the standards of morality. In this context a technological activism displaying a disastrous dialectic has been glorified: the more this activism tries to optimise the conditions of survival, the more it aggravates the crisis itself. In its turn, the aggravated crisis becomes the new turn-table for an even more intensive technological intervention. Morality tends to be denounced as anti-technology. Every attempt to slow the process down, every ethical comment is perceived as a retarding factor, as a disturbance. It is no coincidence that one of the central notions of the new *evolutionary ethics* is the word 'flexibility'. Flexibility is a cardinal virtue in an ethics that knows only *potentialities*, and has no *limitations* or *categorical* rules. The temptation is strong to assail the anthropogenic crisis of nature and – a fortiori – the inadequacy and the shortcomings of human nature with a *permanent interventionism* and with an over-exaggerated *constructivity*.

4. NOTES ON PERCEPTION, BODY AND IDENTITY

In the context of the third paradigm – indicating technomorphia as a cause and effect of the crisis in nature – the issue of identity is *dramatised*. The homeotechnique gave rise to a hyper-nature or epi-nature. As a consequence 'nature' has ceased to be an 'opposite' of the formation of identity. Identity always took a counter-position – against an *unmanageable* nature, against *social* and hence *enduring* role models, against the *passivity* that the factual world in which we live forces upon us. Only in a second phase, and as a result of this positioning, could a new *synthesis* come into existence, a balancing I-identity, which was able to obtain its substance *in opposition, starting from* and *resulting from* these factors, which in the first phase were experienced as barriers to *singularity* and *individuality*.

Identity is the result of a dialogue between what is *immutable*, and what is *modifiable* and *variable*. It is a synthesis of an *acknowledgement* of *what is unavailable* and a *defence* of *what is available*. The choice of interventionist constructivism, as I would like to call it, makes the first condition for identity, the *acknowledgement of what is unavailable*, disappear. Identity, then, seems to be an *optional object* itself. What is chosen holds only *provisionally*. A circular movement begins between the compulsion to *construct*, the temporality of identity as a *provisional* status, and the determination of identity from a *preferential* background. This forces us to consider identity as a *contingent issue*, a *fictional middle position* on the way to other identities. This is connected with the fact that the *'opposite'*, which can force us to acknowledge what is unavailable, no longer has any say in the matter. Perception has become merely *self-perception*. As Helmuth Plessner put it, people are nothing more than "proprioceptive systems" (ed. of 1980: 367, my translation).

By the end of 1970s the sociologist Dieter Claessens had already predicted the consequences of a society excessively dominated by technique and analysis, a culture compelled to analyse, deconstruct and be technically useful. According to Claessens this process of "severing an abstract dimension of the concrete-sensory, which itself no longer attempts the concrete-sensory" is connected with a "loss of the body". With it, the synthetic and integrating powers of humankind would come to an end. He forecast the "end of the conservative reserves" and "an existentially depressive feeling of fear", a fear "of the inherent inability to continue to extract from the analysed, that is the unfolded reality, a lively reality and a convincing, identity-saving and identity-creating meaning" (Claessens 1980: 310).

What changes is the ability to perform an identity-establishing synthesis in reality, which integrates everything we experience and perceive in a (relatively) meaningful unity. Looking at it from the angle of *what is observed*, that is, from the perspective of the world of objects, one could also interpret the emancipation of the abstract as the continuous *shrinking* of relevant existential information. The genetic code, which escapes every sensory representation and which, even when visualised, points at no anthropological profile whatsoever, is considered to be the decisive object for every statement on identity. From the angle of perception, that is, from the subject side of the world, the aforesaid emancipation could also be described as the *disciplining* of perception. This means that our perception is *formatted* on the basis of cultural codes, which we seldom recognise *as such*. This occurrence does not have to be understood as a compulsive, political determination. It is probably much more subtle. Early in his career Merleau-Ponty drew attention to the fact that the process of perception is not something natural. In his posthumous work "The Visible and the Invisible" (*Le visible et l'invisible*), he proposed a terminological distinction between 'wild perception' ("perception sauvage") and 'cultural perception' ("perception culturelle"). This terminological difference lays open a domain in which the *transformations* of perception can be described. In any case the neurological basis of perception is merely a starting point. The "wild perception" is characterised by its holistic, pre-reflexive and poly-sensible nature. When Merleau-Ponty talks about the "zone of transcendence" it is this very human potential he envisages, a potential to broaden and to transform the "wild perception", led by cultural choices (ed. of 1994: 279). What emerges then is a *selective* perception.

It would be a misunderstanding to presume that "natural perception" has a normative status. This is absolutely not the case: "natural perception" is not a 'terminus ad quem'. Mereau-Ponty simply wanted to explain that perception always implies selection, and that this selection is the *formative* basis of our constructions of identity. Identity is always formed in a field of *habituated* perceptions. Just as there is no "innocent eye" (Gombrich 1960: 264), so there is no perception that has not already been filled with cultural schema, with preferences that are habituated in cultural choices. In and through our habits of perception, a *stylised* world emerges, a world style that is in a certain sense *conventional*.

In this context Nelson Goodman referred to "ways of world-making"(1990: 129). The key to the birth of a world style is not a notion or a theoretical concept but an image, a *privileged* image: a metaphorical line of approach exists which *structures* our perception and our attention. "Just like terms, images can generate and present facts, and in doing so participate in world-making [...]. They induce [...] a reorganisation of the world we are used to" (ibid. 129). What we are dealing with here are not so much *actual* images, but privileged cultural *phantasmata*. Nowadays the genetic code is just such a privileged phantasma. The 'image of the genetic code' consists of a mixture of centrally implanted information, literal visualisations, political and moral expectations, promises and fears, fictional illusions and utopias, cultural fantasies, and personal preferences. We are dealing with an *amalgam*. The result is an "obsessive image" (Rorty 1984: 22). In the instance of the 'image of the genetic code' the investment in the phantasmatic is so important precisely because the sensory is only available to us through technological visualisation. *What we miss with our senses is compensated for in the sphere of the phantasma.*

Nevertheless, of even greater importance are the consequences of 'obsessive images' . Time and again studies on metaphorology stress the *formative power* of certain metaphors (see Ricoeur 1975, Schöffel 1987). The classic formulation of this idea is Hans Blumenberg's statement that privileged metaphors act "as starting-points for orientation, influence behaviour. They lend structure to a particular world" (Blumenberg 1960: 20, my translation). Culturally privileged images are thus to a certain extent *constitutive of behaviour.* They work on the basis of what I call the *logic of adaptation* (Wils 1990). The fact is that an interaction originates between the 'image' and what is imagined. If one constantly imagines a computer as a 'brain', in the end one will use the image of the computer to imagine the 'brain'. The metaphor and what is metaphorised begin to interact: what was the metaphor at first (the brain) becomes the metaphorised; what was metaphorised originally (the computer) becomes the metaphor. It is impossible to give a precise account of this logic. The core of the interaction is connected with the fact that metaphors not only *refer* to something, at the same time they present their "way of looking at things" (Sichtweise).[1] The meaning of the metaphor is always partially contained within it, it does not simply *inform* the semantic surplus value of its reference. This means that the image value of the metaphor cannot or can only partially be determined by its 'reference'. In choosing a certain way of speaking metaphors can even be *obtrusive*. They not only extend the meaning of their reference, they can also reduce it.

The image of the genetic code as a culturally enforced metaphor not only absorbs the scientifically innovative approach to the microbiological constituents of humankind: at some point in time it will also come to dominate human self-reflection and self-imagination. In this case the *logic of adaptation* results in a reductionist point of view: people are nothing more than their genetic code. But this reductionism cannot present itself *as* reductionism. Fictional promises and visions, which concentrate on reprogramming the anthropological and phenotypic profile, are brought into play as camouflaging devices. A genetic fantasy, which is utopian and totalitarian at the same time, positions itself alongside genetic research that does not attempt to raise unrealistic expectations, and which is therapeutically motivated.

Against the background of this cultural-philosophical scenario, one of the main tasks of ethics lies in *aesthetic critique*. This critique should describe the ideological mechanisms that originate a gene mythology which cannot easily be checked by means of scientific criteria, and which spreads across the utopian horizon of our culture. As to the problem of identity formation, the task of this critique is to examine the *aesthetisation* of the issue of identity, the design aesthetics that underlies the post modern ideology of continuous self-conceptualisation, in which an aesthetics of commodities takes the place of human subjectivity. Hans-Jürgen Heinrichs memorably described the condition stipulated by this techno-aesthetic gene utopia for the formation of identity: "This practice requires from human beings an extreme self-thingification in exchange for an astounding broadening of their possibilities to act" (in Sloterdijk 2001: 68, my translation).

This 'extreme self-thingification', the total and final subjection of the body to the mega-technology of microbiological self-conceptualisation, is one of the implications we shall have to accept if, in the future, we consider human beings as the 'stages of subject constitution' ("Schauplätze der Subjektkonstitution", Dietmar Mieth 2002). But the price we will have to pay for this point of view, for an interpretation that understands subjects as fictional entities in a continuous auto-poiesis and an equally continuous auto-(de)construction, is high. We will establish a model of identity in which every identity – as I have already stated – will be *conflictual, relativistic, provisional* and *unprotected*. It is this moral protection that is requested by those who refuse to be subject to the logic of permanent change and who fight for the acknowledgement of a profile that includes *passivity, imperfection* and *finitude*. It is this moral protection that cannot find a trustworthy articulation in terms like auto-poiesis, constructivity and optimisation.

Every transformation, each of the three models, is characterised by acceleration. The theomorphic model *imploded*, so to speak, in eternity. And there it rests forever. But the timeless ideal of a posterity in which perfection and inertia were two sides of the same coin, was already effective in the temporal world. The world was encapsulated in a rotating movement, which represented a substantial goal. Life itself was surrounded, as it were, by a boundary that was impossible to surpass. The acknowledgement of slowness, if not outright *fossilisation*, hallmarked our sense of life. Haste was redundant, since the goal had already revealed itself in the *kairos*-charged past. The anthropomorphic model gave rise to changes that cut deeply into existing structures. Humankind, 'set free from the bonds of creation', now had to perceive perfectibility as an unending challenge. Essential instability required techniques of 'stabilisation through perfection'. Morality was not, in the last instance, linked to humankind's ability to *design* itself with a view to a future ideal. The dawn of "The Age of Activism", perceived by Sloterdijk (2001: 71), demands that time be viewed as a road towards self-building. If possible, one should be careful not to waste any time on it, for life is short as it is. At the focal point of the third, technomorphic model stands what we could term a 'void' time. Early Modernism witnessed an attempt at *standardising* time, so to speak: time is neutral and one can project dates on its linearity at one's discretion. The only use for time is to go continuously beyond *what has been*. Since this business of excelling is the object of an expansive race, the new virtue is called 'acceleration'. The only criterion by which the quality of what is novel can be judged is the speed with which it has been attained. The fact that a commodity has a long shelf-life is no longer a quality that makes it stand out; on the contrary, it has now turned into an imperfection.

If we were honest we would admit that the direction of this evolution has become confused. The 'racing standstill' of which Paul Virilio once talked more or less played down this fact. What we are dealing with now is more a 'chaotic chase', an 'orgy of speed' that leaves a trail of burnt out, exhausted and abandoned protagonists. But where can we turn for salvation? What can be done, or rather: what had we better not do? The counterfactual motto should run: "Self-reflection is the gateway to a culture of receptiveness and deceleration". Indeed, every self-reflection is at the outset but an exercise in scepticism, a questioning of the reputedly obvious, a short-circuiting of certainty, a recoiling before the all too valid. What would be the medium in which this reflection could take place? It would be that of aesthetic experience.

References

Bachmeier, Helmuth. "Schöpfung jeden Tag. Über die Notwendigkeit des Neuen." *Du* 2001, 718, 42–43.

Blumenberg, Hans. "Paradigmen zu einer Metaphorologie." In *Archiv für Begriffsgeschichte*. Vol. VI, Bonn, 1960, 9–142.

Claessens, Dieter. *Das Konkrete und das Abstrakte. Soziologische Skizzen zur Anthropologie*. Frankfurt a. Main, 1980.

Feuerbach, Ludwig. *Das Wesen des Christentums* (1841). Werkausgabe by E. Thies (ed.). Vol. 5, Frankfurt a. Main, 1976.

Gombrich, Ernst H. *Art and Illusion. A study in the psychology of pictorial representation*. London, 1960.

Goodman, Nelson. *Weisen der Welterzeugung*. Frankfurt a. Main, 1990.

Merleau-Ponty, Maurice. *Das Sichtbare und das Unsichtbare*. München, 1994, 2nd ed.

Mieth, Dietmar. *Was wollen wir können? Ethik im Zeitalter der Biotechnik*. Freiburg i. Br., 2002.

Plessner, Helmuth. "Anthropologie der Sinne." In *Gesammelte Schriften III. Anthropologie der Sinne*, Frankfurt a. Main, 1980.

Ricoeur, Paul. *La métaphore vive*. Paris, 1975.

Rorty, Richard. *Der Spiegel der Natur. Eine Kritik der Philosophie*. Franfurt a. Main, 1984.

Schöffel, Gerhard. *Denken in Metaphern. Zur Logik sprachlicherBilder*. Opladen, 1987.

Scholem, Gershom. *Die jüdische Mystik in ihren Hauptströmungen*. Frankfurt a. Main, 1976.

Sieferle, Rolf Peter. *Bevölkerungspolitik und Naturhaushalt*. Frankfurt a. Main, 1990.

Sieferle, Rolf Peter. *Die Krise der menschlichen Natur. Zur Geschichte eines Konzepts*. Frankfurt a. Main, 1989.

Sloterdijk, Peter. "Gottes Werk übertreffen. Horizonte der homöotechnischen Wende. Ein Gesprach von Peter Sloterdijk mit Hans-Jürgen Heinrichs." (preprint of "Die Sonne und der Tod", Frankfurt a. Main 2001). *Du* 2001, 718, 68–73.

Virilio, Paul. *Die Eroberung des Körpers. Vom Übermenschen zum überreizten Menschen*. München/Wien, 1994.

Wils, Jean-Pierre. *Ästethische Güte. Philosophisch-theologische Studien zu Mythos und Leiblichkeit im Verhältnis vor Ethik und Ästhetik*. München, 1990, 51–81.

[1.] The best book on the logic of metaphors is undoubtedly the extensive dissertation of Bernhard Debatin, *Die Rationalität der Metapher. Eine sprachphilosophische und kommunikationstheoretische Untersuchung*. Berlin/New York 1995.

Chapter 17

DISABLED EMBODIMENT AND AN ETHIC OF CARE

JACKIE LEACH SCULLY
Basel, Switzerland and UK

1. EMBODIMENT AND CONTEXT IN MORAL EVALUATION

Differently situated people understand things differently. One part of 'situatedness' is contributed by embodiment. Life as a particular embodiment means not only having experiences that are not shared by people with a different body, but also understanding these experiences in a way that is shaped by this bodily reality. The point of standpoint epistemology (Harding 1991, Hartsock 1998) for example, is that by virtue of their common situation, certain groups of individuals – women, blacks, the poor – will share a privileged perspective on the experience of oppression, one that differs from the perspective of a differently situated group – men, whites, the rich. A situated perspective will give rise to characteristic perceptions and interpretations, including perceptions and interpretations of moral issues. Ultimately this will modify a person's moral evaluations and judgements.

In this paper I suggest that a disabled embodiment means having experiences that are not shared by non-disabled people. Disabled embodiment is likely to affect to some degree the structures of imagination and interpretation that we use to make sense of what is going on around us, including our sense of self and of ourselves as moral agents. This might lead to different notions about the status of concepts such as autonomy and independence that are central to much of modernist ethical theory. These ideals, or at least their status, have rightly been criticised by feminist

247

C. Rehmann-Sutter et al. (eds.), Bioethics in Cultural Contexts, 247–261.
© 2006 *Springer. Printed in the Netherlands.*

ethicists for more accurately reflecting the lives of men in western societies (see e.g. Walker 1998); but they may equally fail to reflect the lives of those whose physical vulnerabilities make autonomy and independence, *as commonly understood*, meaningless.

2.   DISABILITY

The idea that 'the disabled' might form a distinct group of people, and that there might be a unifying concept of 'disability', is historically recent. Some accounts suggest that it developed primarily for administrative reasons, in parallel with the establishment of residential institutions for people whose impairments were considered to prevent their families being productive in the industrial workforce (Barnes, Mercer and Shakespeare 1999: 18–20; Oliver 1990). It was also closely connected with the growing standardisation of the body through advances in medical science, and especially in medicine's ability to define standards in quantitative terms (Matthews 1995; Scully 2002). As a result, deviations from the standard bodily format are easier to identify; and eventually the accumulation of identified deviations becomes so large that they must be 'managed' by being lumped together as a single conceptual category.

Until recently the dominant framework for understanding disability, and therefore getting clues as to what to do about it, was provided by a *medical model* by which disability is thought of as a disease, degeneration, defect or deficit located in an individual. In this model, exactly what constitutes disease, degeneration, defect or deficit is decided by reference to a biomedical norm. Dissatisfaction with the limitations of a *purely* medical perspective for comprehending the collective experience of disability generated several alternatives broadly termed the *social model*. The social model's most fundamental criticism of the medical model is that it wrongly locates 'the problem' of disability in biological limits, considering disability only from the point of view of the individual and neglecting the social and systemic frameworks that contribute to concepts of what disability is. A social model of disability sees it not as solely biologically determined, but takes the societal, economic and environmental factors as at least as important as biological ones in the construction of disability. Meanwhile, other theorists have found the social models inadequate as well, as they leave untouched the aspects of social relations (see e.g. Thomas and Corker 2002), and have begun an exploration of more discursive and phenomenological approaches.

3.   GENETIC TECHNOLOGIES

These models are significant because they represent attempts, by disabled people and others, to understand the social and biological phenomenon of disability. Understanding disability has become a matter of urgency with the development of genetic medicine over the last decades of the twentieth century. The coming of gene technologies has led to a surge of academic and public interest in genetic explanations for biomedical phenomena; both the technical abilities and the interest have major implications for disability. It is easy to forget that only a small proportion of impairments have *predominantly* genetic aetiologies: the majority of disabilities worldwide are the consequence of trauma, postnatal disease, or ageing, in which genetic factors may play little or no role. Nevertheless, the ever-increasing amount of genetic information available encourages the search for genetic aetiologies, while media attention to the almost daily genetic 'breakthroughs' creates the impression that genetics provides explanations (and cures) for all forms of disability and disease.

At the moment, genetic medicine is largely restricted to providing early *identification* of the presence of a particular gene. 'Early' here generally means during prenatal life: in the testing of foetuses, for example where there is a family history of cystic fibrosis, or in preimplantation genetic diagnosis (PID) which is used to characterise genes in embryos created by in vitro fertilisation, before a choice is made of which embryo(s) to implant into the host mother. Direct genetic *manipulation* to cure disease or disability (gene therapy) has so far had only limited success and is still considered experimental (Müller and Rehmann-Sutter 2003; Walters and Palmer 1997; Wivel and Walters 1993). Nevertheless, if gene therapy were ever to be safe and efficient enough, it would provide a route to the eradication of at least some disabilities (those with predominantly genetic aetiologies).

Although many disabled people welcome the contemporary developments in genetic medicine, others (and other non-disabled people) are concerned that they encourage a collective fantasy about the potentially unlimited nature of human existence. For these critics, the desire to prevent or ameliorate suffering can and should be distinguished from the desire to control the body and to perfect it, or to eliminate all forms of deviation from local ideals. The Little People of America, for example, have produced a position statement on genetic discoveries in dwarfism that expresses the ambivalence of many within the disability community: "Some members were excited about the developments that led to the understanding of the cause of their conditions, along with the possibility of not having to endure a pregnancy resulting in the infant's death [achondroplastic couples may produce a foetus that is homozygous for achondroplasia, which is always fatal]. Others reacted with fear that the knowledge from genetic tests such as these will be used to terminate affected pregnancies and therefore take the opportunity for life away from children such as ourselves and our children."[1]

Ironically, just at a time when genetic intervention is becoming feasible the cultural climate is beginning to shift as well. Although the change is slow there is growing recognition of the rights of disabled people, that not all disabilities are the same, that disablement is environmentally contingent, and that not all variations from the biomedical norm are tragedies to be eradicated or overcome. The combination of genetic medicine with a changing cultural understanding confronts us with what we really think about embodiments that differ from the norm. When little direct intervention was possible, and when the 'meaning' of disability was theologically or socially defined, the need to be clear about this was less acute. Being able to diagnose genetically before birth, and in the future perhaps to intervene genetically after birth, makes it imperative to evaluate whether such interventions are morally right or not; and this will not be possible while we remain so unclear about the meaning in contemporary culture of bodily variation or disability.

4.   THE ETHIC OF CARE

Traditional justice-based ethics, that "tend to idealize equality [and] transcendence of difference" (Ruddick 1995: 210), have difficulty incorporating those who are flagrantly different from what is taken as the moral norm. Within contemporary ethics, non-mainstream approaches have mounted strong criticisms against the tradition that relies heavily on a Kantian ideal of the moral subject, developing instead the notion that "we can only survive and develop within networks of interdependence with others, and these networks of dependence constitute the 'moral bonds' that continue to bind even as moral adults." (Benhabib 1992: 50). These critiques have come not only from care ethics, or even feminist ethics. Care ethics, however, give a central place to relationships in ethical reasoning, highlighting the relationship of care as the critical one, and have shown themselves especially able to focus more explicitly on *asymmetric* relationships between beings who may be markedly un-equal in terms of abilities.

Having said that, it must not be imagined that care ethics somehow 'solves the problem' of identifying what justice for disabled people would entail. In the context of disability, the care ethic approach has been applied primarily to consider acts of *caring for* or *being cared for*. This is problematic, in that it reinforces the assumption that all disabled people are in need of a greater degree of care, and a fundamentally different kind of care, than non-disabled people. It must not be forgotten that, for disabled people, relationships of care may also be oppressive, and may – especially in societies that denigrate dependency – exploit the carer (Kittay 1999, compare also her chapter "The concept of care ethics in biomedizine" in this volume). Moreover an ethic that prioritises caring relationships can tend towards an oppressive fetishization of particular *kinds* or *qualities* of relationship that may be equally harmful to disabled people's chances of participating in society.

As an ethical theory or approach, care ethics present a number of difficulties that have been well characterized in the literature (Crigger 1997). One is to do with what 'care' actually means. In the discourse of care ethics the concept of care itself is often poorly defined and/or used in a variety of different ways. Some writers use it to refer only to the practical activity of caregiving: caring is more than having a generalised or even a specific benevolent disposition, and many feminist ethicists have insisted that 'real care' must always entail actual hands-on work in real relationships (Noddings 1984). Others refer to the virtue or cluster of virtues that constitute both caring for and caring about. (When this plurality of meanings is used to demonstrate care ethics' inadequacy as a coherent theory compared to justice ethics, it is worth remembering that the term 'justice' is often used in equally ambiguous ways, see Veatch 1998.)

However, I want to suggest here that care ethics nevertheless has a particular advantage for a consideration of moral issues in disability. This is because of care ethics' implicit demand that due attention be paid to the *particularities* of the cared-for and the carer. The question, what kind of care does a person need? cannot be answered without knowing quite a lot about the needs and expectations of that person, both of which depend on biology, biography, and social context; meanwhile the question, how can that care best be provided? cannot be answered without a knowledge of the care-giver's abilities and limitations.

Attention to the particular and concrete, the details of context and embodied subjectivity, *as they are perceived by those most concerned*, is therefore essential to improve the chances that negotiated moral agreements in real life are just. Traditional views of justice, by being avowedly non-relational, non-particular and non-contextual, will often miss the point of why people think what they do is morally right. What is *just* needs to be perceived not by me putting myself into your shoes, but by me trying to understand how you, in your shoes, might perceive things – and accepting that I am and always will be limited in my attempts. If we accept that moral decisions are made by and between persons, it is necessary to begin to understand the features, and positions, of those persons, while abandoning the idea that their features and positions can ever be entirely interchangeable with our own. Here an ethic of care, which makes us pay attention to the specifics of relationality and context, is invaluable in clarifying why particular decisions are being made, and may provide guidance for where and how to apply the traditional conceptual tools of justice, such as universalizable rights.

5.   DEAF 'DESIGNER' BABIES

As an example, I turn now to consider situations in which parents with heritable conditions have wished to use genetic testing to ensure that they have a child with (rather than without) that condition. Several of these cases have involved deaf parents. The deaf example is a particularly complex one. As is now widely recognised, many culturally Deaf people distinguish deafness from other impairments as not being a disability at all (Lane 1999). Within both the disabled and Deaf communities there is considerable variation in opinion about whether Deaf people 'belong' within the disability movement. Thus Ladd and John write, "Labelling us as 'disabled' demonstrates a failure to understand that we are not disabled in any way within our own community ... Many disabled people see Deaf people as belonging, with them, outside the mainstream culture. We, on the other hand, see disabled people as 'hearing' people in that they use a different language to us ..." (Ladd and John 1991: 14–15)

Although the available evidence suggests that the majority of culturally Deaf people express no preference for either deaf or hearing children (Stern et al. 2002; Middleton, Hewison and Mueller 2001), some clearly do. In early 2002 a case involving two Deaf parents was widely reported in North America and much of Europe. A lesbian couple, both with congenital hearing impairment, who wanted a child, decided to increase their chances of having a deaf child by using a sperm donor with a heritable form of deafness. It is important to note that they did not reject the idea of having a hearing child, only that they felt a deaf one would be "a special gift". The couple therefore used a male friend with a genetic deafness as sperm donor, resulting so far in the birth of two children, both hearing impaired (Mundy 2002).

The parents' actions prompted a wide range of responses. The commonest was disapproval. Even reports that were predominantly supportive of their decision carried a sense of incomprehension or disbelief: "It may seem a shocking undertaking ..."(Mundy 2002). At the other extreme some advocates for the Deaf community (both Deaf and hearing) have strongly defended the couple's right, not only to have a child but to choose the *kind* of child they wanted to have.

Shortly after this, an Australian newspaper reported that a deaf Melbourne couple planned to use preimplantation genetic diagnosis to ensure (not, like the American couple, increase their chances) that they would have a child with normal hearing. In this instance, however, there was little or no debate of the ethical grounds for the parents' action – there appeared to be no felt need to justify a decision so much in tune with mainstream opinion. Interestingly, the newspaper report *did* mention that because the legal use of PID in Australia is restricted to preventing the transmission of genetic disease, the local regulatory body, the Infertility Treatment Authority, was asked to decide on the legitimacy of the request because "we have to ask if deafness is a disease … Some people would say deafness is a disease. Others would say it was an unfortunate condition." (Riley 2002). But no mention was made of those who would say that deafness is neither of these, but another way of being. It seemed that this was unimaginable.

Both critics and supporters have drawn their arguments almost exclusively from the language of parental (sometimes societal) rights and obligations. Thus those who were critical of the womens' choice here spoke about the strength of the parental obligation not to harm their child – in this case, the harm of condemning the child to a disability that could have been avoided. In Dena Davies' well known article (Davies 1997) on the situation in which deaf parents ask for prenatal genetic testing to select for a deaf child, she locates the moral problem in the damage to the child's right to what she describes as "an open future"; being hearing impaired will necessarily narrow the range of choices that would eventually be open to the child as he or she grows up. Davies avoids being drawn into a discussion of whether being deaf is itself a harm. She instead concentrates on the contempt being shown for the child's autonomy, either through being forced irreversibly into the parents' notion of what constitutes a good life, or through being treated solely as a means to the end of perpetuating Deaf culture. Davies argues that, for a liberal state, having a diversity of communities (including the Deaf community) is something that *increases* autonomy because it offers a wider variety of ways in which people can choose to live. However, this benefit can only exist while individuals are free to choose which community they wish to join or to leave; and if the existence of the group conflicts with the right of the individual to make such life choices, the liberal state must, Davies believes, support the ethical priority of the individual. Note that these arguments apply to other situations than prenatal genetic testing. They have also been used in the debate over giving cochlear implants to prelingually deaf children. In theory, cochlear implants have the potential to offer (a form of) hearing to profoundly deaf children, an intervention that it is recognised "can *determine* community membership" (Crouch 1997) (and by implication the futures open to the child).

Davies' article has been criticised on a number of points. The basic concept of an "open future" is problematic. Given that parents make decisions about the form and content of a child's life from the moment it is born (and often before), including the education it receives and the company it keeps, any child's future must be seen as substantially constrained by the decisions of its parents. In general, we don't find this problematic, accepting that no child can survive, let alone flourish, in the absence of a framework that guides the child's development and that thereby imposes constraints on choice. Questions must also be asked about the presentation of an oversimplified picture of a community as something that an individual can simply decide to leave or join (see the discussion in Corker 1998: 21–25); and about how far the flourishing of the individual and of the community can actually be separated from each other.

My intention here is not to evaluate these arguments. I want only to point out that these analyses, whether essentially for or against the choice made, focus on rights. The lack of attention paid to the biological and social context within which these decisions are being made, is striking. This approach is entirely in line with traditional models of justice, according to which moral agents are pretty much interchangeable: knowing what to do in any given situation should not require a knowledge of doer, done-to, or the circumstances in which it all takes place. Outside the realm of theory, however, real moral agents don't (can't) behave like this. Real moral agents are embodied, and embodiment – male or female, hearing or deaf – is a biological and material particularity that informs social and environmental relationships, forming a moral terrain within which certain moral judgements become justifiable. And these terrains differ in accordance with the biological, material and social particularity. Without attention to the embodiment of difference, justifications made from within an alien terrain appear incomprehensible or just plain wrong.

If we can "conceive of the Deaf as being members of a linguistic and cultural minority, our moral landscape should be altered" (Crouch 1997: 17), even if it does not (cannot) became the *same* as that of Deaf people. In order to make this imaginative shift however we need to understand the concrete particularities of the Deaf world. For example, most commentators on this case, working from the standpoint of a hearing or, in some cases, deafened person, accepted that the choice of deafness over hearing entailed a reduction in the sum of abilities, such as the ability to communicate, and hence also a reduction in future options, such as job choice or social role. Conversely, for those Deaf people who think of themselves as a cultural or linguistic minority[2], it makes no sense to claim that 'choosing' deafness violates the child's future autonomy. Choosing deafness, they might say, is rather like choosing to practice one's Judaism, or to send your child to a Rudolf Steiner school: a cultural choice that closes down some options, for sure, but opens up others that are equally valuable. Aware that by objective criteria many deaf people perform poorly in terms of education or employment, they may ascribe this to the negative effects of discrimination (inappropriate schooling, lack of sign language interpreters in higher education and the workplace) rather than the impairment itself. Some Deaf people might then choose to avoid deafness in their children, to protect them from these disadvantages, rather as parents might elect not to practise a minority religion so that their children can fit better into the mainstream culture. But others would believe that difficulties caused by societal prejudice do not constitute good grounds for choosing a hearing over a hearing impaired child.

A moral terrain is shaped by the unique features of the culture. Since most forms of deafness are not genetic, the majority of deaf children are born to hearing parents, while in most cases two deaf adults will bear hearing children. Deaf culture is therefore unlike most of the linguistic and ethnic cultures to which it is compared (by both Deaf and hearing commentators) in that it is predominantly maintained and transmitted through peer contacts such as schools for the deaf or deaf clubs. Only in the small minority of cases of genetic deafness is Deaf culture transmitted vertically, that is, through the family. As Paul Preston has described, within the Deaf world the so-called deaf of deaf are peculiarly significant: they are "the crucial link between their Deaf home environments and deaf children from hearing families" (Preston 2001: 71). Deaf children from deaf families therefore often have conferred on them a kind of leadership status (in an article on Deaf activism in the US, a woman is quoted as saying, "I'm Deaf of Deaf. I've always said that I'd get to the top and open as many doors as I could for the whole Deaf community," and another says with pride, "My father and my grandfather went to Lexington [a school for the Deaf]. I am Deaf of Deaf of Deaf" (Solomon 1994).

Paying attention to context immediately foregrounds relationships like these: the social context is formed by a network of relationships and it is, after all, through ties with others that moral difficulties arise at all. By prioritising the relational context (Rehmann-Sutter 1999) it becomes possible to see how the parental choices in these situations are rooted in care: both the physical and emotional labour of caring for the child, and caring in the sense of fostering community with others (Baier 1995: 49). For people whose context is the Deaf world, then, a preference for a deaf child can be seen as expressing care for the continued existence of their community, ensuring that the networks of relationship and concern that constitute the community endure to provide support to its members. On the level of the individual, the preference for a deaf over a hearing child can also be seen as an attempt to ensure the best for him or her: that the child will grow up as a key member of a flourishing community rather than being on its periphery.

Again in the particular context of Deaf culture, wanting a child 'like themselves' can be something more than an egotistical desire to perpetuate a parental identity. It may reflect a concern for the parents' ability to care for the child. Paul Preston quotes a deaf mother describing the birth of her hearing daughter: "When Barbara was born, it wasn't until about three days later that I had this funny feeling about her ... [When I discovered she was hearing] I couldn't believe it! I was really upset. I thought, Oh my God ... what on earth am I going to do with her? I don't even know how to talk to her ... *I wanted to be close to my children ... I worried that we would never connect, or that we would drift apart*" [my italics]. Parents might believe, rightly or wrongly, that they can better care for a child who is more like themselves: better able to anticipate their needs, create strong emotional bonds, and to provide appropriate guidance as they grow, than for a hearing child whose experiences and position in the world is too different from their own. Furthermore, for some parents, wanting a child like themselves will also be a necessary affirmation of their own selfhood: preferring a hearing child over a deaf child would be tantamount to denying the validity of their own lives. And it is conceivable that such a denial could be sufficiently damaging to the parents' sense of self, and self-agency, as to compromise their ability to ensure the wellbeing of any of their children, hearing or not.

Interestingly, Robert Crouch notes analogous reasoning used by hearing parents deciding that their hearing impaired child would benefit from a prelingual cochlear implant: "struck by the otherness of the life that they imagine their child will lead ... parents will usually choose [an implant] ... to prevent a chasm from opening up between them and their child (so that their child is in the same community as they are)" (Crouch 1997: 15). What is interesting is that the desire for parents and child to share a culture is seen as justified in this case, where it is condemned as selfish or egotistical when used by Deaf parents. Since the pros and cons of being hearing versus being deaf are not explicitly raised here, it seems that the deciding factor is that the observer shares the same moral terrain in one case and not in another.

Attention to context means being able to acknowledge features that might be, from one perspective, counterintuitive. Commentators who were critical of the decision to prefer a deaf child (or to have one at all) focused on the difficulties faced by the deaf child in an alien hearing world: as Davies put it, the constraints placed on its right to an open future. They worked from assumptions about the benefits of being able to hear, and the difficulties of being deaf. More intriguingly, they invariably ignored the possibility that a hearing child might also face difficulties growing up in a deaf familial and social world. Both sociological and anecdotal evidence suggests that the life of a hearing child within a deaf family is not always straightforward (see Preston 2001; Singleton and Tittle 2000; Bull 1998). There may be an awareness of difference and isolation; difficulties in generating a distinct individual identity when that is overwhelmed by the sense of categorical difference between hearing and Deaf members of the family[3]; problems of mixed loyalties; deferred stigmatization; and the harm that may result from being used in an age-inappropriate way as an interpreter and 'cultural mediator'. I want to emphasise that these problems are not inevitable, and may in any case not be experienced as *problems* by individuals or families. My point here is rather that these potential difficulties for the hearing child were never articulated, where the potential difficulties of the deaf child in a hearing world were repeatedly rehearsed. A certain form of embodiment (hearing) is assumed not only to be normative, but to be universally unproblematic, irrespective of the context within which it exists. Normative embodiment essentially fades from sight as something that might be a relevant factor in ethical judgement: it becomes no embodiment at all, and hence morally neutral.

This point is particularly important if it is claimed that some person can represent others in negotiations about justice, as for example in John Rawls' model (Rawls 1971) in which a representative behind a so-called veil of ignorance is required to be fair to the interests of those whose positions she has not personally experienced. The idealised Rawlsian representative must understand that certain facts are more or less epistemologically visible to differently situated agents, and – to retain her impartiality – should not know where she herself is placed within this universe of differences. When we consider a real representative in a real moral situation, however, we are faced not only with her limits to her ability to ignore her own position, but with the difficulty of using her imagination to enter into situations so far from her own as to be literally unimaginable. The assumption that being hearing is better than being deaf *in all contexts* reveals precisely that inability to make an imaginative leap, or even to realise that one needs to be made, that I described earlier.

6.   CARING ABOUT WHAT CARE MEANS

I want to emphasise that the above discussion is not intended as an argument for the legitimacy of the decision of the parents in this case. Some people will find one or more of the points made above convincing, others will not be convinced by any or all of them. My intention is rather to bring out some of the contextual and relational elements, unfamiliar to mainstream hearing society, that for those living within it make a relevant difference to the moral terrain and hence to the rationale for their decision. These points also illustrate how rights-based arguments, with their drive to universalization and to abstraction of concepts like harm, damage, wellbeing and so forth, can miss the point of why people believe that what they do is right.

The moral evaluation made by deaf parents preferring a deaf child becomes more understandable, and also more plausible, when viewed from the perspective of relationships of care. But in order to have anything to bring within that perspective, that is to see the relevance of care and to see what caring might actually concretely mean in this situation, requires understanding, as far as is possible, the moral frameworks of those concerned. And that demands attention to the realities of their social experience. Differences in perspective may be hard to articulate, precisely because one's own standpoint is the air that one breathes; for the deaf couple it was reported that "Because they don't view deafness as a disability, they don't see themselves as bringing a disabled child into the world. Why not bring a deaf child into the world? What, exactly, is the problem?" Thus there is a need for moral explorations that try, as far as it can be done, to map unfamiliar terrain and translate some of its features into terms that can be understood by others. As Gilligan writes, "The question of what responses constitute care and what responses lead to hurt draws attention to the fact that one's own terms may differ from those of others. Justice in this context becomes understood as respect for people in their own terms" (Gilligan 1995).

7.   CARING FOR JUSTICE FOR THE DISABLED

Care ethics have an uncertain position within the whole structure of ethical theory, and specifically in relation to theories of justice. Some writers take care and justice as complementary approaches; others suggest that they are polar opposites, generating two incompatible ways of tackling moral problems. What I suggest here is that the approaches to moral judgement that work through the lenses of justice and care respectively are actively interdependent. Neither need take precedence: each is essential for a fuller moral understanding, not in the sense of providing two alternatives, but with each informing and restraining the other. As Elizabeth Bartlett notes, real moral behaviour goes beyond either/or: "At the core of both the ethic of care and the ethic of justice lies a deep and passionate concern for human dignity, for respect of that in us which cannot be reduced to abstract principles ... Justice ... must be embodied in particular persons, their passions, their friendships, their concrete realities" (Bartlett 1992: 87). Similarly, Robin Dillon has distinguished between what she defines as care respect and Kantian respect (Dillon 1992). The former is akin to what I have been describing in this paper. It "involves trying to understand what it is like to be her living her life from her point of view". Dillon sees care, care respect and Kantian respect as forming a continuum that runs from the form of interaction most appropriate to close and personalised relationships, in line with the familiar meaning of care, to the universalised respect appropriate to the traditional model of justice; care respect lies somewhere in between, fostering the attention to context and embodied particularity that is characteristic of the ethic of care, but without the intensity of engagement and response that care, as it is commonly understood, demands.

When non-disabled people think about *doing the right thing* for disabled people, what tends to spring to mind are actions like making doorways accessible, or installing ramps for wheelchair users and visual and audible signals for those with sensory impairments. These actions are important steps towards participatory equality, but they are also in a very real sense superficial: thinking of them does not require much in the way of imaginative exploration. I suggest that to know what justice and care really mean for disabled people, we need to know in much more detail about the lived experience of disability and chronic illness. Then we might find ourselves more open to different kinds of needs: modifying timetables to accommodate differences in energy levels, for example, or exploring the different understanding of touch and proximity that deaf people have.

Even in theory, the ideals of justice cannot be applied to abstractions devoid of all human features and context. Robin Dillon has argued that caring for another is a way of respecting her, and that justice requires "respect [i.e. care] for persons and respect for the basic rights that every person possesses" (Dillon 1992: 69), and for the communities that these persons constitute. The connection between justice and care can, however, also run in the opposite direction. If "justice in its deepest sense can be understood as treating persons in truthful accordance with their own concrete reality,"(Farley 1998) then caring requires an openness to understanding what justice means for a particular individual or community: how they would choose to enact their rights, in their very specific ways. Both justice and care therefore entail knowing what other people care about – not necessarily *caring about the same things*, but letting the fact that these others care about these things in this way, matter to us.

References

Baier, A. "The need for more than justice." In *Justice and care: essential readings in feminist ethics.* V. Held (ed.). Boulder, Colorado: Westview Press, 1995.

Barnes, C.; Mercer, G. and Shakespeare T. *Exploring disability: a sociological introduction.* Cambridge: Polity Press, 1999.

Bartlett, E.A. "Beyond either/or: Justice and care in the ethics of Albert Camus." In *Explorations in feminist ethics: theory and practice.* E. Browning Cole and S. Coultrap-McQuin (eds.). Bloomington and Indianapolis: Indiana University Press, 1992.

Benhabib, S. *Situating the self: gender, community and postmodernism in contemporary ethics.* Routledge: New York, 1992.

Bull, T. *On the edge of deaf culture: hearing children/deaf parents.* [Annotated bibliography.] Alexandria, Virginia: Deaf Family Research press, 1998.

Corker, M. *Deaf and disabled, or deafness disabled?* Buckingham: Open University Press, 1998.

Crigger, N.J. "The trouble with caring: a review of 8 arguments against an ethic of care." *J Prof Nursing* 1997, 13, 217–221.

Crouch, R.A. "Letting the deaf be Deaf: reconsidering the use of cochlear implants in prelingually deaf children." *Hastings Center Report* 1997, 4, 14–21.

Davies, D.S. "Genetic dilemmas and the child's right to an open future." *Hastings Center Report* 1997, 2, 7–15.

Dillon, R.S. "Care and respect." In *Explorations in feminist ethics: theory and practice.* E. Browning Cole and S. Coultrap-McQuin (eds.). Bloomington and Indianapolis: Indiana University Press, 1992, 69–81.

Farley, M.A. "Feminist theology and bioethics." In *On moral medicine: theological perspectives in medical ethics.* S.E. Lammers and A. Verley (eds.). Grand Rapids/Oxford: WM Eerdmans Pub Co, 1998.

Gilligan, C. "Moral orientation and moral development." In *Justice and care: essential readings in feminist ethics.* V. Held (ed.). Boulder, Colorado: Westview Press, 1995 36–37.

Harding, S. *Whose science? Whose knowledge? Thinking from women's lives.* Ithaca, New York: Cornell University Press, 1991.

Hartsock, N. *Feminist standpoint revisited and other essays.* Boulder, Colorado: Westview Press, 1998.

Kittay, E. Feder. *Love's labor: essays on women, equality and dependency.* New York, London: Routledge, 1999.

Ladd, P. and John, M. *Deaf people as a minority group: the political process. Course D251, Issues in Deafness.* Milton Keynes: Open University Press, 1991.

Lane, H. *The mask of benevolence: disabling the Deaf community.* San Diego: DawnSignPress, 1999.

Matthews, J. Rosser. *Quantification and the quest for medical certainty.* Princeton: Princeton University Press, 1995.

Middleton, A.; Hewison, J. and Mueller, R. "Prenatal diagnosis for inherited deafness – what is the potential demand?" *J Genet Couns* 2001, 10, 121–131.

Mundy, L. "A world of their own." *The Washington Post Magazine*, Sunday 31 March 2002, W22.

Müller, Hj. and Rehmann-Sutter, C. *Ethik und Gentherapie.* Tübingen: Francke, 2003.

Noddings, N. *Caring: a feminist approach to ethics and moral education.* Berkeley: University of California Press, 1984.

Oliver, M. *The politics of disablement.* Basingstoke: Macmillan Press, 1990.

Preston, P. *Mother father deaf: living between sound and silence.* Cambridge: Harvard University Press, 2001.

Rawls, J. *A theory of justice.* Cambridge: Harvard University Press, 1971.

Rehmann-Sutter, C. "Contextual bioethics." *Perspektiven* der Philosophie 1999, 25, 315–338.

Riley, R. "Pair seeks IVF deaf gene test." *Herald Sun,* June 30, 2002.

Ruddick, S. "Injustice in families: Assault and domination." In *Justice and care: essential readings in feminist ethics.* V. Held (ed.). Boulder, Colorado: Westview Press, 1995.

Scully, J. Leach. "A postmodern disorder: moral encounters with molecular models of disability." In *Disability/Postmodernity: Embodying disability theory.* M. Corker and T. Shakespeare (eds.). London: Continuum, 2002, 48–61.

Singleton, J.L. and Tittle, M.D. "Deaf parents and their hearing children." *J Deaf Studies Deaf Educ* 2000, 5, 221–236.

Solomon, A. "Defiantly deaf." *New York Times,* August 28, 1994.

Stern, S.J. et al. "The attitudes of deaf and hearing individuals towards genetic testing of hearing loss." *J Med Genet* 2002, 39, 449–453.

Thomas, C. and Corker, M. "A journey around the social model." In *Disability/Postmodernity: Embodying disability theory.* M. Corker and T. Shakespeare (eds.). London: Continuum, 2002, 18–31.

Veatch, R.M. "The place of care in ethical theory." *J Med Philos* 1998, 23, 210–224.

Walker, M. Urban. *Moral understandings: a feminist study in ethics.* New York, London: Routledge, 1998.

Walters, L. and Palmer, J.G. *The ethics of human gene therapy.* New York, Oxford: Oxford University Press, 1997.

Wivel, N. and Walters, L. "Germline gene modifications and disease prevention: some medical and ethical perspectives." *Science* 1993, 262, 533–538.

---

[1] Position Statement on genetic discoveries in dwarfism. Little People of America, January 1996.

[2] Note that not all culturally Deaf people would agree with this position either.

[3] Preston, 1995: 89: "Paradoxically, within a community of shared identity, individual differences can emerge – identities that are not restricted to a single, all-encompassing feature."

Chapter 18

COPING WITH LIMITS
Two Strategies and their Anthropological and Ethical Implications

WALTER LESCH
Louvain-la-Neuve, Belgium

The aim of this paper is to reflect upon some fundamental issues in bioethics and how they may be related to the topic of limits. Of course, one might ask if there is any ethical item that cannot be related to the topic of limits. Ethics could even be defined as the art of setting and justifying limits in order to instil a sense of reasonable, acceptable regulations. Without limits everything and everyone would lack coherence and identity. On the other hand there seems to have been an important cultural change in attitude towards many forms of limitations which are no longer automatically accepted as the lines at which we have to stop, or at least must ask permission to go any further. They are seen more or less as borders that can be crossed in order to discover areas of completely new possibilities, broadening the range of human activities and conferring the power to transform the original structure of nature. As far as I can see the ethical evaluation of limits depends more on assumptions linked to general worldviews and less on the construction of an ethical argument in specific situations. It makes a difference whether the ethicist is fundamentally seen as the border guard between the areas of the permissible and the forbidden, or whether ethics first of all has the task of surveying a partly unknown territory where we are not sure of the precise demarcations. In the modern understanding of nature, normative standards must be justified and can no longer be deduced from the description of a natural framework implying pre-existing moral rules.

I would like to contribute to the clarification of some decisive *contexts* of bioethical debates which have long been organized in a much too isolated way, as if there were clearly defined tools for 'doing bioethics' within a narrow range of scientific and therapeutic challenges. If this were the case all other aspects of medical decisions would be less important. The project of demarcating bioethics from neighbouring disciplines may have been

C. Rehmann-Sutter et al. (eds.), Bioethics in Cultural Contexts, 263–273.
© 2006 *Springer. Printed in the Netherlands.*

reasonable during an early period of transition, in order to establish it as a genuine field of research by underlining differences much more than common ground. But nowadays the most challenging ethical questions are those that cannot be dealt with within the limits of a single discipline.

My question is a methodological one: using the experience of medical limits as a starting point I want to find out how the concrete bioethical norms and background theories interact. It is not evident that background assumptions simply determine the finding of norms. The opposite can be true as well: changes in the application of normative standards will probably have an impact on the plausibility of the general framework of philosophical theories. After some preliminary remarks (1) I want to introduce the idea of *coping strategies* as a well-known topic of research in psychology because I am convinced that it can be helpful in the context of biomedical ethics as well (2). The central part of the paper will outline two major types of coping with limits and corresponding ethical and anthropological theories (3). The strategies and the theories have been developed as clearly separated structures. But they can also be found in new combinations with elements of each other (4). In a short conclusion, attention will be drawn to the hermeneutical profile of a bioethical theory taking into account the necessity of coping with limits (5).

1.   PRELIMINARY REMARKS

Talking about limits in the context of modern philosophy inevitably brings to mind the existentialist jargon of the experience of limits (Jasper's "border situations") such as death, suffering and guilt. In extreme situations individuals have to make a choice that will not always be a rational one. But in doing so they can discover themselves as free and responsible agents. Dealing with these limits is a matter of personal concern, and sometimes even of taste. Some people prefer to construct their moral point of view by focussing on the difficult, extreme or even tragic aspects of life. Others like to focus on ordinary and less spectacular events as real life situations in which the extreme cannot be accepted as a standard for the ethically right decision. The use of the word 'tragic' is only appropriate for the condition of dependency on destiny. So we could put aside the existentialist vocabulary of dramatic encounters with limits and just stick to the more trivial case of limited possibilities. In a general sense this is true because of the economic condition of scarce resources and the difficulty of distributing limited means in a just fashion. That's why bioethics is nearly always research into appropriate criteria for distribution on the macro and the micro level. Whenever we have to define priorities, we exclude a wide range of interesting alternatives that unfortunately never have the opportunity of being tested and proving their validity.

On a more existential level the limits of life are defined by the fact of aging and mortality. Life has an end. That's why we learn to appreciate all we can do within the limits of the span from birth to death. Days count because they are counted. This formulation is not quite correct because we usually do not know the exact number of days we still have. Having this kind of knowledge (which may soon no longer be just another bit of science fiction) would substantially change our relationship to death. In spite of the process of aging and approaching death we are used to seeing life as an open future even if our possibilities diminish. The open nature of this life project makes us amenable to the idea of going beyond the limited possibilities and struggling against anything that looks like the overwhelming power of fate putting an end to our autonomy. Instead of accepting life passively as something that can never be modified, the new paradigm stresses the aspects of invention and transformation. In the long run this may lead to the notion of total control over all the body's performances.

I am aware of the danger of painting a black and white picture when blaming biotechnologies for deliberately creating the myth of the perfect mastery of a nature that loses its enigmatic aspects and increasingly acquires the artificial character of a machine. The cultural project of dominating biological processes indeed aims to turn a maximum of natural functions into technically controllable systems that can be controlled from outside. In doing so money and technology have become the most important instruments of power – in the field of medicine as much as elsewhere. In addition to the medical duties of diagnosis and therapy, biomedical technologies are an integral part of a big business that functions according to its own rules and no longer responds to ethical imperatives and warnings. In this respect bioethics has already reached its limits when it protests against the attempt to identify ethics with cost-benefit analysis.

2. LIMITS AND COPING STRATEGIES

In spite of the impressive achievements of science and technology many uncertainties remain and completely new issues are constantly generated. They mainly concern the beginning and end of human life because there is no consensus on the definition of morally relevant criteria for the domain of life. If bioethical reflection is to show a way through the jungle of difficult decisions, it has to deal with these limits to life and with the limits to medical intervention. But there is no guarantee of finding clear standards for a definition of life that must be protected, or for a justification of the physician's power to intervene in matters of life and death. Uncertainty provokes stress reactions and a wide range of personal attitudes towards limitations and the destruction of life plans. In psychological research there is a well-developed branch of investigation into life events that cause stress and the various ways of dealing with these difficulties. In a way, bioethics itself can sometimes be seen as the nervous activity of experts who are often acting near the total collapse of their own capacities. Such an image is of course far from the ideal of the wise philosopher detached from the urgent decisions of professional practice.

It makes a significant difference whether we accept illness as something that cannot be changed immediately or whether we combat it by all means possible; whether we ask for medical assistance when we encounter the slightest difficulty or whether the barrier against professional intervention is placed rather high. Coping strategies depend on personal notions of the conditions for a good life. These personal resources are primarily based on shared opinions that are part of a traditional worldview. Such a background can be a religious belief, but there are also secular theories about the range of medical influence.

The problem becomes quite obvious when we think about life-sustaining techniques in cases of incurable diseases. The ethical questions about the justification for putting an end to suffering can only partly be answered within the logic of coherent argumentation – which of course remains indispensable. Arguments are nevertheless embedded in a larger context of dispositions we have to take into account. One of these contextual elements is the method of coping with the challenge of a difficult situation. Coping can be defined as a mental procedure of interpreting and giving sense to life events that suddenly disturb the usual ways of thinking and behaving. It is not a totally spontaneous reaction, but a result of learning and arranging former experiences as a helpful network of well-considered convictions, arguments and emotional attitudes.

In a recent study the Dutch bioethicists Marli Huijer and Guy Widdershoven (2001) related models of the physician-patient relationship to models of philosophical theories of desire as they are presented in Martha C. Nussbaums influential book on Hellenistic ethics (Nussbaum 1994). Jackie, a fictitious person with metastatic lung carcinoma, consults five physicians who give five rather different interpretations and recommendations. Let us take a brief and oversimplified look at the five carefully reconstructed approaches that can serve as examples of structuring the field of ethics.

1. *The paternalistic/Epicurean approach:* According to this point of view there should be objective criteria for finding out what is best for the patient. It is first of all up to the physician to take decisions according to his or her

professional competence and according to his or her evaluation of the patient's will to avoid pain.

2. *The informative model:* This second perspective is guided by absolute respect for the patient's choice that must be made on the basis of an informed consent, taking into account all the relevant facts. This approach would favour a strategy of counselling adapted to the complexity of medical possibilities.

3. *The sceptical model:* The sceptical physician emphasises an interpretation of pain and suffering as part of the human condition. From such a point of view the patient should be freed from the illusion of a carefree life and accept the limits of medical intervention.

4. *The interpretative/Stoic model:* The main point of the Stoic model is the clarification of the patient's desires and values. This comprehensive approach takes into consideration the whole story of a patient's life with its changing interpretations and preferences.

5. *The deliberative/Aristotelian model:* In order to reach a consistent truth the multiple aspects of desires are critically discussed as part of a process of defining acceptable norms.

These are five approaches to decision making when curative treatment becomes difficult or even impossible. "What does human wellbeing or human flourishing at the end of life mean? Should we rely on 'shared objective criteria' of what it is to die well? Should we respect each patient's desires and preferences as truths not to be discussed? Should we steer a middle course between objective and subjective criteria for living and dying well, and find ways to discuss general ideas and connect them with personal beliefs and desires?" (Huijer and Widdershoven 2001: 157) Comparison of the patterns of ethical consideration does not lead to a clear-cut solution because we are confronted with different styles of providing good advice, reliable information, comfort or support.

The five models show us five different ways of coping with death – as a patient, as a physician or as any other person consulted. If the views of the persons do not fit together, there will be no trust in this important relationship between the patient and those around him or her, whether they are professionals, relatives or friends. And ethical advice might in any case not provide helpful perspectives. Quite the converse: it might even disturb the patient's equilibrium and destroy the resources that could have been used for coping with the difficult situation.

Coping strategies can be investigated with the instruments of qualitative empirical research. We have to ask a representative number of people how they react in comparable situations of stress. The careful analysis of the interviews will lead to a wide range of patterns of reasoning and behaviour: from outcries of pain and revolt to melancholic forms of withdrawal, from the stoic acceptance of bad news to the creative reinterpretation of a problem that might have a positive side as well. Coping strategies can be essential to preserve one's mental health even if they do not correspond exactly to the medical data. They are interpretations, sometimes illusions or more or less intelligent forms of self-deception. What counts is their mental and physical impact in helping to live or simply to survive.

If we now want to make a link between these coping patterns and patterns of ethical discourse we have to consider an important methodological difficulty. Coping strategies may have implicit moral considerations. But they cannot replace the work of the moral philosopher who wants to know how the moral argument is constructed and not how we manage to present plausible opinions and even useful lies. Nevertheless there is an advantage in taking into account coping strategies at least as parts of a personal moral conviction with the authority of authentic experience and as a first test of application. If a coping strategy does not work at all, the individual will invent new ways of arranging the pattern of reasoning in a sort of new equilibrium of all the aspects that have to be considered.

To cut a long story short I will go directly to making the link between psychologically reconstructed patterns and types of ethical theory. Of course this hypothesis should be supported by more evidence from social research and philosophy in further studies.

3.   TWO BASIC TYPES OF COPING STRATEGIES

While bearing in mind the obvious danger of oversimplification, I suggest that two major attitudes towards the idea of limited length and quality of life are commonly to be encountered within medical ethics (Durand 1999). Both use the reference to the concept of human dignity, but in very different ways. The first emphasises respect of limits and is consequently oriented towards the idea of morally limited medical intervention. The second model primarily understands medical interventions against a background of the moral responsibility of overcoming the limitations of life as far as possible by using technologies in the service of expanding bodily capacities. The description of these differences has often been linked to an apparently fundamental difference between the american and the european ways of conceiving bioethics, in which the american version is represented as technology-friendly, taboo-free and oriented towards the autonomy of the patient, while the European tradition embodies a more sceptical, risk-averse and paternalistic variant. In practice this opposition is rarely found in its purest form. The comparison can be helpful however to clarify some basic differences that may also be detected in the construction of individual coping strategies.

3.1   Protection of vulnerable people

The first position can be considered as the more traditional one, looking for clear standards of human life in the light of finitude and dependence (Marx 1986; MacIntyre 1999). It is based on the personal ideal of self-limitation and is articulated on the collective level in the imperative to treat the risks of new technologies with great caution. The individual and the societal dimension are expressions of the same demand for self-control and moderation. On the individual level such an attitude corresponds to a virtue ethic, in the sense of the cultivation of a modest and controlled lifestyle. On a social level it contains the common vision of ethically informed society taking care of its members' health problems and providing equal access to treatment for all people asking for professional help. Medical interventions must satisfy first of all the basic needs of all patients, not the demands of some privileged clients.

With this model I do not intend an ideological elevation of illness as the occasion of abasement or other anti-life constructions, but the ideal of the freely willed self-limitation out of respect for the natural lifecycle. This attitude, which does not escape a certain paternalism, is based on a number of background assumptions concerning the global vision of life, the individual's obligation to cultivate a modest and controlled way of life, and a theory of a (naturally) given framework of realities that should not be changed. One global theory behind such an approach could be the defensive understanding of human rights as the negative right to protection against state intervention.

3.2   Autonomy and technological assistance

In a sharp contrast to the first model the second type has fewer problems with the new possibilities offered, for example, by genetic diagnosis and the new reproductive technologies, and tends to claim positive rights concerning the access to new treatments (Hottois 1999). Why should it be forbidden to repair defects of nature, if the technological means are to hand? Why should the access of patients to the new treatments not be explicitly defined as a positive right and defended against the sceptics of progress, who with their demands for the protection of human dignity hard-heartedly deny healing to others? If infertility is no longer regarded as an irrevocable fate, the desire to have a child of one's own becomes a reasonable project of autonomy for which the entire palette of reproductive medicine is at the individual's disposal. Personal desires are acceptable in this moral evaluation and do not have to be justified before an ethics commission. Within the model of unlimited progress desires are potentially endless, and the biomedical imagination is morally obliged to discover new horizons without an anxiety driven restraint: the boundaries of the doable are shifted in line with the trust in technical possibilities. In any case, in this understanding a society's readiness to order itself in line with modesty and prudence declines.

As I see it, the two basic models described here, of self-limitation and self-realisation, are implicated in the continuing conflict between two antagonistic traditions of meaning in bioethics that centre around either the protective or the modificatory implications of the conferral of dignity. In another context I have called these the taboo and the identificatory functions of dignity. According to the first paradigm the carrier of dignity takes a role to be acted out in harmony with a superordinate authority. The moral status is coupled to this reference and implies an orientation towards objective values. The second paradigm is something like Pico della Mirandola's licence to attain an independence/autonomy that is expressed in creativity and that mistrusts every form of paternalism. From this stance there develops a selfconsciousness that takes a critical attitude to traditional norms and trusts in its creative power to develop completely new patterns of life and that discards the ballast of old calls to self-abasement. Behind the gesture to the special dignity of humanity there could also be hidden a completely different picture of humankind, which in the interests of ethically concrete debate in situations of conflict needs unpacking to clarify divergencies.

Frequently a bioethical debate heats up around an implicit evaluation of the medical measures taken to avoid or prevent disability. Both sides are mutually distrustful of each other. If the proponents of gene tests are seen as essentially hostile to disabled people, the statements of the disabled are also too quickly discounted as antiprogessive defamations. Here it could be helpful investigate more fully the world views that have come into conflict, in order to ensure that an active solidarity with disabled people need not be compromised by a personal decision against the further transmission of inherited disease. What is particularly interesting is the learning process involved in exposure to disabled people. Disability is apparently primarily a problem for the non-disabled who are unable to cope with otherness or integrate otherness as part of the normality of their world view. The phantasies and obsessions in the area of reproductive medicine to a great extent are centred around the fear of a disabled child. Here some detached and sober work of clarification is needed, before a less helpful exchange of distrust over hidden motives begins.

4. CROSSING THE BORDER BETWEEN TWO CULTURES OF BIOETHICS

In reality the two attitudes outlined here are not to be found very often. The more restrictive one will never be completely separated from legitimate interests in scientific research going beyond the limits and testing the promises of new therapies. And the second point of view would be without any reason if there were not also elements of care, security measures and voluntary renunciation.

The following comparison is often used in order to illustrate the fact that both approaches need each other. "I like the metaphor of the automobile, wisely equipped with brakes and accelerator. The purpose of the car is movement, but to drive safely it is important to be able to slow down and stop, sometimes suddenly. Inevitably, you get drivers who crawl, drivers who speed and drivers who prefer to keep their car up on the blocks in the garage where it will come to no harm. (…) Like all wise motorists, we'll keep moving, but we make sure our brakes are in working order" (Holloway 1999, 148f.).

As we agree to keep moving we also must admit that there will always be difficult roads and slippery slopes. It is a triviality to focus on this dimension of bioethics. Bioethics does not *create* the problem of changing values and uncertain rules. It tries to give answers to the situations we are find ourselves in. It would be much more interesting to discuss how one or the other expression of self-limitation or of transgression can be imposed on others who do not share this opinion. This can only be clarified in an open democratic debate. In other words: the confrontation between the types of coping strategies and of ethics will lead to no result if we do not start to negotiate the items that should be subjected to moral discourse and (later on) to legal regulation. This is one basic difference between the psychologist's task of counselling and of understanding coping strategies and the ethicist's role at the point of transition from private to public affairs.

5. BIOETHICS AND THE HERMENEUTICS OF FINITUDE

How do we know when to accelerate or to slow down? This will always be a matter of controversial interpretation and analysis of the roads we choose. Some roads will not be dangerous even if we drive faster. On others, the rules of prudence will soon impose a number of restrictions to be met. And there may even be good reasons not to use them at all and to prefer public transport or bicycles or to avoid crossing the area because of higher priority ecological interests or for other reasons. In other words: there is no automatic route leading from scientific discovery to practical application. The concrete design of medical therapies has to be explained, discussed and justified. Its uses are a question of priority setting and interpreting the interests of all persons involved in the project.

The German philosopher Odo Marquard has defined hermeneutics as an answer to limited possibilities (Marquard 1981: 117–46). I think that this approach can also be applied in the context of medicine. Bioethics is the place where different intellectual civilisations meet. The interaction between contradictory personal visions of a good life and the claims of justice (limited resources and common good) has to be accompanied by hermeneutical research towards understanding the various attitudes towards finitude. Some comprehensive ethical theories express the hope that the conflicting points of view can be unified in a sort of reflective equilibrium. The emotional dimension of many public discussions about bioethics seems to underline that it will be hard to get beyond the clash of convictions and to reach acceptable compromises (or to do so only at a very basic level).

In any case, we cannot escape from the burden of our limited horizon and our limited life. Our days count because they are counted. Imaginary constructions of immortality will not change this limitation. But they point to the understandable desire to tell a life story from the point of view of its ending, in order to see if the narrative makes sense or not (Bauman 2001). As far as my own life is concerned, I will never be able to say so definitively. What counts more is my capacity or incapacity to see others suffering and dying, coping with their ultimate limits in admirable, courageous, anxious or hypocritical ways. Where ethics could contribute to helping them to keep their horizons open – even close to the most painful limits – it would be an expression of autonomy and dignity.

But our response to limits is not only a matter of existential beliefs. It must also be related to our technical surroundings. Do we have the technologies we need? Do we need the technologies we have? Attempts to answer the second question are often considered as idle talk, because the hard facts of science and research and the economic circle of supply and demand have already decided the role they will play. Legislators who want to set limits must begin by interpreting the pressure of needs and desires that have led to the dilemmas we have to face. An ethics of desire could be a useful supplement to moral philosophy dealing with the validity of normative positions. Otherwise ethical expertise will not be able to communicate with the different backgrounds of coping strategies that already being applied long before the discussion of rational clarification begins.

References

Bauman, Zygmunt. *The Individualized Society.* Cambridge: Polity, 2001.

Durand, Guy. *Introduction générale à la bioéthique. Histoire, concepts et outils.* Montréal/Paris: Fides/Cerf, 1999.

Holloway, Richard. *Godless Morality. Keeping Religion out of Ethics.* Edinburgh: Canongate, 1999.

Hottois, Gilbert. *Essais de philosophie bioéthique et biopolitique.* Paris: Vrin, 1999.

Huijer, Marli and Widdershoven, Guy. "Desires in Palliative Medicine. Five Models of the Physician-Patient Interaction on Palliative Treatments Related to Hellenistic Therapies of Desire." *Ethical Theory and Moral Practice* 2001, 4, 2, 143–159.

MacIntyre, Alasdair. *Dependent Rational Animals. Why Human Beings Need the Virtues.* London: Duckworth, 1999.

Marquard, Odo. *Abschied vom Prinzipiellen. Philosophische Studien.* Stuttgart: Reclam, 1981.

Marx, Werner. *Gibt es auf Erden ein Maß?* Frankfurt: Fischer, 1986.

Nussbaum, Martha C. *The Therapy of Desire. Theory and Practice in Hellenistic Ethics.* Princeton: Princeton University Press, 1994.

V. INNOVATIVE MODES OF ANALYSIS

Chapter 19

WHAT CAN THE SOCIAL SCIENCES CONTRIBUTE TO THE STUDY OF ETHICS?
Theoretical, Empirical and Substantive Considerations[1]

ERICA HAIMES
Newcastle, UK

1.   INTRODUCTION

Since the late twentieth century the Euro-American mass media have given a great deal of coverage to debates over topics such as abortion, euthanasia, fertility treatment, surrogacy, organ donation, genetic screening and access to medical treatment. Topics outside the medical field such as genetically modified crops, investment policies, child labour and environmental issues have also been thoroughly aired. Since the debates have been primarily concerned with the ethics of such practices it could be argued that their prominence represents an increase in awareness of ethical issues. However, the voice of sociology and the other social sciences is rarely heard in these debates. Is this because (i) the social sciences have little to say on these issues, or is it because (ii) though it has much to say, the voice of the social sciences has had little impact, and is this, in turn, because (iii) the social sciences are not usually associated with the study of ethics and ethical issues?

On the first point, some writers claim that the social sciences have not had a major interest in ethics, or in medical ethics and bioethics in particular (for example Osborne 1994a and Zussman 2000). However, although there may not (yet) be a 'sociology of ethics', for example, there is a substantial body of theoretical and empirical work in the field of ethics from sociology and the social sciences in general. It is part of the purpose of this paper to identify that body of work and to suggest ways in which it might contribute to the study of ethics. It has to be acknowledged that, so far, this work has

C. Rehmann-Sutter, M. Düwell and D. Mieth (eds.), Bioethics in Cultural Contexts, 277-298.
© 2002 Blackwell Publishing. Printed by Springer, the Netherlands.
Previously published in Bioethics 16:2, pp. 89-113. Reprinted with permission.

had only limited impact beyond its own disciplinary walls. This is surprising given the increasing use of the acronym, ELSA (a reminder to scientists and policymakers to consider the ethical, legal and social aspects of developments in medicine and science) that would appear openly to invite social scientists to join law and philosophy in contributing to these debates. Several countries, including the United States and Sweden, have nationally funded programmes to encourage research in the ELSA fields. However it appears that the voices of ethicists and lawyers have dominated those of the social scientists, possibly because of their more precise focus: what after all is included in the rather vague term, 'social aspects'? Social scientists are not frequently cited by these other disciplines and they are not often called upon to participate in public debates or to contribute to the policymaking arenas. Thus another purpose of this article is to try to make the voice of the social sciences heard more clearly, though not by drowning out those other voices. The aim is not sociological imperialism but rather a dialogue with other disciplines (Silverman 1985: 194f.). Thus, even though I am writing primarily as a sociologist I shall be using 'sociology' and 'social sciences' as broadly interchangeable terms. The boundaries between them are blurred in much of the relevant work so it is more useful to emphasize their shared interests in the conceptual, cultural, political and practical aspects of the social world, and the contributions that these can make to the study of ethics, than to get caught up in their differences.

On the third point, it probably is true to say that the social sciences are not associated in the lay, scientific or political mind with ethics and ethical debates. Indeed most ethicists probably do not associate social science with ethics other than in the somewhat arbitrary distinction between normative and descriptive ethics. Nelson questions this presumption of a linear relationship between ethicists and social scientists, in which the latter provide the data upon which the former make judgements: "Moral theories, informed by facts, judge practices." He argues instead for an interactive model between the two (Nelson 2000: 12, 16). Another purpose of this paper is to develop Nelson's argument further and to demonstrate that, by virtue of their theoretical as well as their empirical interests, the social sciences have more to contribute than just 'the facts'. The social sciences see legal and ethical issues as primarily social issues and, because of this encompassing perspective can contribute not only to the understanding of ethical issues but also to the understanding of the social processes through which those issues become constituted as ethical concerns.

In brief, therefore, the overall aim of this article is to contribute to the hitherto limited discussion (although see also DeVries and Subedi 1998; Hoffmaster 1992; Spallone et al. 2000) on the contributions of the social sciences to the study of ethics. I shall frame the discussion through a consideration of three broad questions:

(i)   What theoretical work can the social sciences contribute to the understanding of ethics?

(ii) What empirical work can the social sciences contribute to the understanding of ethics?

(iii) How does this theoretical and empirical work enhance the understanding of how ethics, as a field of analysis and of debate, is socially constituted and situated?

It may well be that some colleagues reading this article see no objection to Nelson's description of the relationship between ethicists and social scientists since they see (bio)ethics as a branch of moral philosophy which aims to evaluate ethical arguments in order thereby to eliminate poor quality reasoning. They therefore might have little interest in the above questions, seeing them primarily as matters of social practice, which might be of interest to social scientists but which are of little relevance to ethicists. However I should like to suggest several reasons why these three questions are of direct relevance to bioethicists, beyond the point of making a case for the inclusion of social science in ethical debates.

First, ethics is no longer a purely abstract discipline since there is a growing interest within bioethics in conducting empirical investigations and, within philosophy more broadly, with applied work. This alone suggests that there is potential for a fruitful collaboration between ethicists and social scientists. Therefore a clearer understanding of each other's perspective on the same issues of substantive interest would be of mutual benefit. At the very least these three questions, and their answers, provide an indication of how social scientists approach ethics and ethical issues. Second, whilst there is still a legitimate case to be made for seeing ethics as an abstract formal discipline, distanced from matters of actual social practice, it is not possible to posit the same view of ethicists themselves. As individual and collective practitioners of their discipline, however varied in their approaches and interests, they are members of professional and other social groupings and are thus subject to the influences of, and in turn influence, broader social changes and developments. At the very least I would argue that there is some interest for bioethicists in seeing themselves in a social context, even if there is a resistance to seeing their subject in such a way. All three questions above, but particularly (i) and (iii), suggest ways in which the practitioners of ethics might be understood in terms of broader social patterns and organisations. Third, I suggest that there is more to be gained than mere interest from understanding this social identity of ethicists, since one of the consequences of the growing lay and media interest in ethics and ethical issues is the increased scrutiny, and even criticism, social and political, of ethicists themselves (see Spallone et al. 2000). This is part of what it means to see ethicists in a social context. From a social science point of view, to understand more about ethics is to understand more about ethicists, and vice versa, as the answers to the questions above will indicate. A clearer understanding of how this social identity affects their influence on the conduct of ethical matters is, I would suggest, a matter of practical as well as theoretical interest for ethicists.

I shall take each question in turn though the first will be treated at greater length since it has received the least attention in previous writings. The division between theoretical and empirical work is somewhat artificial (can one conduct good quality empirical research without being informed by a particular theory of knowledge; equally can one understand theoretical claims without an understanding of how the everyday empirical world operates?) but, given the tendency hitherto to see the social sciences as having only a limited, empirical, contribution to make it is important to consider these two inter-linked strands separately for the moment. This both ensures clarity and expands the understanding of the social science contribution.

2.   WHAT THEORETICAL WORK CAN THE SOCIAL SCIENCES CONTRIBUTE TO THE UNDERSTANDING OF ETHICS?

Social theory covers a wide array of approaches to the analysis of the social world. These can be broadly delineated as: theories of the subject and the relationship between the individual and society; theories of social structure, organisation and institutions, including explanations for change in these and for their changing relationship with the individual; critical theory, which examines the relationships between social life, politics and the ordering of society through processes of rationalisation and bureaucracy; the sociology of difference, defined in a number of ways such as multiculturalism, ethnicity, gender and, finally, theories of modernity and postmodernity (Elliot 1999). Within this body of work, a number of theorists (primarily though not exclusively sociologists) have taken an explicit interest in ethics, either as the subject of their investigations or as part of their explanatory repertoire in the analysis of broader social phenomena.

(i) Weber's writings contain two possible lines that might shape a sociology of ethics (Weber 1958 and 1946). First, he demonstrated that a concern with how one ought to be and how one ought to act can influence a broader cultural trend towards the emergence of certain types of society and away from other possible lines of development. He coined the term "economic ethics" in describing the effects of religion on economic affairs (Osborne 1998). Second, he was interested in the practice of scholarly life itself. These two strands (the first a broadly defined notion of ethics as a field of sociological study, and the second, a narrower view of ethics that informs a reflection on how one's values affect how one acts as a sociologist) have varying degrees of prominence in later work (e.g. Dingwall 1980; Silverman 1985).

(ii) Theorists such as Hobhouse and Ginsberg saw ethics as central to their work in the analysis of systems of government. Helmes-Hayes claims that they saw sociology's "real or first purpose" as being "to formulate a general, historically informed theory of social and moral development" (Helmes-Hayes 1992: 340). Ginsberg in particular was interested in the "sociologically related issues of ethics" and saw "no grounds for assuming that [a moral judgement] is not susceptible of investigation by rational methods" even though this went against the prevailing practice of separating ethics from social science (Ginsberg 1942: 127f., as quoted in Helmes-Hayes).

(iii) Foucault's interest in "practices of the self" spanned a concern with individualism, liberalism and modernity: "The heart of this problem being how to construct oneself ethically without recourse to overriding moral norms" (Osborne 1994: 487; Rabinow 1997: 255f.). He identified a continuity from ancient Greece to contemporary Europe and argued that in both societies there is a reluctance to relate ethics to religious beliefs and to accept that legal systems have a part to play in personal, private lives. In contemporary life, ethics becomes an aesthetics of the self, a process in which one crafts oneself into being a 'moral agent'. This approach is expanded in Osborne's own work on 'ethical stylizations'. "This means I want to connect ethics to the issue of person-formation. Ethical styles are to ethics what styles of reasoning are to scientific truth; that is, before one can be ethical or unethical, one needs a style of ethical judgement, or rather a particular stylization of the person that is equipped to make ethical judgements." For example, he claims that Victorian ideals of altruism had less to do with one's conduct towards others and more to do with the need to sustain an acceptable sense of self by avoiding sloth and misery, altruistic acts being one way whereby this was achieved. Turning to contemporary society he suggests that we do not now live in an immoral world, as some argue, but rather that we live in a world of many ethical stylizations but with few rules about 'ethical content' (Osborne 1998: 221, 231).

(iv) Giddens seeks to understand the relationship between the self and society without privileging either concept. His structurationist work on life politics links substantive concerns (the environment, biological reproduction, globalisation, and self-identity) with a view of trends distinctive to societies at a stage of late modernity: "... life politics concerns political issues which flow from processes of self-actualisation in post-traditional contexts, where globalising influences intrude deeply into the reflexive project of the self, and conversely where processes of self-realisation influence global strategies ... Life politics is a politics of life decisions." These decisions are a feature of everyday life for individuals who no longer have the certainties and rules of traditional society to guide their actions and to tell them who they are and what their place is within society. Thus, "how should we live?" is the question at the centre of life politics. "Life political issues ... call for a remoralising of social life and they demand a renewed sensitivity to questions that the institutions of modernity systematically dissolve" (Giddens 1991: 214ff.).

(v) A concern with ethics has been taken up in more recent debates in cultural theory. Lash acknowledges that "ethical debates have become so central in recent years to social and cultural theory that they can no longer be ignored" which implies that though they have been present they have not yet received adequate attention. One failing of early social scientific work on ethics was, according to Lash, the tendency to see ethics as part of the superstructure, "to be explained through a set of social causes and in terms of social interests. The cultural turn itself brought, thankfully, a departure from such positivist notions of causation and problematized the utilitarian assumptions of the notions of interests." He characterises these debates as being between liberal theorists such as Rawls and Habermas, and communitarian writers such as Charles Taylor and Alisdair MacIntyre (Lash 1996).

(vi) This debate has now been expanded by the postmodern ethics of writers such as Bauman whose work is particularly useful for advancing the sort of discussion advocated in this article. He notes how important the notions of universality and foundation have been to philosophers and legislators in modernity as they have sought to eliminate ambiguity and conflict and to achieve order and uniformity, in morality and ethics as in everything else. He observes, "... the moral thought and practice of modernity was animated by the belief in the possibility of a non-ambivalent, non-aporetic ethical code ... It is the disbelief in such a possibility that is postmodern ... The foolproof – universal and unshakeably founded – ethical code will never be found" (Bauman 1993: 2–27). Whilst some have seen this as a precursor of social collapse (Elliott 1999: 27), Bauman is more optimistic, seeing possibilities for individuals and communities to formulate new ideas of responsibility and ethics: "a radically novel understanding of moral phenomena has been opened" (Bauman 1993: 2–27).

(vii) Feminist ethics, based on a confluence of feminism, the social sciences and philosophy, has been developed by a number of writers. As

Bennett observes, "Feminist theory provides an alternative theoretical framework for bioethical decision-making by providing an analysis that is based on interpersonal relations and connections rather than individualised rights. Feminists have argued the need for greater acknowledgement and valuing of the role of caring and nurturing relationships ...". Such work challenges what it argues are male-centred theories of ethics, specifically the communitarian writers who assume shared understandings of the social world and of social order. Such challenges have implications for how central ethical concepts such as autonomy might be understood. For example, Bennett argues that we should talk of "embodied autonomy", which depends on understanding "individuals whose lived realities are mediated through the embodied intersections of race, class, gender, sexuality, disability and fertility. Autonomy is not found in an extra-corporeal individual carrying a bag full of rights as a safeguard against the world. Instead, autonomy is articulated by an embodied self, through relationships with others" (Bennett 1999: 300f.).

As these different strands of work suggest, not only has there been a great deal of social science interest in the concept of ethics, there has also been sufficient space and liveliness within that work for debates and opposition to arise. Just on the one topic of social change, social theory has produced two tales about the transition to modernity and the effects of the growth of science, the market and bureaucracy upon notions of morals and ethics. First, the tale that "modernity was ushered in by forces inimical to ethics, to community, to values" so that there was a fragmentation of moral and ethical life, and, second, the tale that there was "an erosion of all values by the uniform, levelling and monolithic forces of bureaucratic culture and rationalisation" (Osborne 1994: 493). In the transition from modernity to either late modernity or even postmodernity, the focus has changed again. As Lash puts it, "Now no longer were questions of rights and justice necessarily pivotal in ethics. No longer was ethics necessarily a question of moral agency. Now the moral self was a singular ethical subjectivity finding itself in face of the unbounded demands of 'the other'" (Lash 1996: 76).

It would be inappropriate to try to make unified claims about the cumulative contributions of such a diverse body of work. However what is clear is that ethics is a part of what needs to be explained when analysing the key areas of social science interest. From these diverse theories we gain an understanding of ethics as shaping, and being shaped by, major social forces; of ethics being historically and culturally located; of ethics working in association with key social institutions, such as religion, government and the family; of ethics defining particular social groups, such as genders, professions, communities or sexualities; of ethical values being both reflective and constitutive of the self. That is, of ethics as being embedded within those aspects of the social world that tend to be taken as given, or as unproblematic, features of the world, by the normative statements of ethicists in general.

There is a danger that, with such diversity of interests, contexts and uses, the very notion of ethics disappears in a myriad of diverse meanings. However whilst it could be argued that that is true of the field of ethics more generally (how precisely do most writers employ this term?) it is a charge that warrants more serious response. There are at least two possible responses. The first is to say that the key contribution of the social science work considered so far lies in its very diversity, since we can see if our understanding and usage of that term is extended, clarified or even challenged by the uses presented here. In that respect, this work can be said to have made a contribution to rigorous thought.

The second response to the charge of the disappearing concept of ethics is provided in the next section, where examples of empirical research on ethical issues are presented. Much of the work reviewed hitherto, with the exception of feminist sociology, tends to lack an empirical engagement with the practices of ethics or, more precisely, with the practices of what becomes designated as 'ethics', and this makes it easier for the concept to become even more elusive. Several commentators from within the discipline have noted the tendency for the social sciences to become somewhat abstract when dealing with such issues. Ginsberg called for small-scale studies to provide empirical content in order to achieve the goal of understanding social life as a whole; Osborne calls for a "plurality of sober-minded forms of analysis and critique" of ethical issues; Lash, in a somewhat different style, declares, "We want to bring ethics out into the streets" (Osborne 1994: 494; Lash 1996: 76).

3.   WHAT CAN EMPIRICAL SOCIAL SCIENCE CONTRIBUTE TO THE UNDERSTANDING OF ETHICS?

As already mentioned, empirical social science could be seen as providing descriptive 'facts' to go alongside normative statements. And there is an acknowledged need for this, since, as Zussman claims, many ethicists' propositions are based on empirical claims that are usually unsupported by evidence (Zussman 2000: 8). However besides resisting being seen as the 'handmaiden' of ethics, as it was once seen as the handmaiden of medicine, sociology in particular would question this simplistic division. As Nelson points out, empirical inquiry is not just an exercise in "scooping up the facts" (Nelson 2000: 13). Social scientists would argue that those 'facts' are inherently tied to theories of knowing the world and to a repertoire of techniques and skills for accessing that world (or indeed for constructing that world through an ongoing set of social processes, see Velody and Williams 1998).

Therefore in this section it is useful to consider the contributions of empirical work from the social sciences in a number of different ways. First, examples of the empirical work that has been done; second, the potential within that existing work to give rise to a number of further questions and third, in terms of the epistemological and methodological techniques for gaining empirical access to the social world. The purpose in bringing this material together is not to try to impose a false unity on existing work but to establish that such work exists and to identify a web of connections between that work, to assist in the identification of new questions about the relationship between the social sciences and ethics.

3.1   Examples of empirical work

Given the breadth of theoretical work it is perhaps not surprising that there is also a wide range of empirical work within the social sciences that has the potential to enhance our understanding of ethical issues. Just a few examples shall be covered to illustrate this claim. Much of this work did not set out to be an explicit investigation of ethical issues per se, but it nonetheless touches upon and reveals how ethics are 'done' (identified, thought about, acted upon) in everyday life, across a range of substantive examples.

Given the prominence of assisted conception and embryology in ethical debates over the last twenty years this is a familiar and useful place to begin to lodge certain claims about the potential of social science work in this field; one can then move from that point to consider 'medical ethics', 'bioethics' and 'ethics' more generally. Others might want to argue for a different starting point, although it is worth noting Toulmin's claim that "medicine saved the life of ethics" (Toulmin 1982) as an additional reason for focusing on medical contexts.

Commenting on reproduction and on the technological malleability of the body more generally, Giddens remarks, "The more we reflexively 'make ourselves' as persons, the more the very category of what 'a person' or 'human being' is comes to the fore" (Giddens 1991: 217). These and related issues lie at the heart of concerns over the new reproductive technologies and assisted conception. It is usually at this point in a public or policymaking discussion that philosophers are invited in, to explain the different forms that answers to such questions might take. Whilst there is a place for abstract principlism, there is also a place for the social sciences to throw further light both on the questions themselves and on the contexts in which they arise. Examples of empirical investigations of what issues come to the fore in the field of assisted conception, and how they are reasoned through and handled by everyday actors, can be seen in the work of several social scientists.

Franklin conducted an ethnographic study of couples' experiences of the in vitro fertilisation (IVF) treatment process. On the question of what counts as a person Franklin found that some women blurred the boundaries between chemical pregnancies, embryos and actual babies. As one woman said at the point of embryo replacement, "well, they are my babies 'cos they are my embryos" (Franklin 1997: 182). This blurring provided support for couples and gave them the stamina to sustain the arduous treatment. Thus we have here a highly contextualised definition of what is a person: one that is given meaning and consequentially acted upon. This study reveals other aspects of IVF treatment that raise ethical questions. Franklin's couples found that the rigours of IVF treatment dominated their lives but were hard to resist as the promise of resolution (of their infertility or of their difficulties in coming to terms with their infertility) beckoned. However as they proceeded with treatment it became apparent that for most couples there would be no resolution in either sense. The paradox was that the ethical arguments advanced in support of IVF ('helping desperate couples') as a response to ethical concerns about the process (experimentation on human embryos), are shown, by empirical investigation, to be open to challenge, as the evidence from Franklin's study suggests that IVF creates a desperation to succeed that had not been there in the first place. Also IVF is shown to create other ethical difficulties, such as women being sold or pursuing IVF treatment when there are reducing levels of likely success. However the data also show that despite acknowledging these costs most of the women interviewed strongly endorsed the treatment. As Franklin concludes, "... women's experiences of IVF ... are far from simple. They are composed of feelings and perceptions that are equivocal and ambivalent, positive and negative, empowering and disempowering. These paradoxical dimensions of the experience of the IVF procedure are fundamental to many of the ways of making sense of it ..." (Franklin 1997: 194).

Price's work also shows additional ethical dilemmas created by IVF that are rarely discussed: that of treatment succeeding "too well", resulting in multiple births. In her study of parents coping with triplets, quads or quins (1992). Price found a vast range of practical and emotional problems associated with multiple births. The practical problems included: not being able to pick up all babies at any one time; coping with different sleeping patterns amongst the children; isolation because of the difficulties of taking three or more small babies out at any one time; financial difficulties; the sheer amount of work and the lack of formal or informal assistance on a regular basis from family, friends or social services. Emotional problems arose between the parents who had little time or energy to deal with each other's needs and from the difficulties of treating any of the children as individuals given the demands they made as a group. Therefore, the risks of a multiple birth require serious consideration between doctor and patient before IVF is provided. However, this entails other ethical considerations, such as should doctors prioritise the avoidance of a triplet birth by only replacing two embryos (rather than the three allowed in UK legislation). Thus, the impact of this research has been that exposure of the practical difficulties of coping with higher order births has added ethical considerations to the physiological considerations of whether it is more effective to transfer three or only two embryos. This in turn raises other ethical dilemmas for practitioners, such as the question of balancing, on the one hand, their patients' desires for a baby (and their own needs to produce satisfactory IVF success rates) with the now known problems of a multiple birth. A question that Price's study did not attempt to answer but which she raises as part of the discussion is the possibility of using selective reduction as a solution to a multiple pregnancy. The paradox here is that the very success of IVF creates further ethical dilemmas over abortion. Therefore what is learned from Price's study is that ethical questions about assisted conception are rarely confined to single treatment contexts but are likely to impinge on, and be affected by, the broader social context of the interplay between families, medicine, the state and reproductive issues.

Both Price's and Franklin's work point to questions about what counts as informed consent. Should information include the consequences of going through the treatment process itself? Should information include detailed consideration about whether a multiple birth is actually something parents have seriously considered and know that they can cope with? The social science research described here suggests that the answer to both questions is 'yes'. Clearly both sets of data can be challenged but only by other empirical findings.

A different type of study is reported by Haimes. This is an empirical examination of the normative statements (taken from documents and interviews) that comprise the two sides of the debate over whether children conceived through donated gametes should be told information about their genetic parents. The aim of the analysis is to "prise open those unstated assumptions upon which the debate rests, to see what process of reasoning, and what repertoire of concerns, are available to those participating in such discussions." This is another example of the success of assisted conception creating further ethical difficulties, since the process of helping people to have children creates the situation where those children have their genetic origins rendered unknowable. These difficulties are usually hidden from most bioethicists' gaze since they lie within the communities and families rather than within the more visible context of the clinic. The analysis shows that the apparently virulently opposed sides of the debate share a large amount of common ground, including: a view of families as needing to be discrete, self-contained units; uncertainty about the place of genetics and biology in making the family, and uncertainty about prioritising the needs of one member of the family over another. These in turn show that the debate depends upon notions of 'ordinary' and 'extraordinary' families, which are themselves statements of ethical values and are thus open to challenge from other value positions, for example claims about the legitimacy of diverse family forms. This study demonstrates that "the handling of genetic origins is not a simple matter, but can be shown, through a sociological analysis, to have consequences for and connections to, other aspects of the wider society" (Haimes 1992).

A final example of empirical work comes from Edwards who provides valuable insight into how ordinary people, with no direct involvement with assisted conception, think about these ethical and social issues. In her ethnographic study of "Alltown" (Edwards 1998), she discovered that residents identified three related sets of dangers with these practices: psychological (in terms of worrying about the effects on the children conceived), biological (in terms of the risks of incest which was used as a "conceptual brake ... a boundary, a limit which ought not to be traversed") and relational (the impact of reproductive technologies on wider social relationships). Edwards found that her informants located the ethical complexities raised by assisted conception within other complex family relationships that they had either experienced themselves or had heard about, such as remarriage, step-relatives or cohabitations. If those pre-existing frameworks raised no concerns then their parallels within assisted conception were also accepted; if there were concerns in the pre-existing scenarios, such as fears about incest, then the parallel case in assisted conception caused concern. One aspect that worried respondents was the possibility that a child might not be able to make full kinship claims (for example, for unquestioned support throughout life) on his/her relatives because of being conceived through donated gametes. This worry was explained through the examples provided by half-relatives or step-relatives who might feel not fully attached to a family. However there was also the experience in existing families of marginal individuals (such as long-lost fathers or other relatives) wanting to stake unjustified kinship claims on children. This example was used to explain the dangers of a donor wanting to make a claim on a child that s/he had helped to conceive (or indeed the dangers of the child making claims on the donor). These dangers were thought to increase if the donor was already related, for example a sister donating an egg. However, other experiences such as the closeness of sisters, and the need to keep things in the family were also used to justify that practice. Elsewhere, in reference to the same study, Edwards concludes, "people interpret what they see as the implications of NRT not through what they know of the techniques and philosophies of reproductive medicine, but through what they know about the practice and predictabilities of kinship" (Edwards 1993: 63f.).

This study shows that new technical possibilities do not always create new ethical dilemmas but when they do, everyday actors are seen to have other cultural reference points through which they resolve such difficulties. In this case, that which is usually presented as an issue of medical ethics, is seen by everyday actors as an issue of family ethics. That is, the 'same' ethical issues are located in different contexts by different sets of actors and are accordingly seen to have different implications and meanings. It is these differences that are open to empirical investigation.

3.2   Questions raised by empirical investigations

The studies cited here have a contribution to make beyond their own individual empirical findings. They have the potential to stimulate further questions about the ethical context of reproductive technologies since they provide an empirical basis to Giddens' observation that reproduction "is now a field where plurality of choice prevails" (Giddens 1991: 219). This theme is further analysed by the social anthropologist, Marilyn Strathern: "Choice is a significant value to which Euro-Americans give weight in setting up families ... developments in new reproductive technology have made newly explicit the possibility of choosing whom and what one desires to call family. In opening up ways of thinking, they offer a cultural enablement of a kind" (Strathern 1996: 47). This claim can be pursued through further empirical research. For example, taking the aforementioned studies as a guide, one might ask to what extent do different groups (patients, clinicians, policymakers, ordinary people) share a view of the importance of choice in specific reproductive circumstances? In the case of preimplantation genetic diagnosis, for example, in what sense does a woman 'choose' to have her embryos screened: is this perceived to be a choice by the woman or does she see it as an obligation, in order to avoid what are said to be the personal and social costs of a disabled child? Also does choice in such situations include the option of the woman actively choosing to have a child with disabilities, and rejecting those embryos that have been shown not to have disabilities? Equally how are certain views of choice curtailed by other 'ethical' considerations, such as the welfare of the child? (Burbidge and

Haimes 1996). And how much are such choices constrained by the exercise of professional power and authority? (see Freidson 1975; Osborne 1994; Pippin 1996). As we have seen, these questions, when subjected to empirical investigation, expand our repertoire of what counts as an 'ethical' question by alerting us to the possibility of multiple perspectives on ethics.

Therefore, such empirical studies both complement the broader theoretical sweeps outlined in the previous section (by revealing yet further 'ways of thinking' and by showing what 'cultural enablement' might look like in specific contexts) and stimulate further theoretical and empirical lines of investigation. Such an empirical grounding also sheds further light on how the 'politics' of life politics might be enabling in some contexts but also constraining in others.

The findings of these empirical studies and the additional questions and lines of investigation that they give rise to suggest that if we are to understand more clearly how individuals 'act ethically' we have to engage in the detailed, contextualised dilemmas. As Bauman argues, "human reality is messy and ambiguous and so moral decisions, unlike abstract ethical principles, are ambivalent. It is in this sort of world that we must live; and yet, as if defying the worried philosophers who cannot conceive of an 'unprincipled' morality, a morality without foundations, we demonstrate day by day that we can live or learn to live, or manage to live in such a world, though few of us would be ready to spell out, if asked, what the principles that guide us are, and fewer still would have heard about the 'foundations' which we allegedly cannot do without, to be good and kind to each other" (Bauman 1993: 32).

3.3   The contributions of social science methodology

A third feature of the contributions of empirical social science is the epistemological frameworks and the methodological skills and techniques that provide the means for accessing and engaging with messy human reality. As the above studies demonstrate, methodological strategies such as ethnographic research and participant observation, and research methods, such as in-depth interviewing, are particularly powerful in revealing the details of how people think and act ethically in everyday life (Hoffmaster 1992; Holm 1997). Other techniques have also been fruitfully deployed to reveal empirical data on ethical considerations. Also in the field of reproductive technologies these have included the documentary analysis of Hansard debates (Haimes 1990; Franklin 1993; Mulkay 1993), survey research to elicit clinician's strategies for implementing the welfare of the child clause in the UK's Human Fertilisation and Embryology Act, 1990 (Burbidge and Haimes 1996) and historical methods to analyse the debates over the provision of fertility services on the NHS. These techniques provide access to (i) the locations and institutions (for example, clinics, hospitals, courts, government committees, ethics committees) in which ethical dilemmas arise, (ii) the actors (for example, professionals, lay people, policy makers, family members) involved in the activities and discussions around these dilemmas, (iii) the diverse range of definitions and meanings that underpin both the dilemmas themselves and the answers proffered (and that underpin the diversity of responses from different actors in different locations and institutions, at different periods in history), and finally, (iv) the broader socio-cultural frameworks (of family, health, community, economy, religion, law, politics and so on) which both shape the repertoire of possible responses and which are shaped by the actual responses given in actual cases.

It is of course not the case that other disciplines lack either an interest, or skills, in conducting such empirical work: Zussman points to the work done by doctors, nurses and health administrators (Zussman 2000: 11). Similarly, Holm is one of several bioethicists now conducting empirical research, taking as his subject the ethical reasoning of health care professionals (Holm 1997: chs. 3 and 4). However, besides providing many of the methodological tools used by others, the social scientist has the broad range of theoretical and epistemological resources and interests that assist him/her to make further sense of those situated practices, by asking questions that go beyond those that are immediately apparent in the situation itself, to connect these dilemmas to broader debates about social and cultural development, as outlined in the section on social theory. Such broader perspectives enhance the growing empirical research of bioethicists and others, in a number of ways. First, by providing opportunities for fruitful collaboration between the disciplines, and, second, by providing further insights into the nature and provenance of particular situations and into the wider social and cultural vicissitudes that will have an impact on those settings and on the groups and individuals that populate the settings. It is these attempts to connect the empirical data with theoretical explanations that make the social science enterprise distinctive and which take it beyond the handmaiden, 'scooping up' role. It is this disciplinary imperative, to connect the particular to the general, that also leads the social sciences beyond the specific ethical questions of particular practices, to ask further questions about the very designation of certain issues as being 'ethical issues'. The next section examines this.

4.   HOW CAN SOCIAL SCIENCE RESEARCH
       ENHANCE THE UNDERSTANDING OF HOW
       ETHICS AS A FIELD OF STUDY IS SOCIALLY
       CONSTITUTED AND SITUATED?

As the above empirical examples show, assisted conception practices have been culturally, professionally and politically packaged as 'ethical issues' but it is important to ask whether alternative packaging is possible. For example, one study suggests the 'yuk' reaction that greets practices such as the use of ovarian tissue from foetuses or cadavers for egg donation is as much stimulated by a sense of 'social impropriety' (that is, a sense that anyone having to identify this as part of their biography would be seen as drawing undue attention to the more unseemly aspects of procreation) as by any particular ethical stance (Haimes and Williams 1998). Thus, detailed empirical analyses such as these provide insight into how ethical issues are so packaged. However, there is much less work, empirical or theoretical, on this aspect so this section is much more speculative than the previous two sections. The focus is on trying to identify the likely directions that such work might take in the future.

Still within the field of medical ethics, answers to this question might be found through an exploration of the social processes of labelling, whereby certain matters are designated 'ethical', and others 'economic' or 'political'. For example, what is 'being done' when decisions about embryo experimentation are routinely categorised as ethical questions and when issues such as the funding of fertility treatment on the National Health Service are routinely labelled economic or political questions? Where do such labels come from? What are the consequences of such labels? What alternative arguments and actions are constrained by the application of such labels? What would happen if alternative labels were applied?

Other lines of inquiry are likely to be equally fruitful. For example, there is a need to ask what is the relationship between the history of medical ethics and state policies, such that some fields have become highly legislated, for example, abortion, and others, such as euthanasia, are not. The history of medical ethics is important since as Osborne points out, its current focus on patient care is a relatively recent development; its first and long-term concern has been with the etiquette of inter-professional rivalry (Osborne 1994b). Mackenzie, drawing on the work of Hacking, demonstrates how to do an historically informed social constructionist analysis of morality "by 'taking a look' at the local, empirical facts that surround their emergence as problems" (Mackenzie 1998: 201). An exploration of the history of medicine and ethics would also be useful to document the ways in which certain cases have provided a forum for particular formulations of ethical positions, such as the Tony Bland case in the UK. An awareness of the history of medical ethics would also be useful in documenting the transition of certain practices from being "ethical concerns" to "mundane practices", and back again, as has happened to some extent with donor insemination (Daniels and Haimes 1998). Turning to contemporary medicine, the increasingly influential role of 'the four principles' (autonomy, beneficence, non-maleficence and justice) in guiding clinical decisions may shed light on the notion of 'ethics as mantra': some suggest that rather than assisting clinicians in clarifying their possible courses of action these principles simply tend to reproduce "normative frameworks" instead (Chadwick and Levitt 1996, see also the chapter of Schöne-Seifert in this volume). It has already been shown how 'autonomy' can be challenged.

Other questions require systematic theoretical and empirical examination, for example, what, if anything, is distinctive about the different ethical domains? Are medical ethics different from ethics of the environment, or issues around ethical investment, or legal ethics, and so on? Some would argue that there are profound differences (Freidson 1975: 342). For Schenck it is the "radical responsibility" and the "radical threat" of medicine that gives rise to intense ethical debates, mainly because of the bodily concerns of medicine. "Going into discussions of medical ethics attending only to ethical theories, without attention to the phenomenon of embodiment, would be like going into a discussion of the rules governing ball games without paying attention to the differences between football, basketball and baseball, focusing only on abstract discussions of the nature of 'rules' and 'games' as such. Too often ethicists of medicine seem lost in just this way; and that is because they tend to ignore that which distinguishes medical ethics from business ethics or legal ethics which is of course the centrality of the body for the practice of medicine and the texture of embodiment of human life itself" (Schenck 1986: 50).

Such claims are of course open to empirical investigation; for example, returning to Osborne's claims about ethical stylizations, is there evidence of different professionals in different fields (for example, law, medicine, academia) displaying distinctive stylizations? These claims lead on to other questions: what impact does singling out ethical domains in this way have on the practices within those domains; how are the boundaries between different ethical domains established, challenged, penetrated, or changed? What happens when different ethical domains clash (for example, the legal and the medical, see Stacey 1992)? Equally, what happens when it is assumed that two domains are so different that ethical concerns about practices in one field are not reflected to the same extent as in another domain (see e.g. Gofton and Haimes 1999)? Whilst some work is being conducted on these sorts of questions, much of it needs to be more clearly theoretically framed if it is to contribute to broader understandings of social life.

In other words, the task here is to conduct a social science analysis of the social processes, meanings and institutions that frame and produce 'ethics' and ethical problems. Some of these questions might be addressed through a 'genealogy of ethics': how has it become possible to talk of ethical concerns; how have matters become so designated; how do they become attached to particular fields of activity (and not to others); what institutions have formed to deal with ethical issues; what ideologies, knowledges, webs of power have arisen around these institutions; how has the notion of 'ethical expertise' arisen; what is the history of its attachment to particular disciplines? Nelson suggests that the social sciences might mount an implicit challenge to bioethicists in asking why they attend to the questions they do and "even whose interests they think it appropriate to serve" (Nelson 2000: 14).

Work has already begun on other lines of analysis that will help to situate ethics in a socio-historical context, including: the possibility of cross-cultural ethics (see e.g. Gillet 1995), the notion of ethics in developing countries (see e.g. du Jardin 1994) and the multiplicity of ethical positions, including the framing of ethics for particular religions and so on. Each of these in turn raises further questions about the framing of ethical debates in certain conceptual terms (such as lying, truth, trust, autonomy, personhood) and the tendency to ignore or exclude other terms (such as power, politics, authority). Again Nelson is pointed in his commentary: "The social sciences might make a contribution to bioethics by helping the field's practitioners understand better what's behind its deeply installed respect for individual autonomy and whether it has assumed more the character of an ideology than a moral philosophy" (Nelson 2000: 15).

Such a conceptual analysis would allow for the possibility of alternative models emerging (Sharpe 1992). The historical perspective might cause alertness to a consideration of whether the twenty-first century is particularly challenging in terms of ethical dilemmas or whether each period sees itself as confronting new and more difficult dilemmas than previous eras. Such work would lend an empirical base to Bauman's wish that "... the sources of moral power which in modern ethical philosophy and political practice were hidden from sight, may be made visible, while the reasons for their past invisibility can be better understood." It is also likely that such studies would reveal evidence for Bauman's claim about the relativism of ethical codes and of recommended moral practices (Bauman 1993: 3, 14). Clearly, at this point it is only possible to identify some of the questions that might provide analytical purchase: others will arise as the discussions and explorations progress.

5.   CONCLUSION

Silverman suggests that "The social, political and cultural contexts of health communication and ethics is a crucial area which raises central questions about the relevance to society of sociological research" (Silverman 1992). This article has at least outlined the relevance of the social sciences to the study of ethics. A review follows, showing how the four broad purposes proposed in the introduction have been fulfilled, with a brief indication where this leads, in terms of the relationship between social science and ethics.

The first purpose was to identify the various strands in the social science work on ethics. These fall broadly into theoretical and empirical contributions from a wide range of social scientists.

The second purpose was to identify how this body of work contributes to the study of ethics. The theoretical work expands the questions one can ask about the study of ethics by relating it to the dominant themes in the social sciences. The empirical work contributes at several levels. It provides the data on particular substantive issues; that is, the 'facts' about what people say or do on particular topics. This in turn provides data on how individuals and groups think and act ethically; that is the 'how' and the 'why' of what people say and do, that fall outside the frameworks of formal philosophy. In addition, the empirical work demonstrates and refines the techniques for gaining access to the 'what', 'how' and 'why' of ethical behaviour. Together, the theoretical and empirical work provide knowledge on the social processes that lie behind the designation of certain areas of social life as being 'to do with' ethics, and the exclusion of other areas as not being of concern for the study of ethics.

The third purpose was to try to make the voice of the social sciences heard more clearly. However, it is perhaps a weakness of the work that has been done hitherto that its proponents have not made full use of each other's contributions, which might be part of the reason why the voice has not been heard. It is hoped that this identification and delineation of existing work will assist the voice of social science to be heard more clearly, both in communication with each other and in communication with ethicists and other disciplines. Equally the identification of a number of so far unanswered questions should help this dialogue since it both preempts any sociological imperialism and presents a task in common for a number of disciplines to work on together. In that sense there is a dialogue that arises out of the intellectual space created by questions, which in turn generate further questions and collaborative work (Silverman 1985: 194f.).

The cumulative impact of this existing and potential work helps to fulfil the fourth purpose of this article, which is to establish that social science has more to contribute than just the 'facts'. Zussman expresses reservation about sociology being the junior partner to medical ethics yet he does not really make a strong enough case for seeing it as other than this because he fails to embed the empirical work of social scientists within their theoretical roots and hence their analytical framework. It is the theoretical framework that provides the future direction for the relationship between the social sciences and ethics: one that puts ethics centre stage as the subject of social science enquiry. This is a quite different focus and approach to work than that of examining a series of substantive topics (euthanasia, abortion, informed consent etc.) to provide empirical data for ethicists to use to refine their normative analysis. The advantage of putting ethics and ethicists centre stage, as the subjects of sociological research, is that they can then be located in a broader socio-historical context, which enables new questions to be asked. (Such as, why are these issues defined as ethical concerns by these people in these times and these places?) That is, more precise questions that are at the same time more general in so far as they are indicators of broader concerns and trends of particular societies at particular times. It is the recognition of this potential that supports the claim that it is indeed possible to have a 'sociology of ethics'.

One feature this particular article does not lack is questions. It is hoped it too will stimulate a dialogue since it contains the beginnings of an exploration in the sociology of knowledge, in a field where social science perspectives can contribute to the central focus of concerns rather than merely stand on the sidelines.

References

Bauman, Z. *Postmodern Ethics*. Oxford: Blackwell, 1993.

Bennett, B. "Posthumous reproduction and the meanings of autonomy." *Melbourne University Law Review* 1999, 23, 286–307.

Burbidge, R. and Haimes, E. "Infertility treatment and the welfare of the child". Paper presented to the British Sociological Association Medical Sociology Group Annual Conference, 1996.

Chadwick, R. and Levitt, M. "Comment". In *Cultural and social objections to biotechnology*. Biocult Project Team Report. University of Central Lancashire, 1996, 160–173.

Daniels, K. and Haimes, E. *Donor Insemination*. Cambridge: Cambridge University Press, 1998.

Dingwall, R. Ethics and ethnography. *Sociological Review* 1980, 28, 871–891.

DeVries, R. and Subedi, J. (eds.). *Bioethics and Society*. New Jersey: Prentice-Hall, 1998.

Edwards, J. "Explicit connections: ethnographic enquiry in north–west England". In *Technologies of Procreation*. J. Edwards et al. (eds.). Manchester: Manchester University Press, 1993, 42–66.

Edwards, J. "Donor insemination and 'public opinion'". In *Donor Insemination: International Social Science Perspectives*. K. Daniels and E. Haimes (eds.). Cambridge: Cambridge University Press, 1998, 151–172.

Elliot, A. "Introduction". In *Contemporary Social Theory*. A. Elliot (ed.). Oxford: Blackwell, 1999, 1–31.

Franklin, S. "Making representations". In *Technologies of Procreation*. J. Edwards et al. (eds.). Manchester: Manchester University Press, 1993.

Franklin, S. *Embodied Progress*. London: Routledge, 1997.

Freidson, E. *Profession of Medicine*. New York: Dodd, Mead and Co, 1975.

Giddens, A. *Modernity and Self-Identity*. Cambridge: Polity Press, 1991.

Gillet, G.R. "Medical ethics in a multicultural context." *Journal of Medical Ethics* 1995, 238, 531–537.

Ginsberg, M. *Reason and Unreason in Society*. London: Longman's Green und Co., 1942.

Gofton, L. and Haimes, E. "Necessary Evils?" *Sociological Research Online* 1999, 1. http://www.socresonline.org.uk (last visit: June 2005).

Haimes, E. *Family Connections*. Doctoral thesis, University of Newcastle, 1990.

Haimes, E. "Gamete donation and the social management of genetic origins". In *Changing Human Reproduction*. M. Stacey (ed.). London: Sage, 1992, 119–147.

Haimes, E. and Williams, R. "Social constructionism and the new technologies of reproduction". In *The Politics of Constructionism*. I. Velody and R. Williams (eds.). London: Sage, 1998, 132–146.

Helmes-Hayes, R.C. "From universal history to historical sociology: a critical comment." *British Journal of Sociology* 1992, 43, 333–344.

Hoffmaster, B. "Can ethnography save the life of medical ethics?" *Social Science and Medicine* 1992, 35, 1421–1431.

Holm, S. *Ethical Problems in Clinical Practice*. Manchester: Manchester University Press, 1997.

Lash, S. "Introduction to the ethics and difference debate." *Theory, Culture and Society* 1996, 13, 75–78.

du Jardin, B. "Health and human rights." *Social Science and Medicine* 1994, 39, 1261–1274.

Mackenzie, C. "Social constructionist political theory". In *The Politics of Constructionism*. I. Velody and R. Williams (eds.). London: Sage, 1998, 200–220.

Mulkay, M. "Rhetorics of hope and fear in the great embryo debate." *Social Studies of Science* 1993, 23, 721–742.

Nelson, J.L. "Moral teachings from unexpected quarters. Lessons for bioethics from the social sciences and managed care". *Hastings Center Report* 2000, January-February, 12–17.

Osborne, T. "Sociology, liberalism and the historicity of conduct". *Economy and Society* 1994a, 23, 484–501.

Osborne, T. "Power and persons." *Sociology of Health and Illness* 1994b, 16, 514–535.

Osborne, T. "Constructionism, authority and the ethical life". In *The Politics of Constructionism*. I. Velody and R. Williams (eds.). London: Sage, 1998, 221–234.

Pfeffer, N. "From private patients to privatisation". In *Changing Human Reproduction*. M. Stacey (ed.). London: Sage, 1992, 48–74.

Pippin, R.B. "Medical practice and social authority." *Journal of Medicine and Philosophy* 1996, 21, 417–437.

Price, F. "Having triplets, quads or quins: who bears the responsibility?" In *Changing Human Reproduction*. M. Stacey (ed.). London: Sage, 1992, 92–118.

Rabinow, P. (ed.) *Michel Foucault: Ethics, Subjectivity and Truth.* London: Allen Lane, 1997.

Schenck, D. "The texture of embodiment: foundation for medical ethics." *Human Studies* 1986, 9, 43–54.

Sharpe, V. "Justice and care: the implications of the Kohlberg-Gilligan debate for medical ethics." *Theoretical Medicine* 1992, 13, 295–318.

Silverman, D. *Qualitative Methodology and Sociology.* Gower: Aldershot, 1985.

Silverman, D. "Review: social science perspectives on medical ethics." *British Journal of Sociology* 1992, 43, 506.

Spallone, Pat et al. "Putting sociology on the bioethics map." In *For Sociology.* J. Eldridge et al. (eds.). Durham: Sociology Press, 2000, 191–206.

Stacey, M. *Regulating British Medicine.* London: Wiley, 1992.

Strathern, M. "Enabling identity? Biology, choice and the new reproductive technologies". In *Questions of Cultural Identity.* S. Hall and P. du Gay (eds.) London: Sage, 1996, 37–52.

Toulmin, S. "How medicine saved the life of ethics." *Perspectives in Biology and Medicine* 1982, 25, 736–750.

Weber, Max. "Science as a vocation." In *From Max Weber.* H. Gerth and C.W. Mills (eds.). New York: Oxford University Press, 1946.

Weber, Max. *The protestant ethic and the spirit of capitalism.* New York: Free Press, 1958.

Velody, I. and Williams, R. "Introduction." In *The Politics of Constructionism.* I. Velody and R. Williams (eds.). London: Sage, 1998, 1–12.

Weber, Max. *The Protestant Ethic and the Spirit of Capitalism.* New York: Free Press, 1958.

Zussman, R. "The contributions of Sociology to medical ethics". *Hastings Center Report* 2000, January-February, 7–11.

---

[1] Reprinted from *Bioethics*, Vol. 16, No 2, 2002, pp. 89–113; with permission form Blackwell Publishing, Oxford, UK.

Chapter 20

DISCOURSE ETHICS
Apel, Habermas, and Beyond

MATTHIAS KETTNER
Witten/Herdecke, Germany

1.   THE ORIGIN OF DISCOURSE ETHICS IN APELIAN TRANSCENDENTAL PRAGMATICS

The contemporary German philosopher Karl-Otto Apel is best known for his wide-ranging 'transcendental pragmatic' approach to a range of issues in theoretical and practical philosophy. This approach accords argumentative discourse and its essential normative presuppositions a fundamental role in all other philosophical inquiries for which justifiable validity-claims are raised, for example epistemology, normative theories of rationality, critical theory, and ethics. If such presuppositions exist, then any interlocutor's communicative intention to waive them will conflict with the construal of that debate as rationally meaningful, since it involves the interlocutor in the kind of inconsistency that Apel (like Habermas), drawing on speech-act theory, conceptualises as a "performative self-contradiction". Apel (unlike Habermas) develops this concept into the doctrine of rationally definitive justification (*Letztbegründung*). Apel deserves to be better known as the originator of Discourse Ethics (*Diskursethik*), the central contention of which – that some presuppositions of discourse have universally valid moral content – he developed in the mid-1960s.

With regard to Discourse Ethics, Apel (1976) contends that all human needs, as potential claims that can be communicated interpersonally, are ethically relevant and must be acknowledged in so far as they can be justified interpersonally through arguments. More specifically, transcendental pragmatics reveals certain normative proprieties in the practice of argumentation. Anyone who wants all competent participants to

299

C. Rehmann-Sutter et al. (eds.), Bioethics in Cultural Contexts, 299–318.

act as rational evaluators should require every competent participant to recognize these normative proprieties as ideally regulating their discursive commitments. Apelian Discourse Ethics holds that, of the proprieties that ideally govern rational evaluators' communities of argumentation according to the regulative idea of an "ideal communication community", some have a *morally* significant content for all rational evaluators. The understanding of such content is arguably part of how anyone should understand their role as a subject of argumentation, if they are competent and seriously willing to raise questions and provide answers in argumentation. Apel (1988) favours an articulation of the moral relevance of such proprieties in terms of a moral co-responsibility of actual and possible participants for keeping all their actions in accordance with a generic deontic status of free and equal co-subjects. Whoever is involved in argumentation rationally should want interlocutors to accept certain proprieties of mutual recognition and symmetrical situatedness as binding on anyone so involved; this would ideally regulate the discursive commitments. There is mounting controversy between Apel (1998, 2000) and Habermas over how far Discourse Ethics should accommodate mundane constraints and how it can be "applied" – be made practically relevant.

2.   CLASSICAL DISCOURSE ETHICS: APEL AND HABERMAS

Discourse ethics is both a recent paradigm of normative philosophical moral theory (ethics) and a set of beliefs about universally applicable standards for making moral judgments (a postconventional morality or ethos). The original insight that makes the elaboration of discourse ethics a worthwhile pursuit in normative ethics must be credited to Apel,[1] who argued that anyone who takes part in an argument implicitly acknowledges potentially all claims of all members of the communication community if they can be justified by rational arguments (Apel 1980: 277). Apel contended that all human needs, as potential claims that can be communicated interpersonally, are ethically relevant and must be acknowledged in as much as they can be justified interpersonally through arguments.[2] This, together with a substantial normative assumption about the nature of morality – namely that "one should not unnecessarily sacrifice a finite, individual human interest" (ibid.) – led Apel to formulate the following basic normative principle of (Apelian) discourse ethics: "all human *needs* – as potential *claims* – i.e. which can be reconciled with the needs of all the others by argumentation, must be made the concern of the communication community" (ibid.).

It is against this background that Habermas later proclaimed, as "the distinctive idea of an ethics of discourse", the "discourse principle" (D) that "only those norms can claim to be valid that meet (or could meet) with the approval of all affected in their capacity as participants in a *practical discourse*" (Habermas 1990: 66). In its most general reading, D suggests that validity, as claimable of norms regulating ways of acting, depends on consensus building among those who stand to be affected by the norms, provided this is rationally qualified "in their capacity as participants in a practical discourse".

As such, D expresses (1) a general intersubjectivist theoretical view about the nature of validity and validity-claims (not necessarily only in a moral sense, but also generally) concerning norms that regulate ways of acting. It is only when D is read prescriptively, as prescribing how anyone in so far as we are rational *should* govern our recognition of norms, that D expresses (2) a normative principle. And it is only when D is read prescriptively in a specifically moral sense, as prescribing how anyone in so far as we are rational should govern our recognition specifically of moral norms, that D expresses (3) a principle that is normative in a specifically moral sense.[3]

Habermas has recently strengthened the generality of D by making a more abstract discourse principle the centre-piece of a theory that would, if it could be fleshed out successfully, be a normative theory of modern law, democratic governance, and modern morality. In its revised form, D stipulates: "Just those action norms are valid to which all possibly affected persons could agree as participants in *rational* discourses."[4]

Any version of discourse ethics must provide an account of its contention that validity-claims, in so far as they are rational, depend on rationally qualified consensus building. Under which conditions is consensus building a "rational discourse" at all? And under which conditions is a rational discourse a "practical" or any other specifically distinguishable ("pragmatic", "ethical", "moral") kind of "discourse"? Different versions of discourse ethics differ, amongst other things, in how they link consensus building, qualified as discourse, to validity-claims, and both to rationality. For Apel and Habermas alike, the term 'discourse' means roughly that argumentation is conducted under conditions of free and open dialogue. Habermas' more detailed account draws heavily on certain transfigured speech-act theoretical notions (for example "illocutionary obligations") and on a supposedly deep distinction between acting in a strategic way versus acting in a communicative way.[5] Habermas has suggested that a rational discourse is a rational moral discourse if participants adopt as the decisive validity-determining question whether "the foreseeable consequences and side effects of [a norm's] general observance for the interests and value-orientations of *each individual* could be freely accepted *jointly* by *all* concerned."[6]

Apel views the way in which validity-claims depend on argument-driven consensus building as Peirce conceived of how our beliefs about reality depend on an unlimited community of inquiry (Apel 1998b). In Apel's transcendental-pragmatic version, the very idea of a discourse ethics is rooted in some general facts about the praxis of argumentative discourse. More specifically, a transcendental-pragmatic analysis of argumentation (a self-reflexive kind of presupposition-analysis[7]) reveals that there are certain normative proprieties in the practice of argumentation, which anyone who wants all competent participants to act as rational evaluators should want every competent participant to recognize as ideally regulating their discursive commitments.[8]

For Apel, the very idea of a discourse ethics springs from the realization that among the proprieties that should ideally govern a rational evaluator's community of argumentation, there are some proprieties with a *morally* significant content for all rational evaluators. The understanding of such content is arguably part of how anyone should understand his or her role as a subject of argumentation, if they are competent and willing to let "the unforced force of the better argument" (Habermas) take the lead, or to "seriously raise questions and provide answers in argumentation" (Apel). For Apel, discourse ethics is in its central sense a thin morality *in* argumentation – a morality that is ingrained, so to speak, in the informal public system of argumentation, and is as ubiquitous and practically important for us as argumentation. Discourse ethics in the wider sense of a "program of philosophical justification" (Habermas 1993) in moral theory has the tasks of identifying and justifying whatever morally significant content can be derived specifically from or arrived at via the essential non-disclaimable normative elements in the self-understanding of participants in argumentative discourse. Moreover, it has to probe the relevance of such content for questions of moral theory. For Habermas these tasks amount to a "theoretical justification of the moral point of view" (Habermas 1996). In Apel's normatively more ambitious approach they amount to a vindication of a thin but strongly universalistic rational ethos (Apel 1989, 1996).

Presuming there is *moral* (Apel) or at least *morally relevant* (Habermas) content in argumentation amongst rational evaluators of reasons for validity-claims – a content that is (1) always already recognized since it is *implicit* in argumentation, and (2) irrefutable[9] since it is implicit in *argumentation* – what is it? Not surprisingly, such content concerns how anyone with whom we would use, or who would use with us (and think that it is rational to use), argumentation as the sole arbiter of the validity-claims that we associate with our reason-backed judgments, should treat anyone else who could so use argumentation. All involved in such argumentation rationally should want certain proprieties of mutual recognition and symmetrical situatedness to be the norm for everyone actually involved and anyone possibly involved, which would ideally regulate the discursive commitments.

Apel's explanation of the respective (moral) proprieties can be summarized in terms of a moral co-responsibility between actual and possible participants to keep all action in discourse in accordance with a generic deontic status of free and equal co-subjects.[10] Habermas explains the respective (morally relevant) proprieties in terms of rules as follows: "that nobody who could make a relevant contribution may be excluded"; "that all participants are afforded an equal opportunity to make contributions"; "that the participants must mean what they say: only truthful utterances are admissible"; "that communication must be freed from external and internal compulsion so that the *yes/no* stances that participants adopt on criticisable validity-claims are motivated solely by the rational force of the better reasons" (Habermas 1996, 1990: 89).

3.   REALISTIC DISCOURSE ETHICS: RESPONSIBILITY FOR DISCURSIVE POWER

I maintain that Apel's and Habermas' promissory explanations can be improved on if one has a viable concept of what makes a reason for judging rationally controllable behaviour a *moral* reason for a *moral* judgment (cf. section 4). Discourse Ethics, I maintain, is best conceived of as a set of moral responsibilities that conceptually belong to the deployment of a certain form of power, namely discursive power. It is then possible to derive (Kettner 1999) what I call the parameters of idealization that shape a discourse specifically into a "moral discourse": argumentative consensus building driven by discrepant moral judgments (cf. section 7).

In order to see the merits of this view it is necessary to eliminate a deep-seated philosophical prejudice, namely the received view that reason and power are totally antithetical. Contrary to this misguided contemplative idealism, I prefer to take seriously the metaphor of reason as a power of sorts. Argumentation, I maintain, is not only an exercise of rational "capacities" but also of power. We can call this "discursive power". To make the concept of discursive power more palatable, we have to distinguish current interpretations of power that are too immediately causal. Admittedly, the *causality interpretation* of power and its various modes is by no means arbitrary. It follows from the quite sound principle that "if someone has power then it must imply something for others. [...] Yet it is clearly possible to conceive of power as a capacity of a power holder to make a difference without causing any power subject to do anything; in the decisional [and likewise in the discursive] interpretation of power what matters [...] is less the impact that a power holder has upon the behaviour of some power subject, but [...] the capacity to make a difference in decision making, the outcomes of which may or may not cause another actor to behave in certain ways" (Lane and Stenlund 1984: 395). Discursive power, then, is the power to modify via argumentation, to change or to keep from changing the beliefs people have about what is right and wrong in their employment of the authority of good reasons in their actions. Discursive power operates on our interpretations of reasons as better or worse; and in the measure in which our ways of life give importance to how we match our actions to our interpretations of reasons as better or worse, discursive power operates albeit indirectly on our actual conduct. Discourse Ethics, then, is the intrinsic morality of the use of discursive power in the governance of moral or non-moral norms.

In order to clarify the sense in which some intrinsic elements of the normative texture of argumentative discursive practices amount to a minimal *morality* it is helpful first to consider the general notion of a morality.

4.   MORALITY, MORALITIES

Morality in the most general sense of the term can be characterised as a social practice of governing what we do and want to do by a concern for certain aspects of the well-being of certain sorts of beings. In all commonly recognised moralities the set of beings with moral standing includes but need not only include people. Usually, the "bearers" or "subjects" of moral beliefs will be members of the same community of moral concern, one's moral peers. Different moralities draw the boundaries of their respective communities of moral concern in different ways.[11]

The definition common among sociologists, of any particular morality as a particular "normative system" that regulates action by classifying it for "praise and blame", is overly general. For there are many different action-guiding normative systems (law, religion, prudence, mores, etiquette), none of which is identical with, though each is somehow related to morality as we know it (Castaneda 1974).

Moralities are complex socio-evolutionary constructions. A morality is a more or less integrated yet open-textured (Brennan 1977) web of knowledge, emotional dispositions ("moral feelings"), action skills ("virtues"), motivational propensities ("altruism"), and interpretative resources ("moral thinking") (Gibbard 1990). This web provides us with moral reasons (Dancy 1993; Copp 1995; Gert 1998; Kettner 2003), moral judgments and moral norms.

Moral reasons are reasons why certain things may, ought, or ought not to be done, in a sense of those deontic modalities in which failing to do what one ought to do is *morally wrong*. Moral reasons in turn support our practice of moral judgments. Moral judgments are judgments on the moral rightness, wrongness, or praiseworthiness of claims concerning what morally may, ought, or ought not to be done. That something may morally or may not morally be done is usually expressed in the format of a moral norm. A moral norm presents a certain mode of action in certain circumstances as morally required of certain agents. There is, of course, a wide variety of norms, only some of which are moral norms. Legal rules, rules of games, traffic regulations and so on, are all norms. Moral norms, as opposed to other sorts of norms, are norms that are intersubjectively recognised as "categorical" (Gewirth 1978), to the extent that people who are unwilling or unable to respect them in their deliberations about what they do, minimally as negative constraints, maximally as positive ideals, will either be seen to act immorally ("morally wrong") or will not be counted as moral agents by other moral agents.

To clarify the sense in which moral reasons, judgments and norms differ from non-moral reasons, judgments, and norms, it is helpful to consider the notion of *moral responsibility*. Moral agents are persons who bear, and perceive their moral peers to bear moral responsibility (Ladd 1982; Frankfurt 1983; French 1985; Kettner 2001). An agent's moral responsibility is a responsibility that is neither exhausted by that agent's *causal* role in the outcome of actions, nor by the agents' liability that is relevant to a *juridical* assessment of their actions. Rather, the bearing of moral responsibility consists specifically in collectively taking seriously how the outcome of one's conduct – of possible actions or omissions – affects oneself or relevant others for their good or ill. Different moralities, of course, assign different contents to the structure of such responsibility: different ways of acting ("conduct"), different reference groups ("others"), different significant values ("good or ill"). This diversity of interpretations implies moral pluralism.

We have to distinguish moral relativism from moral pluralism. Diversity and pluralism do not at all imply relativism. Like natural languages, moralities are a pervasive feature of human culture. Yet whereas cross-translatability between any two natural languages seems to be feasible in principle without any limitation, the fact of moral diversity and value pluralism has been taken to support the view that it makes sense to speak of justified moral claims or commitments only with reference to particular cultures: cultural moral relativism. However, strong moral relativism (and incommensurability of values across different cultures or across subcultures) is a self-refuting theoretical view, as is the view that understanding across different value-horizons is impossible. I have discussed the difference between relativism and relativity elsewhere (Kettner 2000). Here, I will simply say that the worldwide recognition of those deontic reasons and morally significant values, which we abbreviate as *human rights*, in fact contradicts any strong relativism and incommmensurabilism. At a sufficiently abstract level, where moral deontic reasons take the form of "moral principles", there is considerable overlapping consensus across diverse cultures, despite disagreement over their more refined interpretation and ranking of which real social practices ought to be governed by which principles.[12] A certain range of values (for example freedom, the provision of basic needs, integrity of primary affective bonds, sanity) bear moral significance virtually everywhere (Gert 1990, 1998). Yet their determinate interpretations in terms of moral action requirements or moral norms may vary dramatically across different cultures.

The *fact of moral diversity*, even at the elevated level of principles, should not obscure the fact that the common object of any morality is the facilitation and enhancement of human flourishing (Mackie 1977; Apel 1976). This remains true even in borderline cases where such flourishing is interpreted as a concern for the integrity of inanimate components of nature, as in cosmocentric as opposed to anthropocentric ethical outlooks (for example in "deep ecology").

The fact of moral diversity needs to be sensibly accommodated in any "rational" morality or normative moral theory with universalistic aspirations. To understand why, we should consider that universalistic moral claims make a demand on everyone who properly takes them into account. They purport to command the assent of whoever is a morally responsible rational agent. Yet those who make such claims are always members of a particular community in space, time, and culture. Hence they run the risk of imposing the claims of what they take to be their universalistic morality on others whose moral views, if only they were allowed to express themselves, could be seen to differ from or even defy the claims that they find imposed on them. The normative universalism that is part and parcel of our established notions of rational validity means that any markedly "rational" morality will also aspire to universalism. Yet a sensible universalism in morality, it seems, must not impose rigid moral principles on an unruly moral world of heterogeneous moral views. If it does, it is buying uniformity at the cost of dogmatism or paternalism. Both dogmatism and paternalism, in imposing alienating moral views, create avoidable "moral costs". They are therefore wrong according to the standards of a truly rational morality. To the extent that an allegedly rational morality or moral theory is insensitive to its own impact, or lacks the conceptual resources for the moral assessment of such impact in its application, it is seriously inadequate for the modern pluralist condition.

5.   MORAL HORIZONS, DISCOURSE ETHICS, AND OTHER MORAL PARADIGMS

A moral point of view or, as I would prefer to term it, horizon, is the evaluative stance of someone who identifies with a certain morality M (for example as a result of "having been brought up" in the spirit of M, finding M convincing, etc.) when assessing practical activities and the reasons by which people justify doing what they do. A moral horizon discloses the "moral costs" that accrue to people's transactions, according to the particular value standards that are acknowledged in M. By considering things in a moral horizon one assesses impacts of and reasons for actions, with an eye to minimising whatever moral costs they are seen to have. I am here reformulating the classical moral principle that is expressed in Latin as *primum non nocere*.

A moral horizon, being a *moral* point of view, discloses what is right or wrong as judged by its avoidable moral costs, and what thus ought to be done or not done on moral grounds. Being a *point* of view, no moral horizon is all encompassing. Being a point of *view*, any moral horizon, to the extent that it can be called rational, must admit controversy and consensus, questions and answers, argument and counter-argument with regard to whatever reasons come under rational scrutiny from this point of view.

Naturally, there are many different evaluative perspectives, only some of which are moral horizons. We can assess actions and the reasons for which they are performed from prudential, technical, legal, moral, economic, technical, religious and, no doubt, many other points of view. For instance, a new medical technique may be ingenious and highly recommendable on prudential grounds from a techno-economic point of view because it saves resources, and yet be disreputable and wrong on grounds of its moral costs from a moral point of view because it subverts patient autonomy. A new law might be legitimate from a juridical point of view because it has passed parliament and yet be illegitimate when morally considered because it violates human rights. A new financial instrument, for example derivatives, for all the sense it may make economically, may be found wanting in fairness (morally) when the patterned distribution of risks and benefits engendered by that instrument is assessed.

For the reasons I give in the next section, I assume that Discourse Ethics and its moral horizon (the horizon of taking seriously how the consequences of argumentative uses of discursive power affect peoples' capacities for governing all normative textures that they perceive as having consequences for good or ill) takes centre-stage among moral horizons. That is not to say that Discourse Ethics replaces extant moral horizons. Rather, Discourse Ethics is central in its capacity to synthesize and gauge *more or less different other* moral horizons.

Briefly, four paradigms of normative moral theory and their respective central principles are at present in the foreground of philosophical ethics.[13] These paradigms are the following: (1) Kantian deontologism, with its principle that persons ought to be respected as ends in themselves (O'Neill 1989); (2) utilitarianism, with its principle that utility ought to be impartially maximised (Sen and Williams 1982; Hare 1981); (3) contractualism, with its principle that explicit or tacit agreements for mutual benefits ought to be honoured (Gauthier 1990); and (4) consensualism, with its principle that all normative arrangements ought to be procedurally governed through free and open argumentative dialogue ("discourse"), ideally among all concerned. Of these, consensualism, as developed into a "communicative" or "Discourse Ethics", is the most integrative and flexible position.

Taking up what was said in section I about the transcendental-pragmatic background of Discourse Ethics, the main point can be reformulated as follows: the stimulus behind the philosophical development of Discourse Ethics is the intuition that the reasons on which people claim that something is morally right must be such as to be conceivably acceptable from the first-person-plural perspective ("we") of everyone concerned by the practice, activity or regulation the moral rightness of which is at stake. Moral rightness, then, is a property of action-norms, which is ultimately dependent on the cooperative discursive practice of free and open dialogue between rational evaluators about discordant appreciations of allegedly good reasons.

This is not to say that all moral content is held to be generated in dialogue[14] or that we should devote all our moral life to argumentation. Instead, the discourse principle (the reasons for which people claim that N is morally right must be conceivably acceptable from the first-person-plural perspective of everyone concerned by N) is a problem-driven principle. Its critical force is invoked only when particular issues cannot be satisfactorily handled by the conventional resources that the people concerned take for granted. Hence it operates on subjects that are always pre-interpreted by whatever moral intuitions the participants happen to have. Discourse is the medium to modify and reshape them. In moral discourse, people work through their various moral perplexities in a cooperative effort to reach a maximally value-respecting practical deliberation that everyone can support. This need not totally coincide with what each claimant would judge as the right way to go given only their own moral horizon and supposing that other moral horizons were not part of the problem at hand. In fact, it may deviate considerably from "the" right exclusively within one's own moral horizon. However, there is the possibility of an integrity-preserving and genuinely moral compromise (Benjamin 1990).

A second point should also be mentioned. Consensus building that is constrained by Discourse Ethics allows us to emulate central principles of other moralities. For instance, if all people whose needs and interests are affected by some practice, p, were to agree in a practical discourse to regulating p by, for example, utilitarian standards then the discursively prompted consensus about the morally right way of regulating p will result in the regulation of p in that manner. Yet whatever substantial moral principle people would like to adopt (for example a utilitarian principle of maximising the average satisfaction of individuals' preferences) will become constrained in Discourse Ethics by respect for people's capacity to reach a common understanding about how they want to treat and be treated by others, regardless of egocentric positional differences.

6.   DISPUTING NORMS, VALUES AND FACTS

There are no moral problems independent of people who are morally perplexed when they examine their practices from a point of view that they embrace as moral. Just as substantial interpretations of moral responsibility differ, what is a moral problem to one person is not always a moral problem to another, though both parties view things in moral horizons respectively.[15] We find moral problems when people are unsure whether a course of action is right or wrong. Hence if we want to understand moral problems, we must find out the rationale behind people's perplexity about what is right or wrong.

Dispute about what people take as proper responses to their moral perplexity, to the extent that it is rational, is governed by a logic of discourse. This logic of discourse revolves around our powers of raising and answering questions about: (1) fact (senses in which something to believe can be the case); (2) value (senses in which something to appreciate can be good); (3) norms (senses in which something to do can be required of someone).

Questions of fact and their associated truth-claims can be disputed with reference to the availability and credibility of the evidence for establishing what is the case. Questions of value, and their associated claims of evaluative commitment can be disputed with reference to the appropriateness and importance of the properties by virtue of which something is held to be valuable in an arguable sense of good. Whether the purported good-making or value-giving properties are really present is governed by questions of fact. Questions of norms can be disputed with reference to the values a norm is held to serve or express. Whether the values by virtue of which it is claimed that certain agents ought to do certain things really authorize the norm in an arguable sense of requiredness, is governed by value questions and by factual questions.

Two people in disagreement about what one ought to do must consider whatever other norms they subscribe to and link them with the norm in question. Norms face the tribunal of discourse and experience corporately: commitment to a component normative texture may turn out to mean, on pain of inconsistency, subscription to (or refusal to accept) another component normative texture (Will 1993). Furthermore, people turn to what they take as the relevant values that bear on the norms, on what one ought to do. And two people in disagreement about the sense in which they have reason to take something to be good must be prepared to scrutinise as many other of their evaluations as are found to be somehow related to the one in question. Values, like norms, face the tribunal of discourse and experience corporately, hence a person cherishing a certain value may find him- or herself committed, on pain of inconsistency, to another value. Furthermore, people discuss what they view as the relevant facts and their relations on which they think depends the sense in which they suppose something in question to be good. The unfolding dialogical dynamics of relating factual, evaluative and normative questions, if need be with many repetitions, generates rational inquiry in the perplexity-driven discourse processes.

Using technical terminology, we can sum up this section by saying that normative differences supervene on evaluative differences, which in turn supervene on factual differences. *Supervenience* here is a conceptual relation, such that if properties of kind x supervene on properties of kind y then there can be no difference in x without some relevant difference in y.

7.   FIVE PARAMETERS OF MORAL DISCOURSE

When, in a target domain, a reflective mode of governance like argumentation can be brought to bear on normative change (as applied ethics presumes it can), the corresponding processes of argumentation represent moral discourse if they embody and express a set of parameters that jointly guarantee the moral integrity of the discursive power exercised by the respective community of argumentation.

I will list five normative parameters that are necessary elements in the idea of a moral discourse. Space does not permit me to elaborate on their formulation and vindication. Suffice it to say that each parameter can be introduced as a well-grounded partial answer to a general question. The general question can be framed thus: 'Are there recognizable proprieties such that, if they were not mutually required among co-subjects of argumentation, argumentation in the face of conflicting reasons specifically representing moral responsibility for them would not make sense?'[16]

Parameter 1: Reasonable Articulation of Need-Claims
All participants in a discourse should be capable of articulating any need-claim they take to be morally significant.

Parameter 2: Bracketing of Power Differentials
Differences in all forms of power that exist between participants, both within and outside of argumentation, should not be any participant's good reason in discourse for endorsing any moral judgment.

Parameter 3: Non-Strategic Transparency
All participants should be able to convey their articulations of morally significant need-claims truthfully and without strategic reservations.

Parameter 4: Fusion of Moral Horizons
All participants should be able to sufficiently understand articulated need-claims in the corresponding moral horizons of the participants who articulate them.

Parameter 5: Comprehensive Inclusion
Participants should constrain what their community of discourse can accept as good reasons by the following requirement: that participants must anticipate whether their reasons can be rehearsed by all non-participant others who figure specifically in the content of any moral judgment that results consensually from the participants' discourse.

Note that discourse ethical consensus building is not equivalent to a unanimity requirement, majority vote, or any preference-aggregative decision procedure (for example bargaining). The dynamic of consensus building in moral discourse does not guarantee a unique "solution" to all moral issues. Staking out a *range* of permission is often the best we can achieve. No morality is an algorithm for solving problem cases. To some extent, morality must countenance tragic choices and persistent tensions. Such choices and tensions at best allow alleviation not total resolution, and considerable "moral costs" are bound to remain. However, a consensus that is sufficiently reflective of the parameters of moral discourse, guarantees to all parties who mutually recognize one another as having a credible stake in the outcome of the discourse, that they are mutually aware of all their different "moral costs", and that they are also mutually aware of the right-making reasons from every participant's moral horizon. Realistically, no rational morality can guarantee anything stronger than that. The possibility of reasonable disagreement (dissent) exists alongside the possibility of reasonable agreement (consensus), notwithstanding the conceptual truth that the latter envelops the former.

On this basis, a morally-discursively prompted consensus may well integrate justified dissent. Depending on whether such dissent expresses mutual and omni-laterally justified concessions, a morally-discursively prompted consensus may express a moral compromise. In such a compromise, however, no-one's morally significant need-claims will have been compromised intolerably.

8.   SUMMING UP

Discourse Ethics is a two-tiered normative moral theory. On the first tier, completely general yet morally significant norms of argumentation are identified. This yields a minimal morality ("ethics in discourse") whose claims range over and whose grounds can be ascertained by all subjects of argumentation. On the second tier, moralities are meta-ethically characterised as variations of a common basic structure of moral responsibility. Moral reasons represent how moral communities fill out this basic structure with determinate content. A set of five parameters can be derived by tracing normative requirements that are arguably necessary for argumentation about moral reasons to have a rational point in the face of moral perplexity and moral pluralism. These parameters together define the notion of a moral discourse as a normative ideal type. Moral discourses are reflective modes of governance. If governed by moral discourse, normative textures in transition would not deteriorate and might even progress in their moral qualifications. Moral discourse as specified by Discourse Ethics is a medium in which our moral convictions can face the tribunal of our diverse experience and divergent moral horizons without losing out to cultural relativism and historicism.

The point of the rational ethos of a discourse ethics, as I perceive it, is a concern for the moral integrity in the exercise of discursive power in communities of actual argumentation. Discursive power is the power to modify convictions of rightness concerning reasons regarded as good through argumentation. In moral decision making, for example in the reflective forum of a clinical ethics committee, moral judgements and the reasons that give confidence to those abiding by them, are usually modified. Even in seemingly individual or person-to-person settings, such as reproductive decision making or genetic counselling, institutional and organizational constraints filter into the set of reasons that are typically available for framing particular moral judgements in such contexts. Therefore it is very important that the discursive power mobilized in such contexts is itself morally governed in order for the resulting decision making to aspire to moral integrity.

The discourse ethics approach has already proved helpful with regard to moral perplexities in reproductive medicine, the New Genetics and the human genome project, the reform of "nondirective" counselling for pregnancy conflicts, the discursive design of clinical ethics committees, and a communicative modelling of the doctor-patient-relationship.

References

Apel, K.-O. "The Problem of Philosophical Foundations in Light of a Transcendental Pragmatics of Language." In *After Philosophy. End or Transformation?* K. Baynes, J. Bohman and T. McCarthy (eds.). Cambridge: MIT Press, 1987, 250–290.

Apel, K.-O. *Diskurs und Verantwortung: Das Problem des Übergangs zur postkonventionellen Moral.* Frankfurt: Suhrkamp, 1988.

Apel, K.-O. "Universal Principles and Particular (Incommensurable?) Decisions and Forms of Life – a Problem of Ethics that is both post-Kantian and post-Wittgensteinian." In *Value & Understanding. Essays for Peter Winch.* R. Gaita (ed.). London: Routledge, 1990a, 72–101.

Apel, K.-O. "The Problem of a Universalistic Macroethics of Co-responsibility." In *What Right Does Ethics Have? Public Philosophy in a Pluralistic Culture.* S. Griffioen (ed.). Amsterdam: VU University Press, 1990b, 23–40.

Apel, K.-O. *Transformation der Philosophie.* Vol. 1, Sprachanalytik, Semiotik, Hermeneutik. Vol. 2, Das Apriori der Kommunikationsgemeinschaft, Frankfurt: Suhrkamp Verlag, 1976. (*Toward a Transformation of Philosophy*, London: Routledge, 1980, reprinted by Marquette University Press 1998.)

Apel, K.-O. "Is the Ethics of the Ideal Communication Community a Utopia?" In *The Communicative Ethics Controversy.* S. Benhabib and F. Dallmayr (eds.). Cambridge: MIT Press, 1990, 23–60.

Apel, K.-O. "Do we need universalistic ethics today or is this just eurocentric power ideology?" *Universitas* 1993, 2, 79–86.

Apel, K.-O. *Discussion et responsabilité. L'Étique après Kant.* Paris: Les éditions du Cerf, 1996.

Apel, K.-O. *Auseinandersetzungen in Erprobung des transzendentalpragmatischen Ansatzes.* Frankfurt: Suhrkamp Verlag, 1998.

Apel, K.-O. "Auflösung der Diskursethik? Zur Architektonik der Diskursdifferenzierung in Habermas' *Faktizität und Geltung.*" In *Auseinandersetzungen. Erprobung des transzendentalpragmatischen Ansatzes.* Frankfurt: Suhrkamp, 1998a, 727–838.

Apel, K.-O. "Transcendental semiotics and truth: the relevance of a Peircean consensus theory of truth in the present debate about truth theories." In *Form a transcendental-semiotic point of view.* Manchester: Manchester Univ. Press, 1998b, 64–80.

Apel, K.-O. "First Things First. Der Begriff primordialer Mit-Verantwortung. Zur Begründung einer planetaren Makroethik." In *Angewandte Ethik als Politikum.* M. Kettner (ed.). Frankfurt: Suhrkamp, 2000.

Baynes, K. *The Normative Grounds of Social Criticism. Kant, Rawls, and Habermas.* Albany: SUNY Press, 1992.

Benhabib, S. *Critique, Norm, and Utopia.* New York: Columbia Univ. Press, 1986.

Benjamin, M. *Splitting the Difference.* Kansas: Kansas University Press, 1990.

Brennan, J.M. *The Open-Texture of Moral Concepts.* London: MacMillan, 1977.

Castaneda, H.-N. *The Structure of Morality.* Springfield, Ill.: Charles Thomas, 1974.

Colby, A. and Kohlberg, L. *The Measurement of Moral Judgment.* Vol.1 & 2, Cambridge: Cambridge University Press, 1987.

Cooke, Maeve. *Language and Reason. A Study of Habermas's Pragmatics.* Cambridge, Mass: MIT Press, 1994.

Copp, D. *Morality, Normativity, and Society.* Oxford: Oxford University Press, 1995.

Dancy, J. *Moral Reasons.* Oxford: Blackwell, 1993.

Dewey, J. *Human Nature and Conduct. An Introduction to Social Psychology.* New York: Random House, 1922.

Foucault, M. *Die Ordnung der Dinge. Eine Archäologie der Humanwissenschaften.* Frankfurt: Suhrkamp, 1980.

Frankfurt, H.G. "What we are morally responsible for." In *How many questions?* L.S. Cauman (ed.). Indianapolis: Hackett, 1983.

French, P. "Fishing the red herrings out of the sea of moral responsibility". In *Actions and Events.* E. LePore (ed.). Oxford: Basil Blackwell, 1985.

Gauthier, D. *Moral Dealing. Contract, Ethics, and Reason.* Ithaca: Cornell University Press, 1990.

Gert, B. "Rationality, Human Nature and Lists." *Ethics* 1990, 100, 279–300.

Gert, B. *Morality. Its Nature and Justification.* Oxford: Oxford University Press, 1998.

Gewirth, A. *Reason and Morality.* Chicago: Chicago University Press, 1978.

Gibbard, A. *Wise Choices, Apt Feelings: A Theory of Normative Judgment.* Cambridge: Harvard University Press, 1990.

Habermas, J. "Discourse Ethics: Notes on a Program of Philosophical Justification." In *Moral Consciousness and Communicative Action.* Cambridge, Mass: MIT Press, 1990, 43–115.

Habermas, J. *Justification and Application: Remarks on Discourse Ethics.* Cambridge: MIT Press, 1993.

Habermas, J. "Eine genealogische Betrachtung zum kognitiven Gehalt der Moral." In *Die Einbeziehung des Anderen. Studien zur politischen Theorie.* Frankfurt: Suhrkamp, 1996, 11–64.

Habermas, J. *Between Facts and Norms. Contributions to a Discourse Theory of Law and Democracy.* Cambridge, Mass.: MIT Press, 1998.

Hare, R. *Moral Thinking. Its Levels, Methods and Point.* Oxford: Clarendon Press, 1981.

Harman, G. *The Nature of Morality. An Introduction to Ethics.* Oxford: Oxford University Press, 1977.

Kettner, M. "Nuclear Power, Discourse Ethics, and Consensus Formation in the Public Domain." In *Applied Ethics. A Reader,* E.R. Winkler and J.R. Coombs (eds.). London: Blackwell, 1993a, 28–45.

Kettner, M. "Ansatz zu einer Taxonomie performativer Selbstwidersprüche." In *Transzendentalpragmatik.* A. Dorschel, M. Kettner, W. Kuhlmann and M. Niquet (eds.). Frankfurt: Suhrkamp, 1993b, 187–211.

Kettner, M. "Reasons in a World of Practices. A Reconstruction of Frederick L. Will's Theory of Normative Governance." In *Pragmatism, Reason, and Norms.* K. Westphal (ed.). Chicago: University of Illinois Press, 1998, 255–296.

Kettner, M. "Neue Perspektiven der Diskursethik." In *Ethik technischen Handelns. Praktische Relevanz und Legitimation.* Armin Grunwald and Stephan Saupe (eds.). Heidelberg: Springer, 1999, 153–196.

Kettner, M. "Kulturrelativismus oder Kulturrelativität?" *Dialektik. Zeitschrift für Kulturphilosophie*, 2000, 2, 17–38.

Kettner, M. "Moralische Verantwortung als Grundbegriff der Ethik." In *Diskursethik – Grundlegungen und Anwendungen*. Marcel Niquet, Francisco Javier Herrero and Michael Hanke (eds.). Würzburg: Königshausen & Neumann, 2001, 65–94.

Kettner, M. "Gert's Moral Theory and Discourse Ethics." In *Rationality, Rules, and Ideals. Critical Essays on Bernard Gert's Moral Theory*. Walter Sinnott-Armstrong and Robert Audi (eds.). Lanham: Rowman & Littlefield, 2003, 31–50.

Ladd, J. "Philosophical remarks on professional responsibility in organizations." *Int. J. of applied Philosophy* 1982, vol.1, no.2.

Lane, J.-E. and Stenlund, H. "Power." In *Social Science Concepts. A Systematic Analysis*. G. Sartori (ed.). Beverly Hills: SAGE, 1984, 315–402.

MacIntyre, A. *Whose Justice? Which Rationality?* Indiana: University of Notre Dame Press, 1988.

Mackie, J.L. *Ethics. Inventing Right and Wrong*. London: Penguin, 1977.

O'Neill, O. *Constructions of Reason. Explorations of Kant's Practical Philosophy*. Cambridge: Cambridge University Press, 1989.

Outka, G. and Reeders, J.P. (eds.). *Prospects for a Common Morality*. Princeton: Princeton University Press, 1993.

Rehg, W. *Insight and Solidarity. The Discourse Ethics of Jürgen Habermas*. Berkeley: University of California Press, 1994.

Sen, A. and Williams, B. (eds.). *Utilitarianism and Beyond*. Cambridge: Cambridge University Press, 1982.

Snare, F.E. "The Diversity of Morals." *Mind* 1980, 89, 353–369.

Wallace, J.D. *Ethical norms and particular cases*. Ithaca: Cornell University Press, 1996.

Walzer, M. *Thick and Thin. Moral Argument at Home and Abroad*. Notre Dame: Notre Dame University Press, 1994.

Westphal, K. (ed.). *Frederick L. Will's Pragmatic Realism*. Chicago: University of Illinois Press, 1998.

Will, F. *Beyond Deductivism. Ampliative Aspects of Philosophical Reflection*. London: Routledge, 1988.

Will, F.L. "The Philosophic Governance of Norms." *Jahrbuch für Recht und Ethik* 1993, 1, 329–361.

Wren, T.E. (ed.). *The Moral Domain. Essays in the Ongoing Discussion between Philosophy and the Social Sciences*. Cambridge: MIT Press, 1990.

---

[1] Discourse ethics, though usually associated with Habermas 1990 and 1993 in the American philosophical context (cf. Benhabib 1986, Baynes 1992, Cooke 1994, Rehg 1994), has its *locus classicus* (dating back to a lecture in 1967) in Apel's essay "The A Priori of the Communication Community and the Foundations of Ethics" (Apel 1976, 1980).

[2] Note that where Apel is referring to 'claims' here he is not only referring to what Habermas later called universal validity-claims of speech acts (intelligibility, truth, rightness, truthfulness), but also to human needs generally considered as expressions that make certain demands on (other) human beings as potential fulfillers of these demands.

[3] One important disagreement between Apel and Habermas concerns the additional resources that each of them thinks are necessary and sufficient to move from 2 to 3: a philosophical theory of transcendental arguments (Apel), or a

sociological theory of modernisation (Habermas). Elsewhere I have shown that this disagreement boils down to whether a clear sense can be given to calling some of the constitutive norms of argumentative discourse 'moral norms', and that Apel is right (against Habermas) in that there is such a sense (Kettner 1999).

[4]   Habermas 1998: 107. His reasons for revising D are complex and cannot be described here. Partly, they have to do with the problem of distinguishing different types of discourse for different domains of validity-claims. For a critical discussion of Habermas' distinctions between "theoretical", "pragmatic", "ethical", "moral" discourses, see Kettner 1995a. Habermas also wants to avoid the charge of building into his theory of normative validity the moral content that the theory purports to reveal, a charge to which his concept of communicative action (Habermas 1985) is already susceptible. This is how Habermas views the theoretical status of the revised D: It: " ... expresses the meaning of postconventional requirements of justification. Like the postconventional level of justification itself – the level at which substantial ethical life dissolves into its elements – this principle certainly has a normative content inasmuch as it explicates the meaning of impartiality in practical judgments. However, despite its normative content, it lies at a level of abstraction that is *still neutral* with respect to morality and law, for it refers to action norms in general. (...) The predicate "valid" (*gültig*) pertains to action norms and all the general normative propositions that express the meaning of such norms; it expresses normative validity in a nonspecific sense that is still indifferent to the distinction between morality and legitimacy. (...) '[R]ational discourse' should include *any* attempt to reach an understanding over problematic validity-claims insofar as this takes place under conditions of communication that enable the free processing of topics and contributions, information and reasons in the public space constituted by illocutionary obligations" (Habermas 1998: 107f.). In my view, the proposed revision creates more problems than it solves (Kettner 1999c; cf. also Apel's criticism of Habermas, in Apel 1992a, 1998a), not least among which is that Habermas' recent theory relativizes the concept of a rational morality to modern societies and the psychologically rare cognitive achievement of a "postconventional" level of moral consciousness. Another problem is that Habermas uses the notion of impartiality as if it were co-terminus with moral impartiality. These problems are already present in Habermas 1996.

[5]   For illuminating discussions about how Habermas' "formal pragmatics" ties in with his theory of communicative action and with his discourse ethics, see Baynes 1992: 88–115; Cooke 1994: 1–29 and 150–162; Rehg 1994: 23–36. Briefly, *communicative action* covers both social action in which, from the perspective of the agent(s), the aim is to reach or maintain a common understanding (=*Einverständnis*, consensus) about validity-claims, and any social action, *a*, linguistic or other, which is a consensually validated way of acting and in which it is more important for the agent(s) to keep in the track of this consensual validation than to attain any further goals they may want to attain by *a* that are not in keeping with the consensus that validates *a*. "Strategic action" is social action in which it is exclusively important for the agent(s) to attain whatever goals they want to attain by doing the action. This typological distinction within a theory of social action is not without difficulties (cf. Baynes 1992: 80).

[6]   Habermas 1996. Habermas refers to this question as "the principle of universalization 'U'", cf. Habermas 1990: 65. Typically, U is portrayed as "a rule

for the impartial testing of norms for their moral worthiness" (Rehg 1994: 38). Understandably, Habermas has been criticised for reducing moral worthiness to (a somewhat contractual concept of) justice. For a response, see Habermas 1996. Apel accepts U heuristically but points out (cf. Apel 1998a: esp. 789–793) that it is a mistake to equate U (or any further determinant of moral worthiness in argumentative discourse) with the rational ethos of a discourse ethics as Habermas does, since U would be unobjectionable only in a world in which it would be normatively natural for everyone to govern all controversial moral judgments discursively. In the actual world, however, the rational ethos of a discourse ethics confronts, and must be supplemented in order to cope with, problems of application in all contexts where it is not, or not as yet, normatively natural to govern all controversial moral judgments discursively. For an early statement of Apel's position concerning application cf. Apel 1980: 282; for Habermas critique of Apel, see section 10 in Habermas 1993.

[7] A transcendental-pragmatic analysis of a practice P, or of a feature of P, is an analysis of P's significance for, or role in enabling us to fix jointly beliefs about the validity of reasons by using the informal public system of argumentation. More precisely, it is an analysis of how P contributes to what we find is necessary for everyone to acknowledge if anyone is to be permitted to give, take, or reject reasons specifically for claims which we intend to be universally acceptable (acceptable from anyone's point of view), provided the claims and reasons are assessed by rational evaluators.

[8] An example of a normative propriety of the kind described is the norm that logical contradiction be avoided. This norm, like all other norms of logical well-formedness of propositional contents, is not the kind of norm whose violation would normally count as a violation of a commitment of moral responsibility; it is not a "moral" norm.

[9] For Apel's claim that the recognition of moral constraints in argumentation cannot be denied, see the introduction of his notion of a "performative self-contradiction" as a predicament that rational evaluators would basically want to avoid in argumentation. We attribute a performative self-contradiction to someone, S, if S intends to claim validity for a suitable content, c, but c is such that if it is valid (in the sense intended by S) then we cannot attribute to S the intention to claim validity for c. For Apel, a validity-claim for c is rationally definitely grounded if arguing for its negation involves one in a performative self-contradiction, and if justifying the validity-claim by representing c as the conclusion of a deductive argument involves one in the logical fallacy of begging the question (cf. Apel 1987).

[10] Cf. Apel 1990a, 1990b, 1992b: esp. 265f., 2000.

[11] Whereas "particularist" moralities draw the boundary narrowly, it is also possible to extend the community of one's moral peers in a more universal manner so as to encompass all human beings indifferent of cultural differences and spatial or temporal distances between them. Such universalism, far from being a lofty idealism, is a built-in opportunity for moral-cognitive development. This development is patterned into a number of stages in an invariant sequential order. There is ample cross-cultural evidence for this in cognitive developmental psychology (Colby & Kohlberg 1987; Wren 1990).

[12] Outka & Reeders 1993. For a less sanguine view, see Snare 1989.

[13] For the first three, see Harman (1977).

[14] For this misunderstanding see esp. Walzer 1994: 7–15.

[15] E.g., an atheist will have no moral qualms about sacrilege, because for him or her the concept of the holy (on which the characterization of a certain transaction as sacrilege depends) will be an altogether empty concept. A Roman Catholic woman's belief that one ought not engage in sexual activity unless the two values of possible procreation and marital affective solidarity are jointly served will bear heavily on her moral evaluation of the impact of contraception practices, as these ply apart ("de-naturalise") procreation and the pursuit of sexual happiness, two endeavours whose natural nexus is a morally significant fact when considered from the Catholic faith perspective.

[16] I discuss the notion of a moral discourse more fully in Kettner (1999).

Chapter 21

THE CONCEPT OF CARE ETHICS IN BIOMEDICINE
The Case of Disability

EVA FEDER KITTAY
Stony Brook NY, USA

My aim in this paper is to offer an oblique approach to the question of biomedicine and the limits of human existence by discussing the role of a care-based ethic in contemporary discussions of disability. Contemporary discussions of disability have resisted the notion that disability is essentially a matter of biology and medicine – that biomedicine has any exclusive right to define, or even to redress, the adverse living conditions that physiological impairments can impose on individuals. In this paper I endorse this critique, but at the same time want to urge caution in a concomitant rejection, which is also found in the disability literature, of the conception of care. Care addresses the limits and limitations of human existence, and disability is a condition in which humans at once encounter and challenge those limits. In this respect, disability shares with many issues of biomedicine questions of vulnerability and dependency. An ethics based on care offers distinct resources for discussions of biomedicine, but I will confine my remarks to exploring the importance of these for disability.

1. DISABILITY, INDEPENDENCE, DEPENDENCE, AND CARE

Care and disability are topics very close to me, both professionally and personally, as I have spent much of my philosophical career developing a care ethics and as I have been a caregiver for most of my adult years, a parent of a young woman with disabilities. Sesha is always part of my discussions on disability, both because it is through her that I have encountered questions surrounding disability, and so I feel it is important to

C. Rehmann-Sutter et al. (eds.), Bioethics in Cultural Contexts, 319–339.

situate my own position in these discussions, and because as she is a member of a group of disabled persons who cannot speak for themselves, I feel compelled to speak on her behalf. I speak then – not for her, because that is first impossible and second presumptuous – but from the lessons I have learned through her and those who have helped me care for her. My daughter is a sparkling young woman, with a beautiful face and an even lovelier disposition. She is very significantly incapacitated, incapable of uttering speech, of reading or writing, of walking without assistance, or in fact doing anything for herself without assistance. She possesses a condition which is clumsily identified as mental retardation or developmental delay. I say clumsily because it is not clear that her problem is really the backwardness or slowness suggested by retardation or the delayed development that suggests that the development will come only a bit later. Although her cognitive functioning is limited, she loves music, water, good food, people, attention, love and life. (And so one might say that there is nothing amiss is her taste for the best life has to offer!) She is fully dependent and while, at the age of 34 she (like us all) is still capable of growth and development, it is quite certain that her total dependence will not alter much. She has lived at home with us till the age of 32, and now she has moved to a home with five other multiply and developmentally disabled young adults in a community in a rural setting. There is a way in which this move may be seen as isolating her, but in fact when we made the move, we discovered that she actually was more isolated while living at home with us "in the community." But the care and level of activity she receives in her new home is exceptional, and I fear not the norm for such communities in the US and in most parts of the world. Many are, in fact, isolated in institutions in which they are supposed to receive care. My daughter's disabilities always threaten to isolate her. And it is only with care, much of it and of the highest quality that she can be included, loved, and allowed to live a full and rich life. When I speak of disability, I think a great deal about the cognitive disability that marks her life and my concern is that persons who have these sorts of disabilities, as well as those who are involved in their care not be left out of not be left out of discussions of justice and moral personhood.

1.1   Dependence, Deviance, and Disability

Disability and care have a long and uncomfortable relation with one another. The same may be said for disability and dependence. While for some a physical dependence on caregivers enable them to carry on the activities of daily living, for many there is an economic dependence created by an inability to earn an adequate income given prejudice, discrimination and lack of access to public spaces. There is also a dependence on social services, sometimes blamed as creating needs and thus sustaining the very dependence that these services were intended to relieve, a critique reminiscent of denunciations of welfare provisions more broadly conceived. When it is taken up by disability scholars and activists, the claim is that dependency is "created amongst disabled people, not because of the effects of the functional limitations on their capacities for self-care, but because their lives are shaped by a variety of economic, political and social forces which produces this dependency" (Oliver 1989: 17).

In a book entitled, *'Cabbage Syndrome': The Social Construction of Dependence*, the author writes:

> The relationship between disability and deviance can be understood with reference to the freedom from social obligations and responsibility, explicit in the sick role model … in the negative views of illness, disease and impairment that continue to hold sway throughout modern industrial capitalist societies. Because such societies are founded upon an ideology of personal responsibility, competition, and paid employment, any positive associations with sickness, such as the exemptions outlined above must be discouraged … (Barnes 1990: 6)

As these two passages suggest, disability, particularly as it is cast as a "personal tragedy" is a concept that links dependency and deviance.

The default assumption is that a disabled person is a dependent person. Similarly, the disabled person is identified as deviant – deviant from a norm of typical species functioning (or form), which negatively effects self-sufficiency and social integration. The two presumptions come together, particularly within the Western industrialized nations, for the deviance that is perhaps especially salient is the deviation from one particular norm, that of independence, and hence is a deviance that renders one dependent. In a world in which independence is normative, the person with impairments comes to be isolated through a stigma which is linked to dependence and the need for care.

It is no accident then that the challenge disabled people in the US in the late 1960's and early 1970's mustered against their deviant status was entitled the Movement for Independent Living. This movement, created by people who were young, intellectually capable, white and largely male, did not interrogate the norm of independence, but affirmed it for a group that had previously been excluded. Their aims were inscribed in the important US antidiscrimination legislation, Americans with Disabilities Act (enacted in July 1990). That act states: "the Nation's proper goals regarding individuals with disabilities are to assure equality of opportunity, full participation, independent living, and economic self-sufficiency for such individuals" (*ADA:* (a), (8)). It is noteworthy that it goes on to say "the continuing existence of unfair and unnecessary discrimination and prejudice ... costs the United States billions of dollars in unnecessary expenses resulting from dependency and nonproductivity" (*ADA*: (a), (9)).

1.2   Care instrumentalized or repudiated

I do not think it is unfair to say that group of disabled individuals who so successfully lobbied for the ADA legislation viewed the provision of care in an essentially instrumental manner – much as most of us view our dependence on farmers. But the social dependence on farmers is taken by most to be relatively innocuous. In the case of care, the dependence has seemed for many with disabilities less benign. Provision that imposes itself on the individual and intrudes into his or her life may be identified with the oppressive forces that have the power to turn the disabled person into a suppliant.

Thus, it is not just the state and other institutions responsible for the lack of public access, the persistence of discrimination, the prevalence of need-based services, the labeling of persons as deviant and the exclusion of disabled people in decision-making that are excoriated for their part in the "creation of dependency." Professional providers of services and care-givers share the blame, as in the following passage from the British Council of Organisations of Disabled People: "... [T]he need to be 'looked after' may well adequately describe the way potentially physically disabled candidates for 'community care' are perceived by people who are not disabled ... which has led to large numbers of us becoming passive recipients of a wide range of professional and other interventions. But, however good passivity and the creation of dependency may be for the careers of service providers, it is bad news for disabled people and the public purse (BCODP 1987: 3.2, cited in Oliver 1989: 13).

Yet coming to the question of disability from the position (or "role") of a resolute carer of a disabled person, my daughter, I am invested in the idea that care is indispensable, and even central, to a good life for people with certain sorts of disabilities. (The claim is stated as it is to make it clear that I acknowledge that not all people with disabilities require care different – in manner or extent – from that of those not characterized as disabled and that due our human dependence, we each have required and are likely at some future time to require extensive care in order to survive and thrive. More of this in my concluding remarks.) Given that people with disabilities are attempting to cast off the perception of the disabled individual as hapless, in need of "looking after", and are working to retrieve independence in the face of practices and persons who reinforce and heighten the sense of dependence, how is care to be regarded in the face of those limitations exacerbated by impairment.

2. BASIC CONCEPTS OF A CARE ETHIC

When I speak about an ethics of care I am speaking primarily of the conception of ethics that has been developed by feminists wanting to render visible and valuable activities that women have traditionally been charged with, namely the care and nurture of children, the ill, those with impairments who require assistance, and the frail elderly. A number of analytic philosophers who do not necessarily align themselves with feminist philosophy have also, of late, taken up work in an ethics of care. There has been a parallel development among some Continental philosophers, beginning with Emmanuel Levinas, who focus on care, although their work has been less influenced by feminist work. Most of my remarks will be limited to the feminist scholarship.

A care ethics as a feminist ethics challenges the univocity of male voices in ethical inquiry. The starting point for much feminist ethics of care, Carol Gilligan's empirically based claim that the abortion debate, structured as a conflict of rights – the rights of the fetus v. the rights of the women – fails to reflect decision-making of women who are faced with an unexpected pregnancy. Rather than ask if the fetus was a rights-bearing person, the women in Gilligan's study asked questions such as: Is it responsible to give birth at this time of my life? Am I prepared to take care of a child? How will giving birth to a baby now affect my relationship to my lover/my spouse/ my parents/my children? How will my own vision of my possibilities be affected and can I be true to the person that others and I expect me to be? What harm will I do if I carry this pregnancy to conception, or if I abort this conception?

Rather than ask about rights, these women asked about responsibilities. Rather than frame the dilemma as a conflict between oneself and the unborn, they tended to think in terms of their relationships to a future child, current children, a spouse or lover, and other family. Rather than frame their concerns as matters of right, they were concerned about their ability to give care.

The term care can denote a labor, an attitude, or a virtue. As labor, it is the work of maintaining others and ourselves when we are in a condition of need. It is most noticed in its absence, most appreciated when it can be least reciprocated. As an attitude, caring denotes a positive, affective bond and investment in another's well-being. That labor can be done without the appropriate attitude. Yet without the attitude of care, the open responsiveness to another that is so essential to understanding what another requires is not possible. That is, the labor unaccompanied by the attitude of care cannot be good care (see Kittay 1999).

Care, as a virtue, is a disposition manifested in caring behavior (the labor and attitude) in which "a shift takes place from the interest in our life situation to the situation of the other, the one in need of care" (Gastmans, Dierckx de Casterlé and Schotsmans 1998: 53). Relations of affection facilitate care, but the disposition can be directed at strangers as well as intimates.

2.1   A comparison of care- and justice-based ethics

The characteristics of care orientation to ethics are frequently expounded by offering a contrast to some more traditional justice-based approaches, especially Kantian-deontological and (to a lesser extent) utilitarian/consequentialist theories. (One might also wish to contrast a care ethics with a virtue-based ethics, although on some accounts, a care ethics is a variety of virtue ethics, see Slote 2001.)

Comparison of Care and Justice:

	Justice	*Care*
1. Moral Agent	Independent, autonomous self, equal or potentially so	Relational, dependent self, unequal in age, capacities, and/or powers
2. Moral Relations	Rights, relations of equality	Responsibilities, relations of trust
3. Deliberative Process	Principled, reason-based calculations, formal contexts	Contextual, narrative, emotion taken seriously
4. Scope of Decisions	Impartiality required, universal applicability	Partiality respected, applicability context-dependent
5. Moral Aim	Protect against conflict, adjudicate competing claims	Maintain connection, avoid violence
6. Moral Harm	Harm when clash between persons	Harm when connections are broken

It will be helpful to explore the contrast with reference to six questions to which a care-based and justice-based ethics offer different answers:

1. Who is the moral agent and what is the nature of moral agency?
2. What is the nature of moral relations?

3. What skills and processes are involved in moral deliberation and action?
4. What is the scope of moral decisions?
5. What are the aims of moral relations?
6. What constitutes moral harm?

Table 1 offers a rough summary of the responses. The following sections amplify these brief replies.

(1) The nature of the moral agency
Standard theories of justice begin with the autonomous individual moral agent who pursues his (sic!) desires and the fulfillment of needs in the context of a social situation in which there are other moral agents who do likewise. On a care ethics, each self is inextricably related to other selves. Their relationships play a constitutive role in the formation of their desires and in their identity (Tab. 1). Furthermore the self in theories of justice is a self-determining adult who is an independent agent. A care ethics does not presume that all agents in a situation demanding moral action are adults who are capable of self-determination and independence. Instead, the fact of our dependence on one another is seen as a part of our inevitable dependency and connectedness with one another. In my own elaboration of an ethic of care, I stress the nested dependencies in which we all find one another, as well as the interdependencies in which we all are engaged.

Tab. 1: Moral Agency

Justice	Care
Autonomous self	Self a self-in-relation (transparent self)
Self as bounded	Self as vulnerable
Independence, self determination valued	Dependence accepted as connection
Presumption of mature adult agents	Different ages capacities, abilities

(2) The nature of moral relations
Within traditional justice-based theories, moral relations are presumed to be among self-determining independent persons, equally situated and empowered, with whom they form associations that are voluntarily insofar as they are either chosen or affirmed. Moral interactions are bound by contractual, law-like sets of obligations or duties and a concomitant set of rights. We are bound to respect the rights of others as we expect others to

respect those we possess. The binding nature of these interactions is fixed in the form of a contract or understanding that I can only hope to have others respect my rights if I respect theirs in turn.

A care-based ethic does not presume that our ethical relations are self-chosen ones among equals, but maintains that we find ourselves in certain relationships to others, some of whom are better situated and have greater powers, others of which are with those who lack our capacities, who are not as well-situated as we are, and over whom we may have power. In engaging with others morally in situations of care, we need to assume a self that is transparent to the needs of another, rather than a self in which our actions are essentially self-directed as autonomy would generally require (See section 3 below). Moral relations are imbued with trust and trustworthiness, and willy nilly, we are beholden to the responsiveness and responsibility of those with greater power or capacity.

Tab. 2: Nature of Moral Relations

Justice	Care
Equality of relations	Inequality of relations
Between generalized others	Between concrete particular others
Emphasis on rights	Stress on responsibilities
Noninterference	Responsiveness
Bound by (voluntary) contractual relations	Bound by trust and dependence
Reciprocity	Asymmetrical giving and receiving

(3) Moral deliberation

How we do or should deliberate about the actions we take as moral agents is also quite different given the two ethical orientations. Within most theories of justice, deliberation is based on our ability to use an algorithm or procedure that will ensure a morally correct solution regardless of who deliberates and about whom one deliberates. These theories utilize a hierarchy of values defined by a set of principles. The categorical imperative and the attempt to maximize marginal utility are standard examples of such methods of moral deliberation. The idea is that given truthful and accurate inputs, a sound procedure carried out according to sound principles, we can each reason so as to come up with *the correct* solution to a moral problem. We need only rely on our capacity to reason, and depending on the theory, a sense of duty, a sense of justice, or a prudential understanding of our own good. We are not called upon to be empathetic or sympathetic, to be kindly inclined, to have a significant range of emotional capacities, or even emotional responses to others within the moral sphere. These are either viewed as morally irrelevant or a hindrance to morally responsible judgement or action.

Coming to a moral decision or judgment within a care-based ethics is less guided by principle than on a justice based account, although principles can still play a role. Instead, a sensitivity to context, to the needs and capacities of those involved are often set within a narrative account that replaces or supplements context independent principles and hierarchical values. The requisite moral skills include a capacity to be responsive to need and an understanding of the specificity of the good for the affected persons. A recognition of the ways in which inequality of power or situation can turn from benevolence to abuse, and an ability to emotionally connect to another and their welfare are equally valued moral skills. Elsewhere I have spoken of the need of a carer to be transparent to the needs of the one in need of care. (Kittay 1999). By this I mean that one needs to be able to bracket one's own needs and wants and not to have these cloud one's perceptions of the needs of the one who is dependent on the carer. This transparency of self is a possibility of a self that views itself as relational.

Tab. 3: Deliberative Process

Justice	*Care*
Role of reason elevated	Value of emotion and shaped inclinations
Autonomous decision making	Transparent self-apparent heteronomy
Principles emphasized	Emphasis on contextual reasoning
Calculation of moral rights and wrong based on hierarchy of values	Narrative, specificity of context, culture, historical factors

(4) What is the scope of morality?

It is generally noted that justice-based theories tend to be applicable to settings governing relations between strangers, or acquaintances in non-intimate settings. Moral judgments, on this account, need to be universal in scope. Care is thought best reserved for private life and more intimate, less rule-governed contexts. Within a justice tradition, proximity of those affected to the moral agent is irrelevant.

A care-based ethics remains sensitive to proximity, whether it be the relational proximity of family or friends or the geographical contiguity of neighbors and fellow citizens. What may be morally appropriate in dealing with strangers is not necessarily deemed morally appropriate in dealing with those close at hand, and what may be justifiable in the case of a neighbor may not justifiable in the case of a child or parent. Valid moral decisions may well be partial, not impartial. It is deemed morally justifiable, and sometimes morally required to care more about those close at hand than those with whom we are more distant.

Tab. 4: Scope and focus of moral deliberations

Justice	*Care*
Formal contexts	Informal contexts
Public (in dispute)	Private (in dispute)
Universality	Context specific
Impartiality valued	Partiality accepted as deemed appropriate

(5) What are the aims of moral relations?

Perhaps the most radical and most relevant differences are directed to telos or point and purpose of moral relations. A justice-based ethics stresses the importance for people to be able to live their lives according to their own lights, free of unnecessary inference from others. Ethics is about the limits of that pursuit, insofar as others have the equal right to pursue their desires. The role of ethics is to avoid conflicts that arise from each self attempting to pursue their own desires.

These points are often encapsulated in the idea that the individual has certain claims upon others in the form of *rights*, rights which also protect one from the unwarranted interference of others. Each person is thought to have the full measure of rights that are compatible with others having the same full measure of rights. (That, at least, has been the aim of theories based on a *liberal* conception of justice.)

A care-based ethics stresses, first of all, the concern for the well-being of a person, and some, such as Stephen Darwall would add, the well-being of an individual *for their own sake*. That is, when we care for another we are concerned with the well-being of that person as it serves that person's welfare for the purposes of that person's flourishing, not for the sake of the larger community or some abstract conception of goodness. It is this concern with the other's well-being for his own sake that places *responsibilities* on us for the other's care. The other's care is, however, not external to our own well-being for these affiliative relationships by which we care for another for their own sake are themselves constitutive of who we understand ourselves to be, for the self of a care-based ethic is a "self-in-relationship." (See also MacIntyre 1997; Gilligan 1982).

Tab. 5: Aim of Moral Relations

Justice	*Care*
Individuals live according to own lights, noninterference in rational life plans	Foster well-being (flourishing) of person for persons sake
Protect rights	Respond to need
Protect against and adjudicate conflict among persons	Foster, preserve connections, serve "progress of affiliative relationship" (Gilligan 1982: 170)

The aim of a care ethics becomes then the maintenance of relationship fostered through attention to and concern for the other's well-being. Self-sacrifice is often viewed as the ideal of a caring self, but as Gilligan importantly points out, when the self-sacrifice is complete, there is no self left and so there is no longer the possibility of relationship. The sacrifice of self may be spoken of as a "temptation of care" (as Sarah Ruddick 1989 puts it), rather than a virtue of care. This brings us at last to the final question.

(6) What constitutes moral harm?

Within a justice-based morality, moral harm is identified as the violation of rights or unwarranted intrusion in the form of paternalism, domination, or violence. It is to be treated unequally, discriminated against in employment, educational opportunities, political and social life, etc. for morally irrelevant reasons, and treated as an inferior with regard to the distribution of the benefits and burdens of social cooperation. Moral harm is seen as resulting from the clash of rights or interests among individuals.

In contrast, in a care-based ethic, moral harm results when important needs, especially of vulnerable persons are unmet, when our concerns elicit only indifference, when vulnerability arouses distain and abuse rather than care, and when human connections are broken through exploitation, domination, hurt, neglect, detachment or abandonment.

Tab. 6: Nature of Moral Harm

Justice	*Care*
When persons are interfered with unnecessarily, rights violated	When vulnerability and need are met with indifference, detachment, or abuse
Unequal treatment, discrimination	When persons are abandoned
When there is a clash of rights and interests among individuals	When connections between persons are broken

2.2 Temptations of virtues of care and virtues of justice

While one can argue that care and justice are both virtues that can serve as the basis of a moral theory, both can be seen as subject to certain temptations and limitations. Temptations are failures of a particular ethical stance, not merely the violation of that ethical ideal. A temptation of justice for example is a failure to be merciful. But to lack mercy is not yet to be unjust. It is to aim for justice but fall short in a manner characteristic of such an aim.

In the case of justice these temptations would also include being overly rule-bound, placing undue reliance on impersonal principles or institutions. An argument that justice fails us in an over-reliance on impersonal institutions is made by Alasdair MacIntyre, when he argues for the importance of the virtues with regard to a need for a standard of care. While legal enforcement of a standard of care is necessary, MacIntyre writes, "the networks of giving and receiving in which we participate can only be sustained by a shared recognition of each other's needs and a shared allegiance to a standard of care … [without which] laws will often be observed from fear of the consequences of doing otherwise, sometimes grudgingly and always in a way that has regard to the letter rather than to the spirit of the laws" (1997: 84–85). Interestingly, this suggests the need for the practice of care, even to enable the proper functioning of justice.

Temptations that undermine the practice of care include the sacrifice of self of which we spoke earlier. The contrary temptation, one that people with disabilities who need care are especially wary of, is the potential for the carer to lose sight of the separateness of the person for whom she cares. The danger is that she will impose her own conception of the good, or alternatively an abstract notion of what is good for the other without sufficient attention to the subjectivity of the cared for. Here the separation of the self that justice-based ethics underscores is important to make caring work well.

2.3 Limitations of an ethic of care with special attention to disability

Having outlined features of an ethic of care, I want to redirect our attention to relevance of a care ethic for disability.

On the face of it, this ethic has a number of serious limitations for people with disabilities. First, as feminists have commented, if women doing the traditional work of caring do in fact exhibit an ethics based on care, should we not say that this is labor that women have been constrained to perform and so an ethics based on it is one borne of subjection. Is it not, as Nietzsche would have it, a "slave morality"? Is this the morality that is usefully adopted by a group of people who are struggling to emerge from a subordinate status?

Moreover, a care-based ethic, as we have seen, has been thought most suitable to informal and private domains. Applying it to the situation of disability would appear to favor the more individual, medical model of disability that is out of favor. Even if it is useful to people with disabilities in the informal, private contexts, why suppose that it can address the structural problems that a social model of disability highlights.

And finally, care, has been taken to be too closely tied to the very image of dependency that disabled people have in large measure tried to shed. Dependency implies power inequalities and a care-based ethics appears to embrace rather than challenge these inequalities.

3. WHAT A CARE-BASED ETHIC HAS TO OFFER DISABILITY THEORY: LIMITATIONS TURNED INTO ADVANTAGES

Although these limitations appear serious, I think they can be answered, and sometimes the limitations may be turned into advantages.

3.1 Slave morality?

To the charge that a care ethics is a "slave morality" we can reply that an ethic that springs out of practices arising from a subjugated position reveals that the subordinated do have a voice, and that it is one that needs to be heard because it can inject new values into a society that does not treat some of its people well. To aspire to the values of an ethics as practiced by the dominant group may be to collude with the very values that subordinate some persons. For example, in talking about the idea of independence for physically disabled people who require aides to assist them with daily tasks, people with disabilities can inadvertently fall into morally questionable habits that mimic those of privileged groups who have taken for granted caring work, relegating it to unpaid or the worst paid labor.

Wanting to show how problematic the linkage of care, dependence and deviance is, Mike Oliver writes, "professionals tend to define independence in terms of self-care activities such as washing, dressing, toileting, cooking, and eating without assistance" (Oliver 1989: 14). Yet, he points out, "Disabled people, ... define independence differently, seeing it as the ability to be in control of and make decisions about one's life, rather than doing things alone or without help". I am suggesting that we still need to ask: "What about those who do the washing, dressing, toileting?"

Judy Heumann, one of the founders of the Independent Living Movement, wrote influentially: "To us, independence does not mean doing things physically alone. It means being able to make independent decisions. It is a mind process not contingent upon a normal body" (Heumann 1977).

This suggests that care, if it can be dissociated from the stigma of dependence, is not only compatible with independence of the sort that Heumann alludes to, but is in fact indispensable to it. But at the same time, we also need to consider that at least conceptually, if not strategically, de-stigmatizing dependency, or rendering it a value-neutral feature of the human situation and utilizing the resources of a care ethics will serve both the disability community and the larger community better than an emphasis on independence. For "independence" as the aim of a movement to include disabled people as full citizens of the human community, and with it the justice-based morality which here has been contrasted with a care-based one, only perpetuates the pernicious effects of the fiction that we can be independent. I suggest that the exploitative nature of care labor is likely to be exacerbated when viewed in the highly instrumental manner indicated by insisting that independence has to do with control and decision-making and nothing to do with needing assistance in carrying out daily tasks. (Also see my discussion of Olmstead v. L. C. and E. W. in Kittay 2000). The stress on independence makes it appear as if it would be preferable to have an aide replaced by a machine. Concomitantly, the person providing care comes to seen as a pure instrument to the achievement of the independence of the disabled person. The fact that there is any relationship of dependency to another person appears as regrettable, insignificant, if inevitable fact. Annette Baier, addressing the absence of the concerns of domestics and care workers within a theory focused on rights, speaks of these persons as "the moral proletariat" (Baier 1995: 55). Where is the independence and control of those who are mere instruments of another's independence and control? What are we to presume of relationships between the person who gives care and the disabled person in need of that care? Elsewhere (Kittay 2001b) I argue, when caregivers are devalued, treated instrumentally, they in turn are more susceptible to devaluing those for whom they give care, particularly but not only, in the case of those with developmental and mental impairments. Is it not better to acknowledge one's dependency on an other, and to examine ways in which there can be a mutually respectful relationship, based on a genuinely caring and respectful attitude. Is it not better to insure that relationships of dependency be replete with the requisite affective bonds, ones which can transform otherwise unpleasant intimate tasks into times of trust, and demonstrations of trustworthiness, gratifying and dignifying to both the caregiver and the recipient of care. Is it not preferable to understand relationships of care to be genuine relationships involving labor that is due just compensation and recognition. Here care and justice support rather than oppose one another.

Moreover, if by appealing to the nature of moral relationships envisioned within a care ethics we conceive of all persons as moving in and out of various relationships of dependence, through different life-stages and different conditions of health and functionings, the person with an impairment who requires the assistance of a caregiver is not the exception, the special case, but a person occupying what is surely a moment in each of our lives, and also a possibility that is inherent in being human, that is, the possibility of inevitable dependence. We see that we need to structure our societies so that such inevitable dependence is met with the care, resources and dignity required for a flourishing life. We again recognize that we need social arrangements enabling those who provide care to be similarly provided with the care, resources and dignity they require for their own flourishing and for the possibility of doing the work of caring well.

Finally, if we see ourselves as always selves-in-relation, we understand that our own sense of well-being is tied to the adequate care and well-being of another. Caregiving work is the realization of this conception of self, both when we give care generously and when we graciously receive the care we require.

3.2   Inequalities?

Critics of care ethics have often pointed out that the paradigm used has often been the mother and child relationship. Clearly this is not the sort of relationship that one wants to model adult relationships of disabled people and care providers. We can at once grant this but insist that an ethics that acknowledges inequalities in situation and power are important if we are to avoid turning these inequalities into occasions for domination and abuse on the one hand, and paternalism on the other. Baier addressing the limitations of a rights approach to morality speaks of the sham in the "'promotion' of the weaker so that an appearance of virtual equality is achieved ... children are treated as adults-to-be, the ill and dying are treated as continuers of their earlier more potent selves ... " She remarks, "This pretence of an equality that is in fact absent may often lead to a desirable protection of the weaker or more dependent. But it somewhat masks the question of what our moral relationships *are* to those who are our superiors or our inferiors in power" (1995: 55). She goes on to suggest that a morality that invokes this pretense of equality and independence, if not supplemented, may well "unfit people to be anything other than what its justifying theories suppose them to be, ones who have no interest in each others' interests" (1995). That is, it may leave us without adequate moral resources to deal with genuine inequalities of power and situation that we face daily, and which not infrequently are conditions that certain impairments (apart from social arrangements) impose on us.

To deal with the inequalities that emerge out of the needs that are a consequence of certain impairments we require an ethic that can guide relationships between different sorts of care providers (family members, hands-on care assistants, medical personal) and people with different sort of care needs. The urgencies of need, whether they arise from medical emergencies, a breakdown in equipment needed for functioning, disabling conditions not addressable by accommodation, are ones that render disabled persons, (*and* not infrequently those who care for a disabled person whose welfare is part of the carer's own sense of wellbeing) vulnerable. This is of course true of each of us, whether or not we are disabled. For instance, we generally come to medical professionals at a vulnerable moment. While paternalism is an inappropriate response on the part of professionals insofar as we may well be able to make or participate in important decisions about our lives, we are likely to require responsiveness to our need and to the particularity of our situation. It is precisely situations such as these that call for an ethic of care and responsibility (on the part of those with greater power and capacity toward those with less), rather than an ethic based on the reciprocity of rights of two equally empowered moral actors.

In raising the issue of vulnerability, it is worthwhile to point to a moral problem within the sphere of genetics and reproduction. We need to be aware of the vulnerability of prospective parents, as well as patients in the face of the presumed expertise of the physician operating in the arena of biomedicine. Issues such as genetic testing, selective abortion for disability, surgical interventions of young children involve physicians, counselors, and bioethicists whose expertise render their relations to parents unequal at a time when parents are exceptionally vulnerable. We need to ask if a care-ethics, particularly its distinctive forms of ethical inquiry, for example the attentiveness to context and narrative (rather than hierarchical principles) can be helpful in providing better guidance – and better health care – for the disabled person and her family. I contend that the paradigm of justice-based ethics, the contractual model between equals is of less value in these situations.

A final point bears on distributive issues. In a model where equal parties participate in a fair system of social cooperation, the ruling conceptions are reciprocity, a level playing field, and fair equality of opportunity. On the assumption that all are equally situated and empowered, a conception of negative rights goes a long way to permit individuals to realize their own good. But differences in powers and situation require a more positive conception of rights and responsibilities toward those less well-situated or powered. Positive provisions are critical if people with disabilities are to be able to flourish – whether these are ramps, Braille in public areas or wheelchairs, help making one's home accessible, the service of home-care attendants, or a safe, enriching, stimulating environment in a protective setting. An ethic of care, if and to the extent that it can be made serviceable in the public domain becomes a stronger justification for positive rights insofar as care is seen as carrying out responsibilities we have for another's flourishing, not only the protection against undue interference or the mere assurance of equal opportunity. Ensuring equal opportunity to people is admirable when people are in a position to take advantage of the opportunities on offer, but many with significant disabilities are not in a position to take advantage of such opportunities even when accommodations are made. For persons with severe mental retardations, such as my daughter, Sesha, no accommodations can make her self-supporting regardless of antidiscrimination laws and every equal opportunity that may be legally available to her. Mental retardation poses a special challenge to the justice approaches that have predominated disability discourse. But even for those who are impaired in ways that are less disabling in our society, positive provision of attendants, equipment, appropriate housing and nonpublic sources of transportation require an attitude of care and concern that either is not well-captured in legal structures that enshrine principles of justice or must, as MacIntyre suggests, undergird formal systems in order for them to function properly.

3.3   Taking care ethics public

The above point concerning distributive justice should direct our attention to the claim that a care ethics is best suited for the private sphere of intimate relations and is not appropriate in the larger realm of public policy. This charge, if it is in fact a criticism, has been addressed by a number of authors. However, Joan Tronto (1989), for instance, argues that a care ethics, suitably developed, is the appropriate one to justify and guide welfare policy; Sarah Ruddick utilizes the ethical basis of "maternal thinking" to develop a peace politics (2001); Michael Slote (1987) defends the use of a care ethics to cover the ground usually reserved for justice, including ethical behaviour to those who are in different parts of the globe; and Virginia Held invites us to imagine what a society that governed social policy on a care paradigm might actually look like. These are only a few of the more prominent examples of efforts to show that an ethics of care need not be confined to intimate relationships. I have argued for a public ethic of care in which care and justice are both transformed in the accommodations a just society must make to be caring and caring relations must make to be just (Kittay 1999, 2001a, 2001b). Critical to my conception is the idea that we are all embedded in nested dependencies, and that a justly caring society must be one in which care of dependents is seen as central to the point and purpose of social organization. It is the obligation and responsibility of the larger society to enable and support relations of dependency work that takes place in the more intimate settings. (Gilligan 1987: 31) A society that makes adequate provisions for a flourishing life for people with disabilities will be one in which the fundamentals of a care ethic, such as our interrelationships and inextricable connectedness, our vulnerability and dependencies, our requirement of responsiveness to and responsibility for one another are recognized and valued along with our needs for respect and self-determination. These values will be reflected in public policies and in institutions, and there is nothing in these values and conceptions that inherently restrict them to the private sphere, even if that is where they are most apparent.

3.4   "The virtues of acknowledged dependence"

Rather than see the emphasis on dependence and connection as limitations, I have suggested that we see the emphasis of these in a care ethics as resources. Carol Gilligan talks about the ways in which a conception of relationship from a perspective of care and a perspective of justice may overshadow one another, citing two definitions of dependency offered by high-school girls she studied. One arises, "from the opposition between dependence and independence, and the other from the opposition of dependence to isolation" (1987: 31–32). She develops the opposition: "As the word 'dependence' connotes the experience of relationship, this shift in the implied opposite of dependence indicates how the valence of relationship changes when connection with others is experienced as an impediment to autonomy or independence, and when it is experienced as a source of comfort and pleasure, and a protection against isolation" (1989: 14). We began the discussion with the question of the relation between care, dependence, and disability and in the definitions offered by these high school girls we see that where an ideology of independence is dominant, the positive experience of connectedness we can experience through dependence is eclipsed. To the extent that disability discourse aligns itself with that of independence, the understanding of dependence as a contrast to isolation is hard to fathom.

Acknowledging the inevitable dependency of certain forms of disability and setting them in the context of inevitable dependencies of all sorts, is another way to reintegrate disability into the species norm, for it is part of our species typicality to be vulnerable to disability, to have periods of dependency, and to be responsible to care for dependent individuals. We as a species are nearly unique in the extent to which we attend to the dependencies not only of our extended immaturity, but also of illness, impairment and frail old age. I propose highlighting the commonalities between different conditions of "inevitable dependencies", so that we can recognize that dependency is an aspect of what it is to be the sorts of beings we are. In this recognition, I hope we can begin, as a society, to end our fear and loathing of dependency.

4.   CONCLUSION: REVISITING THE IDEAL OF INDEPENDENCE

When we see our dependency and our vulnerability to dependency as species' typical, we can recall that sense of dependence that is a respite from isolation. This is not an easy insight, but one articulated in a recent interview given by the comedian Richard Pryor (Gross 2000). The interviewer asked Pryor to speak about the Multiple Sclerosis that has incapacitated many of his bodily functions and will accelerate his death. Pryor said that as he lost old capacities, he had to learn new ones. Indeed, he maintained that the Multiple Sclerosis was "the best thing that had ever happened to me"; that his disease has been the occasion for the most important lessons he has had to learn about himself. He said that when, in order to walk from one end of a room to the other, a person *must* depend on another, he learns how to trust.

Learning to trust when he was vulnerable was the most valuable lesson he learned. This is a knowing that can alter us profoundly, especially when independence is touted as the hallmark of personhood.

As persons, in fact, spend a considerable portion of their lives either as dependents, caring for dependents or in relations where they have responsibility for dependents, the trust that Pryor had to learn when he became disabled – and the need for trustworthiness that warrants such trust – ought to be a feature of all our lives. The fiction of independence, and a fiction it is regardless of our abilities or disabilities, will not help us acquire the necessary moral skills – and may even as Baier suggested "unfit us" for the task.

Dependence may, in various ways be socially constructed, and unjust and oppressive institutions and practices create many sorts of dependence. But if dependency is constructed, independence is still more constructed. We cannot turn away from that fact and sufficiently rid ourselves of prejudices against disability, and certainly not for those whose disability cannot be disentangled from the need for care. We currently make resources needed by disabled people available on the supposition that such social "investments" will be cost effective, for these newly "independent" disabled will now be productive. Recall that the last finding that prefaces the ADA reads: "[T]he continuing existence of unfair and unnecessary discrimination and prejudice ... costs the United States billions of dollars in unnecessary expenses resulting from dependency and nonproductivity" (*ADA*: (a), (9)). The commitment to the flourishing and maintenance of connection is absent in these provisions.

To mask inevitable dependency and valorize only a particular segment of human possibility strengthens the hand of those who refuse our collective responsibility to take care of one another and helps perpetuate the isolation of those with disabilities.

Among the many precious gifts I have received from my daughter Sesha has been to learn, as Alasdair MacIntyre puts it, "the virtues of acknowledged dependency" and the extraordinary possibilities inherent in relationships of care with one who reciprocates but not in the same coin, one who cannot be independent, but repays with her joy and her love.

References

ADA. *Americans with Disabilities Act of 1990.* 101st Congress, *PUBLIC LAW* 1990, 101–336, 933.

Baier, Annette C. "The Need for More Than Justice." In *Justice and Care.* V. Held. (ed.). Boulder, CO: Westview Press, 1995, 47–58.

Barnes, Colin. *The Cabbage Syndrome.* London, New York, Philadelphia: The Falmer Press, 1990.

BCODP, British Council of Organisations of Disabled People. *Comment on the Report of the Audit Commission*. London: BCODP, 1987.

Gilligan, Carol. *In a Different Voice*. Cambridge, Mass: Harvard Unversity Press, 1982.

Gilligan, Carol. "Moral Orientation and Moral Development." In *Women and Moral Theory*. E. F. Kittay and D. T. Meyers (eds.). Lanham: Rowman and Littlefield, 1987, 19–33

Gross, Terry. "Interview of Richard Pryor." *Fresh Air*, 2000.

Held, Virginia. "Non-Contractual Society: A Feminist View." *Canadian Journal of Philosophy* Supplementary 1987, Volume 13, 111–135.

Heumann, Judy. Berkeley, CA, August 1977.
http://www.disabilityexchange.org/newsletter/article.php?n=15&a=134
(last visit: June 05).

Kittay, Eva Feder. *Love's Labor: Essays on Women, Equality and Dependency*. New York: Routledge, 1999.

Kittay, Eva Feder. "At Home with My Daughter." In *Americans with Disabilities*. F. Leslie Pickering and S. Anita (eds.). New York, NY: Routledge, 2000, 64–80.

Kittay, Eva Feder. "A Feminist Public Ethic of Care Meets the New Communitarian Family Policy". *Ethics* 2001a, 111, 3, 523–547.

Kittay, Eva Feder. "When Care Is Just and Justice Is Caring: The Case of the Care for the Mentally Retarded." *Public Culture* 2001b, 13, 3, Special Issue: The Critical Limits of Embodiment: Reflections on Disability Criticism, 557–579.

MacIntyre, Alasdair. *Dependent Rational Animals: The Virtue of Dependence and the Virtue of Acknowledging Dependence*. Berkeley CA, 1997.

Oliver, Mike. "Disability and Dependency: A Creation of Industrial Societies." In *Disability and Dependency*. L. Barton (ed.). London, New York, Philadelphia: The Falmer Press, 1989, 6–22.

Ruddick, Sara. *Maternal Thinking*. New York: Beacon Press, 1989.

Slote, Michael. *Morals from Motives*. Oxford, New York: Oxford University Press, 2001.

Tronto, Joan. *Moral Boundaries: A Political Argument for an Ethic of Care*. New York: Routledge, 1993.

Chapter 22

THE THICK SOCIAL MATRIX FOR BIOETHICS
Anthropological Approaches

RAYNA RAPP
New York, USA

1.   INTRODUCTION

Anthropologists working in the fields of health, illness, and medicine share a focus on norms, values and practice with scholars and practitioners of bioethics. Yet there is also a substantial anthropological mistrust of the field. This mistrust is based on what anthropologists often take to be the ahistoricity and ethnocentricity of much bioethical discourse. By "ahistoricity", I intend to signal two concerns: first, the foundational premises of classical bioethics developed in relation to a specific moment in Anglo-American analytic philosophical traditions. Yet these premises are often incorporated into bioethical discussions in universalizing terms, without sufficient acknowledgement of their embeddedness in time and place (e.g. relevant critiques include Kleinman 1999; Lock 2001; Marshall 1992; Rosenberg 1999). Other perspectives and methods, for example, those drawn from the reflexive and phenomenological perspectives that are the trademark of Continental philosophy and critical social science, have held far less sway in US bioethics. Yet these are the discursive traditions with which many US anthropologists now work. Second, the social movements and institutional changes which spawned bioethics are rarely analyzed in relation to its accomplishments. Yet the field developed under the influence of health reform movements in the USA. In attempting to protect the subjects of medical interventions and experiments, activists made powerful regulatory claims which were rapidly institutionalized in American hospitals and the governmental research apparatus. Ironically, recognition of this activist success was quickly muted in relation to the more successful

C. Rehmann-Sutter et al. (eds.), Bioethics in Cultural Contexts, 341–351.
© 2006 *Springer. Printed in the Netherlands.*

rise of bureaucratic rationality within which bioethical practices are now embedded (e.g. Das 1999; Rosenberg 1999; Kaufman 2001; Stevens 2000). Understanding bioethical discourse as an institutional process is rarely part of the bioethical conversation.

This leads to a second concern, anthropological dismay at ethnocentrism. Anthropologists worry that the norms of the dominant culture from which most health care providers and their bioethical colleagues are drawn are too easily and unconsciously valorized in bioethical theory and practice. Yet there is too rarely sufficient recognition of both positive values and negative barriers affecting how the concerns of patient populations outside dominant social strata are articulated (or dis-articulated) in relation to bioethical discourse and institutional structures (e.g. critiques of Cohen 1999; Das 1999; Farmer 1999; Hoffmaster 2001; Kleinman 1999; Lock 2002). These structural constraints become more visible when "the culture of biomedicine" itself is also placed under the social scientific microscope, rather than being taken for granted as the highest form of rationality. Increasingly, medical anthropologists and other social scientists also study bioethical decision-making as a part of their analysis. Social scientists using grounded methodologies have shared their work with bioethicists, hoping to deploy these varied perspectives and knowledge production processes in complementary fashion (e.g. special issues of Daedalus on "Social Suffering" 1996 and "Bioethics and Beyond" 1999; Brodwin 2002; Hoffmaster 2001). My own work is located in this emergent tradition.

Medical anthropologists have long been deeply involved with documenting and engaging ethical issues in the contexts of illness, healing, and health care provision. They have tried to illustrate how local, national, and global forces mutually construct the varied and often contradictory situations in which health care decision-making and healing practices come to "make sense" for their diverse participants. Their rich and unruly case studies often reveal a complexity of norms, values, and practices not easily constrained within classic bioethical categories. To illustrate this problem of methodological disjuncture, I draw on two studies of my own. The first, reported in "Testing Women, Testing the Fetus: the Social Impact of Amniocentesis in America" (Rapp 1999), examined what pregnant women and their supporters from diverse socioeconomic, racial-ethnic, religious, and national backgrounds understood about the offer of prenatal diagnosis, and how they expressed their aspirations for a healthy pregnancy. In this book, I was particularly concerned about what counted for diverse pregnant women as a fetal or childhood disability, and which disabilities justified abortion. And I examined the communication and mis-communication among a multicultural patient population and their health service providers.

Thus ethical decision-making in complex social context stands at the center of this ethnography.

In the second study, "Mapping Genetic Knowledge", my colleagues Deborah Heath, Karen Sue Taussig and I examine how new genetic knowledge has been made in the discovery process when genes for three heritable connective tissue disorders were found (Rapp and Health et al. 2001; Taussig and Health et al. 2003; Health and Rapp et al. 2004). The conditions under study are achondroplasia, Marfan syndrome, and Epidermolysis Bullosa, a family of blistering skin diseases. In all cases, new genetic knowledge has been made collaboratively, if unevenly, by three groups of interested parties: research scientists, clinician-physicians treating patients with these disorders; and the patient support groups who often provide the "blood and stories" (in science historian Susan Lindee's felicitous phrase, Lindee 2005) on which the discoveries are based. Genes for these disorders were found in the 1990s; issues of prenatal diagnosis and aspirations for gene therapy are very much ongoing and discussed among relevant constituencies. Our fieldwork takes us into genetics laboratories, clinical consultations, and the national and local meetings of lay advocacy health groups, many militating for health policy changes and increased research funds. Advocacy groups are becoming important actors in claims on health care resources, and they increasingly participate in national issues of bioethical concern, e.g. the recent, ongoing debates on stem cell research. It is important to stress the particular US national context of my research: Discussions of medical rationing have a different resonance in the US than in the UK, for example, or the North European countries with longstanding commitments to social democracy, or the post-Soviet regimes of East/Central Europe. As readers surely know, our powerful NIH biomedical research structure is not articulated with a national commitment to universal health care, and access issues undergird most American health policy debates. US newspapers are quick to point out the tragically bureaucratic long waiting lines for a CT scan in Scandinavia. But they are less likely to report the 20,000 women who give birth annually in New York City hospitals without ever having received any prenatal care until they present in the Emergency Room in labor. These conditions form the background of my ethnographic research.

2.   ETHICAL CONUNDRUMS IN ACTION

Here are four stories drawn from anthropological field work concerned with genetic testing and genetic disorders. Each usefully complicates for me any discussion of the limits of bioethical understanding and decision-making by filling in the social context within which multiple constituencies share dilemmas, and act to resolve them.

(1) A forty-three year old social worker from a Dominican background received an ambiguous diagnosis of broken fetal chromosomes. Such a finding might indicate a rare underlying genetic syndrome including a fatal anemia. But it might also well indicate that a transient infection had broken the chromosomes; once the infection passed, the next sample should reveal a growing percentage of unbroken chromosomes. In that case, the fetus might well be normal. The genetics staff discussed the case extensively, understanding that a pregnant woman might be panicked by such an anomalous situation, and abort because she could not know what the outcome was going to be. They therefore decided to recommend a second amniocentesis, in the hopes that it would lend weight to the "transient infection" hypothesis. They also considered and set up a (then) more experimental (and hence, riskier) procedure, PUBS (percutaneous umbilical blood sampling), which would draw fetal umbilical cord blood directly, in the hopes of securing a faster and more definitive diagnosis for the mother, who was already eighteen weeks pregnant. They set in motion the cumbersome bureaucracy, both at the pregnant women's HMO and at the hospital to get rapid approval for the procedure. They additionally recommended high-resolution sonography, and a karyotyping of a maternal blood sample, to rule out other less likely anomalies that might be responsible for the breakage.

With these multiple technological strategies in mind to preserve the pregnancy in the face of an ambiguous diagnosis, the woman was called in. She had already spoken once with a genetic counselor, and knew that there was something anomalous in her fetal results. I was present at the counseling session, and spoke with her many times over the course of the next several weeks. The initial counseling session was dramatic. After the head geneticist had gone over the findings and offered both a second amniocentesis and PUBS, the woman expressed dismay and anger at having to wait for additional results. She noted the importance of her job and maternity leave; fourteen year old son's response, and, above all, her fiancé's emotional distance from a problem pregnancy. Despite the best efforts of the genetics team to preserve the possibility of a normal outcome through additional interventions, she said,

> I must be honest with you. If, after the sonogram, I chose to end this pregnancy, are you saying you will still want to study the fetus? I was very excited about this pregnancy, I really wanted this baby. But I'm 43, I'm reconsidering ... I work with learning disabled, with emotionally disabled students, this is taking its toll. I simply cannot wait another ten days. I'm a

professional woman, and I need to know. My fiancé and I are pulling apart, we may well not be together by the time this situation is resolved.

The chief geneticist then interrupted the consultation, ostensibly to consult with the radiologist about the impending sonogram. But he asked the head genetic counselor and me to accompany him. In the hall, he organized an emergency meeting, saying,

> I do not think this woman wants to continue this pregnancy at all. Things have changed in her life, the boyfriend is easing out of the picture, she has a right to end it if she wants to. Aren't we imposing our values on her, keeping her pregnant for longer and longer, to study a case?

Our team was thrown into consternation; counseling needs for the woman clearly superseded fetal diagnosis. We re-entered the room, repeating that the sonography team was ready to examine the fetus if she wanted. But the woman immediately asked for abortion information. Over the next several weeks, I then spoke with the woman about her abortion experiences, her recovery, her support system. I learned a great deal about her mother, son, and women friends, and a bit about her (by now ex-) boyfriend. What I never learned was the salience of broken chromosomes: the ambiguous diagnosis, so central to the work of a conscientious genetics team dedicated to setting up cutting-edge services to diagnose this fetus, had receded into the mists. It had kicked off a decision-making process in which social ambiguity – of appearance, professional commitments, and, above all, the status of a love affair – far exceeded the weight of genetic material. From a bioethical perspective, surely the genetics team conducted itself appropriately, attempting to increase patient autonomy and knowledge under conditions of scientific ambiguity. But socially speaking, those concerns were tangential to the abortion decision the pregnant subject ultimately made.

(2) A de novo anomaly found on prenatal diagnosis involved additional chromosomal material on the top, short arm of the #9s. After extensive re-checking, the head geneticist labeled it "9P+", 9 for the pair of chromosomes on which it was located, P to designate the short arm, and plus to indicate additional chromosomal material. This diagnosis was later stabilized as a "trisomy nine", the closest approximation to other rare reports in the clinical literature. In all those cases, babies born with trisomy 9 had physical anomalies and were mentally retarded. Armed with a provisional diagnosis, the geneticist met with the genetic counselor who then counseled the mother. When the mother decided to keep the pregnancy, the genetic counselor asked the geneticist to meet with the mother, as well. After a second consultative counseling session, the mother remained quite firm in her decision to continue the pregnancy.

The baby was born in early June, and in late July, the geneticist contacted the new mother through her obstetrician, asking if she would be willing to bring her child to the genetics laboratory, for a consultation. The mother agreed. The "trisomy 9" turned into a six-week old Haitian boy named Etienne St-Croix (pseudonyms). His mother, Veronique spoke reasonable English, and good French. His grandmother, Marie-Lucie, spoke Creole, and some French. The two geneticists spoke English, Polish, Hebrew, and Chinese between them. I translated into French, ostensibly for the grandmother and mother. Here is what happened:

Two geneticists, both trained in pediatrics, handled the newborn with confidence and interest. The counselor took notes as the geneticists measured and discussed the baby. "Note the oblique palpebral fissure and micrognathia", one called out. "Yes", answered Veronique in perfect time to the conversation, "he has the nose of my Uncle Herve' and the ears of Aunt Mathilde". As the geneticists pathologized, the mother genealogized, the genetic counselor remained silent, furiously taking notes, and the anthropologist tried to keep score. When the examination was over, the geneticists asked the mother one direct question. "I notice you haven't circumcised your baby. Are you planning to?" "Yes", Veronique replied, "we'll do it in about another week". "May we have the foreskin?", the geneticist queried. "With the foreskin, we can keep growing trisomy 9 cells for research, and study the tissue as your baby develops". Veronique gave her a firm and determined "yes", and the consultation was over.

Walking Veronique and Marie-Lucie to the subway to direct them home to Brooklyn, I asked what she had thought about the experience: from the amniocentesis to the diagnosis to the genetic consultation. She replied,

> At first, I was very frightened. I am 37 I wanted a baby, it is my husband's second marriage, my mother-in-law is for me, not the first wife, she wanted me to have a baby, too. If it had been Down's, maybe, just maybe I would have had an abortion. Once I had an abortion, but now I am a Seventh Day Adventist, and I don't believe in abortion anymore. Maybe for Down's, just maybe. But when they told me this, who knows? I was so scared, but the more they talked, the less they said. They do not know what this is. And I do not know, either. So now, it's my baby. We'll just have to wait and see what happens. And so will they.

Here, marital and kinship relations clearly influence the decision to continue a pregnancy after positive diagnosis; so does religious conversion. But at the center of this narrative lies another important theme: diagnostic ambiguity. Biomedical scientists work from precedent, matching new findings with old. When presented with an atypical case, they build a diagnosis in the same fashion, comparing the present case to the closest available prior knowledge in clinical archives. While the geneticists are confident that this child will share the developmental pattern reported in the literature for other children with very similar chromosomal patterns, the mother was quite aware of the idiosyncratic nature of the case, its lack of clear-cut label and known syndrome. She therefore decided that the contest for interpretation was still an open one. This is a dramatic instance of interpretive stand-off between representatives of biomedical discourse and representatives of family life. Once again, the autonomy of the pregnant woman to make "an informed consent" takes place on a social terrain in which anomalous chromosomes play only a small part. "Risk-benefit ratios" do not speak to the meaning of a child in this "blended family" and kinship lineage.

(3) In 1994, John Wasmuth and his laboratory colleagues published an account of the discovery of FGFR3, the gene for achondroplasia, the most common form of heritable dwarfism, in the journal Cell (Shiang and Thompson et al. 1994). Hailed soon after in The Scientist as the article most frequently cited during 1995, Wasmuth's publication revealed that 98% of those affected with achondroplasia have the very same mutation in the molecule FGFR3, a receptor for what is called a growth factor. Among other things, the discovery opened the possibility for routinized prenatal screening for this condition. One year after Wasmuth published his article, Clair Francomano, Chief of Medical Genetics at the National Human Genome Research Institute at the NIH, attended the national convention of the Little People of America (LPA), the U.S. national organization for people of short stature. Dr. Francomano is a longstanding researcher and health service provider for people with heritable dwarfism and a member of the LPA Medical Advisory Board. As she tells her story,

> The first thing I saw when I came to this convention last year (after the discovery of the gene was publicized) was one of the people wearing that "Endangered Species" tee shirt. It really made a very big impact on me. And I really worry about it. I worry about what we're doing and about how it's going to be used and what it means to the people here.

Dr. Francomano's response was to chair several workshops for LPA members on the Human Genome Project. There, she explained genetic technologies and programs, listening attentively to the fears and hopes of short statured people. She also expressed her own aspirations for the possibilities opened up by genetic research, and her dismay that new discoveries might be eugenically deployed. Her aspirations centered around gene therapy for specific problems associated with dwarfism such as ear or breathing issues, back pain, and skeletal dysfunction. Her ethically responsive and responsible behavior is greatly appreciated by LPA members. In addition, Dr. Francomano collaborated in designing a membership-wide survey on attitudes toward prenatal testing for the LPA.

Inside the community of the short-statured, prenatal diagnosis is widely valued because this autosomal dominant condition includes a 25% chance that a fetus conceived by two parents with achondroplasia will be born with a lethal condition, double dominance. Yet at the same time, they are acutely aware that public discussions of the fast-evolving high flow-through diagnostic chip technology often include their FGFR3 gene. In this more general discussion, it is not the birth of dying babies in dwarf couples that is being discussed, it is the elimination of otherwise-healthy dwarf fetuses that drives the discussion. The anger and fear of LPA members is considerable, for they recognize that longstanding prejudice against dwarfs will influence future uses of genetic knowledge. This case opens up for me the difficult problem of the social fund of genetic knowledge: those intimately affected by a diagnosable condition are differently located than the general public, where bias against the condition abounds. Thus, issues of public education, not only individual choice, require bioethical discussion informed by insider knowledge. Because the principles of bioethics so insistently focus on the autonomous decision-making individual, larger social issues of eugenic discrimination are too-often lost to general debate. Yet they reside in the community of stigmatized experts whose life-experiences we presently have little capacity or volition to incorporate into bioethical practice.

(4) An emergent case involves the possibility of experimental gene therapy for a lethal form of EB, a genetic blistering skin disease. My ethnographic investigation of this case has just begun, so I have no detailed descriptions to offer. But the bioethical problem is this: There are at least a dozen forms of EB, several of which are lethal in infants (although other forms are "compatible with life", despite the range of problems from mild to slowly fatal that they present). More than a decade's work on gene hunting, mouse modeling, and experimental bioengineering of artificial skin has brought scientists from at least five major US research institutions into collaboration and competition. One such team has recently successfully submitted an application for Institutional Review Board approval (IRBs are roughly analogous in the USA to ethics commissions in Europe, governing hospital-based clinical and research experiments). The team hopes to begin experimental gene therapy on a lethal form of EB that kills infants within six months of birth. Their protocol will enroll twelve newborns on whom topical gene therapy will be attempted. Some scientists in other EB research teams expressed serious criticism of this study informally to me, for they believe that families will now prolong the suffering and lives of dying infants in order to participate in Phase I research. The research team that is currently enrolling human subjects has, of course, conscientiously explained the lack of individual therapeutic gain which is likely to accompany participation in the study to the parents of these EB-affected newborns. Yet

scientists and clinicians outside the team are concerned that the standard bioethical model of conducting the riskiest and least beneficial experimental trials on dying patients is inappropriate when dealing with these newborns. In their eyes, the "normal" trajectory of family grief after the birth of a dying EB baby includes recognition of both the child and family's suffering, and reconciliation with a decision to end heroic life-sustaining procedures so that the baby can die as peacefully as possible. They believe that these desperate families cannot appreciate how little their infants will gain and how much they will suffer. They fear that the families will agree to life-sustaining interventions in order to grasp at a frivolous hope of gene therapy. These scientists and clinicians believe that gene therapy should be attempted first on young adult/ adult forms of EB, where the rationality of individual decision-making is more plausibly established, and family knowledge of the condition is longstanding. Yet there are good scientific reasons to believe that this lethal infant phenotype of the disease is the most accessible form on which to experiment. Are these criticisms instances of insider knowledge? Paternalism? Professional envy? I hope to understand this bioethical controversy ethnographically as it unfolds. In the process, ethnographic attention should put in question the multiple and contested authorities — medical, scientific, ethical commission, or parental – on which a decision to proceed with a highly controversial experimental therapy rests.

In these cases, health care providers and researchers of good intention confront both individual and social aspirations and decision-making which challenge their basic understandings of their roles. These cases index the thick social matrix into which normal/ abnormal babies are born, rejected, or accepted and the relative value placed on scientific literacy when bioethical decision-making plays out in its social landscape. They also suggest that "scientific literacy" and "standards of care" regularly confront widespread public prejudice against some relatively common alternative phenotypes like achondroplasia, and that the ethics involved in judging insider/outsider decision-making are murky, indeed. These examples suggest that in daily practice "real world" diversity of cultural and expert background continually challenges straightforward application of bioethical principles. Yet the literature is relatively silent on how such a hierarchy of dilemmas is routinely addressed. I hope that such dense ethnographic nuggets contain the elements to stimulate a complex discussion of how concrete ethical dilemmas play out in "real time", rather than in the hypothetical spaces which characterize much bioethical discourse. I hope, too, that experts involved with bioethical discourse and practice will develop more self-reflexive and institutionalized practices where these daily practical challenges can be not so much "resolved" as "discussed".

At the same time, when I have presented such cases to bioethics audiences, I often find a troubling response that I do not know how to interpret. Despite the "real world" complexity of my data, professionals trained in bioethical methods often quickly return to "society" as a unified construct, arguing about the correct response as if "one" existed where I often see interpretive plurality. Or members of the audience quickly substitute "hypothetical" cases in place of the less tidy ones I have dredged up through observation and interviewing outside the formal frame of biomedicine. This gap between lived practice and theoretically-derived norms cannot be back-filled by recourse to the standard critique of "moral relativism", for that is not the position that I (or other medical anthropologists) espouse. Rather, the theory/praxis gap signals for me and my intellectual "tribe" a commitment to reflexivity: how are bioethicists, heir to a hybrid theoretical/ applied tradition that is increasingly centered in the realm of social and quite instrumental policy, to understand the limits of their own knowledge-making process? How might a more "critical pluralist" perspective on the recent history of their field (cf. Pickstone 2000) be incorporated into bioethical certainty? This is a problem that anthropological fieldworkers regularly confront in our own work, and we believe it is valuable for other disciplines to engage it, as well. If our mode of knowledge production is to enter into useful conversation with bioethics, then an approach, rather than an avoidance, of the messiness of praxis in full social context (including the contexts in which bioethics as a field of inquiry has come to have utility) will benefit from exploration. Over time, I believe a recognition of this "thick social matrix" will serve both the rapidly-expanding field of bioethics, as well as the multicultural, always-changing patients on whose lives it reflects. Yet principled certainty is continually challenged when the messiness of praxis intervenes: to this, the field of anthropology offers no risk-benefit ratio.

References

Brodwin, P. "Genetics, Identity, and the Anthropology of Essentialism." *Anthropological Quarterly* 2002, 75, 2, 323–330.

Cohen, L. "Where It Hurts: Indian Materials for an Ethics of Organ Transplantation." *Daedalus* 1999, 128, 4, 135–166.

Daedalus special issue, "Social Suffering" 125, 1, Winter, 1996.

Daedalus special issue, "Bioethics and Beyond" 128, 4, Fall, 1999.

Das, V. "Public Good, Ethics, and Everyday Life: Beyond the Boundaries of Bioethics." *Daedalus* 1999, 128, 4, 99–134.

Farmer, P. *Infections and Inequalities: the Modern Plagues.* Berkeley: University of California Press, 1999.

Heath, D.; Rapp R. et al. "Genetic Citizenship." In *Companion to the Handbook of Political Anthropology.* D. Nugent and J. Vincent (eds.). London: Blackwells, 2004.

Hoffmaster, B. (ed.). *Bioethics in Social Context*. Philadelphia: Temple University Press, 2001.

Kaufman, S. "Clinical Narratives and Ethical Dilemmas in Geriatrics." In *Bioethics in Social Context*. B. Hoffmaster (ed.). Philadelphia: Temple University Press, 2001, 12–38.

Kleinman, "A. Moral Experience and Ethical Reflection: Can Ethnogrraphy Reconcile Them? A Quandray for the 'New Bioethics'." *Daedalus* 1999, 128, 4, 69–98.

Lindee, S. *Moments of Truth: a History of Medical Genetics in the U.S.A.* Baltimore: The Johns Hopkins Press, 2005.

Lock, M. "Situated Ethics, Culture and the Brain Death 'Problem' in Japan." In *Bioethics in Social Context*. B. Hoffmaster (ed.). Philadelphia: Temple University Press, 2001, 39–68.

Lock, M. *Twice Dead*. Berkeley: University of California Press, 2002.

Marshall, P. "Anthropology and Bioethics." *Medical Anthropology Quarterly* 1992, 6, 1, 49–73.

Pickstone, J.V. *Ways of Knowing: a New History of Science, Technology and Medicine.* Chicago: University of Chicago Press, 2000.

Rapp, R. *Testing Women, Testing the Fetus: the Social Impact of Amniocentesis in America.* New York: Routledge, 1999.

Rapp, R.; Heath D. et al. "Genealogical Dis–Ease: Where Hereditary Abnormality, Biomedical Explanation, and Family Responsibility Meet." In *Relative Matters: Reconfiguring Kinship Studies.* S. Franklin and S. McKinnon (eds.). Durham: Duke University Press, 2001, 384–409.

Rosenberg, C. "Meanings, Policies, and Medicine: On the Bioethical Enterprise and History." *Daedalus* 1999, 128, 4, 27–46.

Shiang, R.; Thompson L.M. et al. "Mutations in the Transmembrane Domain of FGFR3 Cause the Most Common Genetic Form of Dwarfism, Achondroplasia." *Cell* 1994, 78, 2, 335–342.

Stevens, L. *Bioethics in America: Origins and Cultural Politics.* Baltimore: The Johns Hopkins University Press, 2000.

Taussig, K.S.; Heath D. et al. "Flexible Eugenics: Technologies of the Self in the Age of Genetic." In *NatureCulture: Anthropology in the Age of Genetics.* A. Goodman, S. Lindee and D. Heath (eds.). Berkeley: University of California Press, 2003.

Chapter 23

NARRATIVE BIOETHICS[1]

HILLE HAKER

Cambridge MA, USA and Frankfurt am Main, Germany

"The field of bioethics is beginning to take its own narrative turn. Long dominated by the aspirations to objectivity and universality as embodied in its dominant 'Principlist' paradigm, bioethics is now witnessing an explosion of interest in narrative and storytelling as alternative ways of structuring and evaluating the experiences of patients, physicians, and other health care professionals" (Arras 1997: 66).

In this paper I want to throw a little light on the recent trend in bioethics described by Arras in the passage above. I intend to explore the question of what significance and what place within the system should be given to a *narrative* bioethics, taking bioethics primarily as medical and biomedical ethics.

The first part of the paper therefore deals with the broader concept of ethics and narration, or more specifically, with identity and "narrativity", since it is with the question of moral selfhood that narrativity primarily emerges.

1. PHENOMENOLOGY, HERMENEUTICS AND MORALITY: THE FUNDAMENTAL HERMENEUTICAL SITUATION 'ENTANGLED IN STORIES"

1.1 Ethics and hermeneutics

Inasmuch as ethics is related to the actions and attitudes of men and women, and to social contexts, it is always *partially* a hermeneutical enterprise. Actions cannot be described independently of those who carry

C. Rehmann-Sutter et al. (eds.), Bioethics in Cultural Contexts, 353–376.
© 2006 *Springer. Printed in the Netherlands.*

them out or the contexts in which they are carried out. Actions point to their originators who, to the extent that they act intentionally and consciously, pursue goals with their actions to which they ascribe positive value and which they consider to be right or at least appropriate. Actions cannot be abstracted, either from the attitudes, preferences or evaluations on the basis of which they are carried out, or from their social and cultural contexts. Therefore actions and attitudes only become significant by means of their connection to the biographies or life histories that, in turn, remain incomprehensible without their historical, cultural and social contexts. Both dimensions can thus be viewed as 'con-texts' for the quasi-text of the action itself.

This hermeneutical, semantic, interpretative dimension of action and, more generally, of forms of human praxis, is both phenomenologically and existence-philosophically relevant, in that with this "entanglement in stories" (Schapp 1985) a "fundamental situation", a fact of the *conditio humana* itself, is given a hermeneutically mediated reality, behind which no other ontological or metaphysical reality lies hidden.[2]

We can therefore say that actions are bound as tightly as possible with a person's identity, with his or her life history, and with the meaning a person gives his or her actions in the context of this life history and the reality in which it takes place. Actions are one expression of what a person wants to be, how he or she wants to live, and what he or she considers to be significant (Lenk 1993). To this extent, actions, the expressions of people's values and attitudes to values, are an indication of their notions of what it means to live a good life, and indeed of their notion of what is "good" in general. Since people are individuals, with their own individual experiences and life histories, their notions of what is good, even if these are socially mediated, are also individual and different (Taylor 1995).

Nevertheless, people do not move like atoms in the universe. They are connected with each other in various ways, and this is also true for their values. These are formed primarily within society (Lenk 1994; Haker 1999), and are acquired in processes of individuation and socialisation, which allow the individual self to appear simultaneously as the social self. The insights of modern psychology and psychoanalysis suggest that the individual self can emerge only slowly and with difficulty from the social self, from the "me" (Mead 1967), that is to say, that it is always also a social self. If we take these insights seriously, then the "autonomous self" of whom ethics, and especially bioethics, speak is a *relatively* autonomous self: a self that relies on others (relational autonomy) and which is in constant tension with heteronomy. In bioethics, autonomy is understood as an elementary right to the freedom of self-determined action, and thus as a normative principle. This limitation, or rather linking of autonomy with social formation and heteronomies, must therefore be taken into account as much as the fact that autonomy is less a capacity than a reflective attitude of the moral subject with regard to his or her actions and being in history.

The notions of moral identity or of the moral self point to the narrative structure on which this reflective attitude to autonomy is based. Paul Ricœur has shown the extent to which narrative identity is woven into time (Ricoeur 1984-88). Here it is significant that the moral subject is understood as the idea of condensed narrative identity in its moral-ethical sense, as a responsible subject. The moral subject, that is the individuated autonomous subject, is responsible for his or her actions in the past, present and future (compare Liebsch 1999). In order to grasp the content of this formal structure of responsibility, it is necessary to have recourse to the narrative hermeneutical constitution of the moral subject as a self with an identity, which consists of a life history mediated in narrative.[3] But as I have already indicated with respect to relative autonomy, responsibility is also relative – and this is as much the case with respect to perception and bias of perspective as it is for the real possibilities of taking responsibility.

1.2   Ethical experience

The hermeneutics of ethics thus point to the connection of morality with identity, and so with experience. The ethical experience of always having been responsible cannot be seen independently of one's individual life history, which in turn is connected in many different ways with one's social, cultural and political history. Actions can open up experiences in the same way as the passivity of experiencing and/or suffering the actions of others on one's own body. Moral experiences are primarily evaluative experiences: experiences of meaningfulness, injustice, humiliation, or sympathy for others, which themselves provoke actions (Mieth 1999 and 1998; Haker 2001). They are "self-binding, self-obligatory, autonomous in Kant's sense, because they place the self for its own sake under the law which it unavoidably carries in it" (Mieth 1999, my translation).

Experiences are first and foremost things that happen to us – we cannot predict the experiences we will have, and we cannot create them. Nonetheless, experiences are also reflective, that is, they become conscious experiences only through the articulation of what we have experienced. This is particularly the case with respect to moral experiences. Through experiences, values are formed which constitute a background to actions, and are ideally manifested in actions. Moral experiences therefore represent, in their own form of reflection, a bridge between the sensations caused by the things that happen to us and our rational evaluation of those things.

Summarising what I have said so far, it is necessary to keep in mind firstly the connection between agency and identity. Agency is a mode of expressing what a person wants to be, but it is also, to use a Foucauldian term, the concretion of the socially formed (disciplined) self. The self, expressing him- or herself through agency, can be called a self of "relative autonomy". The relativism of autonomy must be understood to be, as well as relationalism, the specific tension between autonomy and heteronomy, which is present in all actions. Speaking of a narrative identity means reflecting upon the reflexive structure of identity and the entanglement of the self in stories. What is impossible, then, is to regard the moral self as an atomistic, individualistic, sovereign self – although it is not possible to rule out autonomy altogether. Furthermore, the moral self is characterised by its capacity to take over responsibility, which is no longer an abstract concept but refers to the "situated", historical self. Morally or ethically relevant experiences befall us, and even when they leave us passive they nevertheless have an evaluative structure: as experiences of injustice, of compassion, of a 'sense' – shedding its light on our life. However, only if these (passive) experiences are articulated – and interpreted – do they turn into reflexive experiences.

In order to understand what narrative bioethics is about and could be about, I will now turn to my own approach: literary ethics. In doing so, I will narrow my perspective for a moment to literary narration. Valuable insights can be gained from literary theory and aesthetics, which should be used in narrative bioethics primarily but not exclusively in dealing with autobiographical or "authentic" narratives.[4] This move seems crucial, because in the (bio-) ethical context, we read stories in the same way as we read literary narratives. Ricœur's theory of overlapping reference of fact and fiction in historical and literary narratives seems pivotal in dealing with different narrative forms.

2.   LITERATURE AND ETHICS

Literary ethics directs the reader's attention to ethical questions posed in and by means of a literary text. It analyses them, in that it attempts to understand ethical questions as a function of the text, but above all it holds up the narrated and evoked world of the work in contrast to the existential and practical reality to which the ethical judgments and perspectives refer. In the difference between the two modes of reasoning, ethical concepts generated mostly in the one mode, namely that of argumentative reasoning and practical rationality, are as much questioned as the aesthetic functionality of the narrative ethics within a work of art. Following Walter

Benjamin we can label the two aspects of literary-ethical interpretation "commentary" and "critique".

Ethics and literature can primarily be regarded as two modes of reflection, each with its own rules and purposes. Ethics assumes the function of orienting personal identity or concepts of character formation, of reflecting upon individual actions and social practices, and upon social and political institutions or institutional structures. It has become common in recent years, and indeed it is crucial, to distinguish the two different areas of ethics: the *ethics of pursuit* (both the self's pursuit of happiness or perfection and the social striving towards social goods), and *normative ethics* (both an ethics of individual rights and obligations, and of the social responsibility to advance social and political justice). Each of these areas relates to "narrativity", or more specifically, to narrative as a genre. Literature, on the other hand, is art characterised by its distance from action and praxis. It is a realm of reflection in its own right: it is reflexive, insofar as its form of expression is concerned. Nevertheless, it is also experimental and creative in a literary sense, insofar as it articulates types of experiences that did not exist prior to articulation, or at least were not cognitively available. In this way literature is *mimetic poiesis*: it "creates" reality and affects the reality of practice and social experience in this specific and extremely powerful communicative fashion.

The connection between ethics and aesthetics is the concept of experience. Aesthetic experience is certainly primarily an experience of freedom from constraints, of a certain autonomy of art in Adorno's sense. But it is also "self-indulgence by way of indulgence to the strange" (Jauss 1982). Seen from the 'reception-aesthetic' perspective, aesthetic experience is also a form of contemplative and corresponding experience that adds to the experience of imagination (Düwell 1999; Jauss 1982; Seel 1991). Moral experience, according to Dietmar Mieth, has its roots in the sensation of outrage (the experience of contrast), in the experience of sense (happiness, success), or perhaps even in being "touched" by someone or their story – compassion or affirmative passion that leads to action (the experience of motivation).

The concept of narrative ethics, which was developed by Mieth in the 1970s and early 1980s, stems from the *ethics* inherent in narrative or other literary genres being expressed in both the form and content of a text. For Mieth, who worked during his early career on the epos "Tristan and Isolde" (G. v. Strassburg), the novel "Joseph and his brothers" (Thomas Mann), and "The Man without Character" (Robert Musil), literature is one form of an "ethics of models", close to a modern ethics of virtue which must complement normative ethics.[5] Narrative ethics as an ethics of models attempts to take seriously the individual and his or her sometimes bewildering reality. It is based on the view that ethics cannot be reduced to a language of rational concepts or deductive arguments. In contrast, ethical reasoning must remain connected to sensations, feelings and emotions. It is even more the case then, that bioethics as an "applied" ethics would be far too narrowly constructed if it did not complement its necessarily *empirical* approach with an *experiential* approach.

For Martha Nussbaum, who has pursued a related concern in the field of narrative ethics, narration is the appropriate and possibly the only form that expresses the anthropological conceptions at the heart of every ethic. Unlike Mieth, she also claims to develop from narrative sources the building blocks of a theory of ethical emotions – the latter being the central allusions to a theory of the Good.[6] Moral learning, according to Nussbaum, oscillates between the two poles of practical reasoning. On the one hand we have the particularity of literature, the incommensurability of the goods, and the value of feelings which literary theory interprets. On the other we have ethical universality, the hierarchy of the goods and the moral judgment, which ethical theory attempts to ground. European theorists appear to emphasise the aspect of aesthetic experience and therefore the form of experience. Nussbaum emphasises the power of the work of art itself: its representation and articulation of human experience as opposed to "rational" ethical theory with its abstraction from concreteness and individuality.

My own approach tries yet another way. In contrast to a narrative ethic directed at a particular experience without reflecting its impact on normative concepts that seem relatively untouched by narrative ethics, and in contrast to a universally understood anthropology with almost direct, normative implications, I am attempting to constitute literature as a medium *of* ethical reflection and *for* ethical reflection. Therefore, I understand this approach no longer as a narrative ethic (which is certainly a part, but not the whole of it), but rather as a literary ethic with more of a commentarial and critical perspective.

Indeed, the value of art in my view lies in its resistance to abstraction and the subjection of experience to an apparently logically consistent order. Literature achieves its own mode of articulation by strengthening the voice of an alternative, experiential reality, that not only has its own rationality, but is characterised by its bewildering concreteness, its orientation towards emotion and not consistency, and its contradictory qualities (particularity). Art is therefore to be taken seriously as both a medium of ethical reflection and a medium by which the world is revealed. However, the alienation of reality through art is a point worth considering. For through alienation and the experience of imagination, the reader is distanced from his or her everyday practical experience, which is set in a context that grabs him or her emotionally, sensually and cognitively. Insofar as literature entails models that stimulate reflection, and insofar as these reflect the question of the pursuit of individual and social life as much as the question of rights and social justice, it is a worthy participant in ethical reflection. Last but not least, literature stems from its own time, it relates to historical events, to the language of a certain time, to its metaphors and forms – even though its mimesis of reality is of a 'poietic' nature. Thus, commitment to literary commentary and critical interpretation involves expert hermeneutic skill on the part of the interpreter.

As I have said, the task of ethical theory in relation to literature – for which I use the term literary ethics – consists of both a commentary on the text (similar to that which Nussbaum accomplishes in her interpretation of literature), and of a critique of the ethics depicted in literature. The concepts of commentary and critique stem from the literary theory of Walter Benjamin, who used them to describe both contemplative and sympathetic interpretation on the one hand, and on the other the "dissecting-destructive" aspect of interpretation – close to the aim of deconstruction in current literary theory. Both aspects are important, not only for literary theory, but also for ethical reasoning, and they should not be played off against one another. Both literature and ethics alienate from moral practice by way of their critical distance in reflection. But in order to know which ethical questions are posed at a particular time, which judgments are questionable, and which contexts change, ethics must turn to literature as one of the sources of interpretation and articulation, and of beliefs from which ethical reflections cannot and should not be separated. People sometimes ask: "why not turn to social studies instead? This source is empirically valid and representative". Yes, we could do this, but literature's specificity lies exactly in its 'poietic' character, in its relatedness to practice which must not be mistaken for practice itself. Ricœur's idea of mimesis, extending from pre-narrative praxis through literary narration to a return to praxis, is useful here. Mimesis is not imitation but representation. By posing questions in its own way, literature participates in articulating the sense and meaning of social and historical reality – even when it denies any meaning at all.[7]

Literature plays with the world of ethical values. It can depict people and actions that would be ostracised or condemned in the moral world. Literature frequently and intentionally transgresses the boundaries of the moral world in order to bring ethical questions indirectly to light. The ethical question then becomes whether we would wish to live in such a way, if it were thinkable and explainable. Literature shows not only the internal realm of the right, but also experimentally leads the self to its limits or even beyond them. Literature does this, however, behind the veil of aesthetic appearance. Ethics does not have this freedom, but it can profit from literature's imaginative rationality, if it recognises the light that is shed upon the sincerity of practical moral life in the articulation of value systems, moral quandaries, and even the lack of moral knowledge or sensitivity.

Notwithstanding these specific features, there are limits that also need to be recognised. Literature concerns itself with concreteness and particularities. It makes no claims to universality, but shows what is meaningful about human reality. It is clear that literature makes reference to general human problems and questions. It does so, however, by posing questions about the appropriateness of desires, about the range of activity and living, about the interaction between the individual and the social sphere, beings or nature, about the place of humans in history or in the cosmos, etc. But its answers are necessarily embedded within the aesthetic form. By means of the narration of a story literature attempts to display to the reader its view of reality and make it appear plausible, and this happens even though an author has a wide range of possibilities for dissemination of the plot, interaction with the reader, or interruption of one perspective by expanding a story to multiple perspectives.[8] The very structure of narration leaves judgment about the narrated story or plot (or fragments of these) to the reader. Accordingly, a literary text does not *argue* on behalf of the values or the hierarchy of goods represented within it – for example, by posing a truth claim. A literary text asserts no claim to the complete representation of an ethically relevant perspective. Rather, a literary text focuses on one perspective. Once again, ethics is subsumed under an aesthetic conception and form, even when it contributes to both the form and content of the text. Morality or the "ethical" in literature is necessarily a function of the aesthetic.

If "freedom" then is the basic concept of aesthetic experience, and "responsibility" that of morality, it may be appropriate to describe the interaction of both forms of experience. Depictions of limitless freedom (insofar as this is possible to express) quickly become insipid, vapid; it is for this reason that the mimesis concept is upheld for literature. Within this structure, the figures in a text struggle with all the existential questions that people face in their practical lives. On the other hand, morality is perverted into force, even violence, if it is not connected to freedom and creativity. In some ways, it may indeed be true that literature plays a more important role for an ethic of pursuit than for a normative ethic. However, considering the role of freedom and compulsion in normative ethics, this may not be wholly true. It is literature's internal or inherent relation to freedom and compulsion (in the end, narrative figures are puppets in the hand of the author!) that makes literature as form an indispensable medium for normative ethics too. Because of this fact, it is not a material universal anthropology that Martha Nussbaum has in mind, but rather the foundation of all morality –

vulnerability and harm that people experience as a result of other people, and the relation of autonomy and heteronomy – which could serve as the normative vanishing point of a literary ethic.

If we now move a step further, we can ask what role narratives play in bioethics and for bioethics, in the afore-mentioned narrower sense of medical and biomedical ethics.

3. THE ROLE OF STORIES IN CLINICAL MEDICAL ETHICS, AND NARRATIVE ETHICS AS A THEORY OF LITERARY HERMENEUTICAL REFLECTION

3.1 Stories and aesthetic experience

Stories are forms of the expression and experience of a person or cultural group. They are a medium of ethical experience and self-reflection, for those who tell the story as much as for those who listen to it or read it. In the telling of a story, something that has happened to someone – such as a serious illness – begins to take on contours: the experience requires articulation and reflective grappling with the process initiated by the event. Illness is crucially determined by the way in which a person deals with physical and psychological distress. The narration of the different biological, psychological and emotional aspects of an illness and its contextualisation in one's life history and in the socio-cultural environment, can affect a change in the understanding of the illness itself, of one's conception of oneself, and of the way in which the illness was coped with.[9]

On the other hand, through the imagination and the emotions set free by stories, readers are drawn into the world of the concrete life experience of other human beings, even if the stories told by the narrator about them are completely fictional and far from practical reality. There is a sense in which stories impose their view of the world and thus their value judgments on their readers, although the value judgments of the characters and those of the narrator can be quite different. As mentioned above, Martin Seel describes the aesthetic experience offered by a story as one of *contemplation*, *correspondence* (that is, the recognition and identification of the listener/reader with what is heard/read) and *imagination*.

3.2   Stories and ethical experience

These three dimensions of aesthetic experience also play a major role in ethical experience: without submerging oneself in the story, without becoming receptive to it, the quality of actual experience cannot be attained. The experience of correspondence draws the story in a sense into one's own practice, and examines it – in its specific form of emotional rationality – against this background of one's own value judgements, ethical and moral ideas or insights. The imagination, finally, allows this imagined reality to become dynamic, releases new images, perspectives and ultimately experiences, without the pressure to act.

> "The pleasure we take in following the fate of the characters implies, to be sure, that we suspend all real moral judgment at the same time that we suspend action itself. But in the unreal sphere of fiction we never tire of exploring new ways of evaluating actions and characters. The thought experiments we conduct in the great laboratory of the imaginary are also explorations in the real of good and evil. Transvaluing, even devaluing, is still evaluation. Moral judgment has not been abolished; it is rather itself subjected to the imaginative variations proper to fiction" (Ricoeur 1992: 164).

This is certainly more the case for literary narratives than for "authentic" tales of the sort that are found in so-called illness narratives. Experiences of illness are, for those who suffer them, "disruptive experiences". They represent a disturbance of conceptions of identity, they demoralise their subjects. Illness stories often bear witness to this shock of disorientation, the loss of the "centre" of identity. But stories also have, among other things, the function of restoring orientation by means of narrative, of recovering balance and autonomy, or re-moralising.

3.3   Functions of narrative in medicine

In their attempt to recover autonomy, illness narratives have a *therapeutic* function for their narrators. In addition, it is precisely authentic stories which are a means of *creating community,* and in so doing they contribute to constituting or stabilising communities. As an articulation of particular experiences, authentic stories *foster knowledge about specific complexes of experience,* and in certain circumstances they offer insight into the *differences between perceptions of reality* and the differences between various perspectives.

The above-mentioned functions of narrative are important not only for illness narratives, but for all of bioethics. Communication about disease and illness, but even more about being ill itself, is a hermeneutical act that arises from the interaction between objective, empirically verifiable data and the subjective feeling of being ill. However, our conception of an illness is not absorbed by this epistemological function; rather, it also guides our actions. If, for example, in the bioethical discussion of pre-implantation diagnosis, a decision has to be made about whether to destroy an embryo with a specific genetic feature (let us say, carrier status for cystic fibrosis that will not affect the health status of the person concerned) or transfer it to the woman's uterus, then we deal with the implicit normative assumption that this selection is permissible. Although not a disease or illness itself, carrier status for cystic fibrosis is *interpreted* as a reason for negative selection. In the context of pregnancy, for example following PND (prenatal diagnosis), this interpretation becomes the *shared* interpretation of doctor and couple as soon as the doctor explains to the mother that the termination of the pregnancy is a possible option.

3.4 The function of working with narrative forms

Narrative bioethics takes up this hermeneutical situation, that is, the necessity of reaching an understanding with oneself, of communication between doctor and patient, and of educating society about semantic arrangements (myths in the sense of Roland Barthes), in order to demonstrate what stories can achieve in this context. Frequently, however, it appears that the primary concern is to give the group of the sick, the patients, or those who have been declared ill, a voice:

> "Medical voices relegate ill people to patienthood and render their stories into fragments of a larger medical story ... they tend to silence 'lay' experience by reducing it to medical relevancies, i.e. the accounts of treatment, history, and prognosis of the disease" (Frank 2000: 354).

> "To Ivan Ilych only one question was important: was his case serious or not? But the doctor ignored that inappropriate question. From his point of view it was not the one under consideration. The real question was to decide between a floating kidney, chronic catarrh, or appendicitis. It was not a question of Ivan Ilych's life or death, but one between a floating kidney and appendicitis. And that question the doctor solved brilliantly, as it seemed to Ivan Ilych ... All the way home [Ivan] was going over what the doctor had said, trying to translate those complicated, obscure, scientific phrases into plain language and find in them an answer to the question: 'is my condition bad? Is it very bad? Or is there as yet nothing much wrong?'" (L. Tolstoy: *The Death of Ivan Ilych*, quoted in Leder 1994).

When authentic stories of illness are absorbed into medical reflection, it must not be in order to seek a demonstration of the validity of ethical criteria by means of concrete stories – as often happens in the medical use of case studies. An instrumentalisation of this kind would destroy the potential of the aesthetic-ethical experience. Arthur Frank and others point in this context to the *dialogical* character of stories. With regard to illness narratives, with which Frank has primarily concerned himself, paying attention to the story itself is already an ethical act. In listening and taking in what is told, the hearer or reader shows the storyteller respect. In this way, the aesthetic act of reading becomes an equally ethical act. Illness stories or the stories of patients are therefore, according to this first view of what narrative bioethics can be about, like literary narratives, to be taken seriously as a source of ethical experience. They can communicate insights, create or stabilise community, but they can also disrupt or disturb moral systems and call them into question. Beyond this, they stimulate further narrative, particularly when they deal with the making of ethical judgments in concrete situations, that is in clinical bioethics.

The narrative forms used in bioethics differ greatly. The cases or case studies to which bioethics refers can be real or made-up; they can be authentic stories in oral or written form, but they can also be poems, fictional stories, parables, or novels which train and deepen a "sensitivity to the particular", and evoke emotions which belong to a comprehensive notion of morality. If reading is detached from its instrumental use as the illustration of a previous opinion or judgement made independently of the reading, then the texts open themselves up as sources of ethical experience.

Bioethics has recourse to stories according to the following interpretation. First, to give a voice to those who are too little heard in general medical debate. Second, to establish a reflective practice for doctors and nursing staff that allows them to call their own positions and prejudices into question. Third, to gain knowledge and insights about the subjective dimension of being ill – in other words, in order to give ethics and medical judgement as deep and wide-reaching hermeneutical basis as possible.

3.5   Narrative bioethics as a theory of reflection

With this meta-narrative and reflexive-ethical approach, the interpretation and use of texts in a specific practice has already undergone a shift in perspective: to that of bioethics as a theory of reflection. The immediacy of aesthetic experience is left behind and instead, as is necessary in literary studies and art theory, the level of interpretation itself becomes the focus of attention.

This means that bioethics is, at least to a certain extent, a *literary hermeneutical ethics* in the sense that it interprets texts or stories which are told in the various fields of practice. Questions related to literary studies become important here: who is telling the story, is the narrator trustworthy, from whose perspective is the story being told, what does the narrator leave out? Which voices, which positions are mentioned, which are not, what images and metaphors are used, what meanings are evoked, how do the images shape the perception of what is going on, and so on. Here, as opposed to literary studies or to hermeneutics in general, the concern of narrative analysis is not (only) to understand the situations and contexts, and the intentions of the actors better. Rather, analysis which uses the methods of literary studies leads to a practical ethical commentary and a normative ethical critique of what is told. Mieth speaks of a method of integration that ideally contributes to the constituting of ethical models as means of fostering moral sensitivity.

"Despite all the differences between them, hermeneutical methods distinguish themselves by three essential characteristics: the *loyalty* of interpretation to that which is interpreted, the continuing critical *examination* of the presuppositions connected with that loyalty, and the *openness* and open-endedness of the interpretation based on the ultimate unavailability of contexts of meaning ... For this reason, the method of critical reflection is very important for it. It must place divergent interpretations in a mutually critical relationship to each other. The resulting overlapping of perspectives which prove themselves in the critical relationship allows comprehensible priorities to emerge as ethical models"(Mieth 1999: 148).

Anna Hudson Jones speaks in a comparable fashion of the 'abductive' method of narrative medical ethics, which consists in the mediation between the single case and the paradigm, and interprets the achievement of understanding between doctor and patient as the creation of a common story.

"In an ideal form, narrative ethics recognises the primacy of the patient's story but encourages multiple voices to be heard and multiple stories to be brought forth by those whose lives will be involved in the resolution of a case" (Hudson Jones 1999: 253).[10]

4. THE BEWILDERING CONCRETENESS OF STORIES

I would like to consider one example. It is the story of Caroline Stoller, which she published in 1996 at a time when the problem of late-term abortion after prenatal diagnostic testing was scarcely known in Germany or Switzerland (the country where Stoller lives), and therefore was not widely discussed. It was subsumed under the more general abortion debate, and was accordingly mined territory. Within this context Stoller's autobiographical report becomes a political report that gives women a voice, that considers case histories and consequently provokes judgments on a case-by-case basis, and that in my opinion best expresses the dimension of narrative bioethics we have been considering.

Caroline Stoller describes the beginning of her pregnancy, which followed a previous pregnancy that ended in spontaneous miscarriage in the 8th week, as a time of excitement and hope, a time of corporeal and psychological change, of the dreams about "this" child.[11]

> "From the first moment on this child existed in his individuality and is already a personality. It may sound somewhat strange to speak of a personality that is comprised at first only of a few cells. But at the very least even in these chromosomes is a part of this personality already predetermined. Thoughts of a hopeful and expectant mother."

> "It is a happy feeling to tell oneself this. To have a child that we have given life to. A child born of our chromosomes. A child that may perhaps be a little like me?"

In the 16th week of pregnancy, ultrasound and an AFP (alpha foeto-protein) examination were performed. Stoller describes how she unwillingly consented to these tests. As a nurse she was well informed about diagnostic techniques, but she was also aware of how suffering affects people when they are sick.

> "This AFP test looked to me to be an assault upon the personality of this little person. What could the blood levels possibly say about this child? In comparison to the miracle of life, to the perfection of a person? Isn't it already a bit arrogant of us to take on such a right? The right to ascertain the 'quality' of a person?"

> "And yet I naturally went back and forth about this. It really is no big deal, only a blood draw. And when I look at this rationally and without feelings of motherhood, I find this test quite acceptable ... In the end, understanding and rationality won out, and above all the need to have an affirmation. An affirmation of the health of the child."

After the test and ultrasound began a one-week waiting period:

> "These were days that allowed once again all dreams, wishes and visions, without my hopes being dashed. Afterwards, it would be different."

On the basis of the AFP test the doctors diagnosed "spina bifida", a severe malformation that brings little or no hope of life for the child. The pregnant woman reacted with aggression and anger, but also with sadness and great sympathy:

"I feel humanity, the inviolable, the miraculous, that no one must change and must be accepted as it comes. But there is my reason, my fear of suffering, of pain. Both have their justification, each side. And yet I have to decide for one side. One side must die. I must say goodbye to one side. One must be stronger than the other. But I still feel both inside of me. It is as though a part of me must die in this decision. If I choose reason, the other side screams out."

The decision process in this situation was extremely difficult, and a heavy burden for the woman, the couple. Both options pointed to a moral dilemma that always arises when a moral conflict cannot be clearly resolved, because there are two conflicting duties that stand irreconcilably next to one another. The option to continue with the pregnancy meant a decision on behalf of a child whose future would be a life which our society would view as not "normal". The child would be permanently dependent upon another person, above all its mother or parents, for aid and provision. Stoller, who because of her occupation was familiar with the care of children, and familiar with her child's diagnosis, considered this position carefully:

"[He] won't ever be able to do even the most fundamental and simple things by himself. Never going to the toilet alone, never able to play around with other kids. Never jump around in the snow or fall on his face laughing. Never be able to jump after a cat. Never race around on a tricycle. None of these things will he be able to do, and I know that now even before he's born ... I think, if I got to know my child, it would be completely different. Then the child would stand in front of me with his own personality. The condition would become somehow secondary. It would belong to him, to his personality."

In such cases, the decision to have the child is made under uncertain conditions: no one can say how the genetic "defect" will affect the child, the prognoses are in any event only relevant as statistical probabilities. In this situation of uncertainty the pregnant woman is completely dependent upon medical explanation and the sensibility of the counsellor (insofar as she has access to counselling that derives from the explanation). But she is additionally dependent upon social support and her own experience and subjective judgment of her abilities and strength to care for a handicapped child until her own death or the death of the child (regardless of its age). However, from a moral perspective the decision must also take account of the wellbeing of the child, and not just her own wellbeing. In her account, Stoller places her vision of doubt in question and contrasts it with another:

"What does it mean, 'quality of life'? My perspective on quality of life is very subjective, and what is important for me doesn't necessarily apply to others. They are given laughter, joy, the possibility to feel and sense. Eyes with which they can see the multiplicity of our world. Ears, with which they can ear marvellous sounds … Hands with which they can touch, play and explore."

Stoller next attempts to bring the perspective of the child into harmony with her own:

"Would it be important for our life to accomplish such a task? Important for the child?"

But this harmonisation does not work. There still remains the hopelessness and fear: paralysed fear in face of the inability to make a decision and the impossibility of escaping it.

"No one says what is good and what is not. The decision and the responsibility lies alone with us. Sure, one discusses with doctors about whether it would be reasonable in our circumstance to abort the pregnancy. Still, we are the ones who must accept the moral responsibility for the decision."

"It seems to me to be a responsibility beyond my strength and ability to conceptualise. I can't take on this responsibility. Somehow this decision leads to a superhuman responsibility."

"All of a sudden, the personal philosophy seems to falter. It appears false. There is no new perspective/orientation. Somehow I feel uprooted. Self-deceived. In a certain fashion one could say that the character of personal conviction is revealed. In a terrifying way it appears that ideals cannot withstand reality, real life. They are only good in theory, but in reality they are worthless. Utopia. Not doable."

Stoller describes not only the overwhelming hours of decision; she also details how she was confronted with a medical apparatus that measured the child in minute detail ("each individual finger, every single organ") and brought it under control. The fact that she thereby learned more about her child than ever before was a great burden:

"For me ultrasound is extremely awful. It hurts to have to look at the child, knowing that it will soon be dead."

The decision against the child destroyed life plans – at least with respect to the hoped-for future of this child. It is easy to see from Stoller's autobiographical account how much even individual ideals and moral convictions waver. Stoller had the ideals of a caring mother, she was convinced of the sanctity of life. But she did not find a place in life for these convictions ("utopia"). And therefore she can – or must – say:

"The personal worldview is turned upside-down. What once had validity no longer does."

The experience of the deep disturbance of bodily sensation ("the child lives") and disorientation of the moral ability to judge leads to a crisis that is experienced as an extreme challenge. Stoller describes how she finally gave into the pressure that appeared to hold sway over this situation. Both doctors independently advised her to terminate the pregnancy, and this seemed to be the "rational" solution, while the desire to keep the child seemed irrational and emotional. But because the emotions themselves were divided, it appeared to be clear that the solution of the conflict could not rest upon them.

The birth of a dead child is a traumatic experience. However, the intentional and informed induction of the birth of a dead child is an almost indescribable situation. Nevertheless Stoller – certainly also in order to offer a contrast with an early-stage abortion – describes it in great detail.

After the birth Stoller says:

"We should not have allowed it. But it wasn't our decision alone. Dr. Ammann stands behind it, the physicians in the birth clinic and the ancillary personnel. Society, our friends, relatives understand us and assure us that our decision was the right one … It offers no consolation."

No consolation. I suppose this is what many patients or relatives feel when having made decisions in dilemmatic situations. This alone should prevent us from conceptualising bioethics as a tool like a mechanism. But it should also prevent us from expecting too many answers from a narrative. In the final part, I will turn to the question of the status of narratives within bioethics.

5. THE STATUS OF NARRATIVE WITHIN A CONCEPT OF BIOETHICS

5.1 Narrative ethics as a supplement to an ethics of principles

John Arras, whose evaluation of current approaches to narrative bioethics I follow to a large extent, would call the conception which many bioethicists now favour, and which I have just sketched out, the "narrative as supplement" approach. That is, narrative ethics as a supplement to an ethics of principles which leaves the latter to a great extent untouched. Alongside the more narrowly biomedical ethical examples, which in terms of content conform to my discussion up until now, Arras also includes the reflective equilibrium of Rawls, to which Martha Nussbaum also refers. For Rawls, it is the reflective intuitions which represent a source of moral insight and moral experience, and which need to be asserted as an opposing pole to the principles. The coherentist position of Beauchamp and Childress (see also Schöne-Seifert in this volume), which can easily be introduced here, is also based on this image of complementarity:

> "advocates of this coherentist approach to moral justification would have us view narrative and stories as intimately bound up with the most sophisticated renderings of principle- and theory-driven moral reasoning. For no matter how far we progress toward the ethereal realms of principle and theory, we ought never to lose sight of the fact that all of our abstract norms are in fact distillations (and, yes, refinements) of our most fundamental intuitive responses to stories about human behaviour" (Arras 71).

I would rather call this approach a 'complementarity approach' than a 'narrative as supplement approach'. As I have argued, an ethics which attempts to deal with values and virtues is unthinkable without recourse to narrative. Since a modern ethics of virtue cannot be conceived as a normative concept (as it certainly is in the case of an Aristotelian ethics), it but must be conceived as a complement to principlism. In my own understanding, this is not coherentist, however, since this approach would be a too harmonious a version of moral reasoning. On the contrary, I would hold that literary ethics and normative ethics are *dialectically* related: they are not to be commensurate but rather controversial. Basic concepts of morality are indeed questioned by narratives and literary ethics. If aesthetic experience is closely connected to 'receptionary' aesthetics, so ethical experience should be connected to a kind of receptionary ethics. Only if moral principles are confronted with the "bewildering" experiences of literary ethics, can the dialectical dynamics be upheld – or initiated.

5.2   Narrative bioethics as a substitute for an ethics of principles

However, there is another approach which differs in a decisive way from the line of thinking we have just discussed. Its most prominent representative is perhaps Alasdair MacIntyre. Principles, for him, are nothing other than symbols for historical narratives whose relativity has been demonstrated throughout history. Reason has always been historical reason, and even bioethics is bound, in its normative statements, to historical traditions or foundational stories.

"Far from being justified before some court of abstract reason, our actions are ultimately sanctioned by appeal to the norms, traditions, and social roles of a particular social group. Obviously, according to MacIntyre and Hauerwas, to lack such a distinctive story is to lack a rationale for one's actions, character, and life" (ibid. 74).

Arras, however, correctly points out that the problem with such a concept is already contained in the notion of narrative. Instead of being "the" story one is entangled with, narratives, by means of their dialogical and open character, entail not only the force of disruption but also the possibility of radical change. However, this means that bioethics, in its attempt to base itself on relatively closed narratives, traditions, roles and norms, would lose this precision once again. If, on the contrary, narratives are "open" in this sense, then the point of view from which moral judgements are made cannot be found within them, but must be sought outside them in the dialectic of narrative concreteness and normative abstraction.[12] The contribution of a narrative ethics to the affirmation of moral norms would then be less than either MacIntyre or Hauerwas suppose.

5.3 Narratives and the post-modern theory of truth

The third variation on the relationship between narrative ethics and normative ethics mentioned by Arras is post-modern ethics, particularly the version of R. Rorty. According to this theory, seeking or attempting to maintain meta-narratives is doomed to failure, at least in the sense that it might be possible to find an objective truth in them. This is the case not only with respect to epistemology, but equally for morality and its claims to validity. Beyond this, meta-narratives only function with the help of exclusion, force, and subjection, as Lyotard has emphasised. They must therefore themselves be subjected to ideological criticism.

If, however, as post-modern ethics might suggest, normative questions are banned from ethics, this raises the question of whether narrative ethics then becomes a theory of reflecting on tradition or on social roles. Indeed, this is exactly what Arthur Kleinman, for example, envisions. Kleinman holds that, if it does not wish receive the verdict of "ethnocentricity, mediocentricity and psychocentricity", the "new bioethics" will have to be constituted as a kind of ethnography. In this context, narratives play a central role, inasmuch as they represent the source of agreement about a specific situation (Kleinmann 1999). Here, however, the question of the normative status of the social-scientific findings remains systematically unanswered. Daniel Callahan observes with regard to Kleinman:

"How do we incorporate what we have learned into the making of good moral judgments? How do we get from the 'is' of 'lived experience' to the 'ought' of those judgments that require us to act in some justifiable manner?" (Callahan 1999: 288).

The alternative to a Weberian ethnology of moral forms of praxis in different cultural contexts consists in the role of the ironic observer, in critical distance, such as appears to suggest itself to Rorty. Nevertheless, the ethical attitude of sympathy and compassion for concrete persons plays an extremely important role for Rorty. In this it becomes clear that even a postmodern approach cannot ultimately avoid the question of normative status.

6.   CONCLUDING REMARKS

This is not the place to determine the interaction and the complementarity of narrative and moral principles or moral norms in an appropriate or comprehensive fashion. But if we need to hold on to the general claim to normativity of moral obligations and moral rights – and I see no alternative approach ready to deal appropriately with injustice and the violation of rights – then the relationship between the justification of moral claims by means of arguments, the establishment and review of "low-range" principles and norms of action in different contexts of praxis, and the narratives which address these contexts must be clarified. With respect to clinical bio- or medical ethics, I would like to offer the following suggestions.

Firstly, since in the context of medicine our notion or understanding of illness is constitutive for the practice itself, and in that the notion of illness has both an objective and a subjective side, narrative (or the self-understanding of a person) is, in this area, a *necessary* component of bioethical reflection. Although a number of works on narrative medical ethics exist, this issue still needs to be outlined more exactly.

Secondly, the normative interaction between principles and norms, and the evaluative terms with which a person interprets his or her experiences in the way I have described above, must be as much a part of the bioethical reflection as of ethical reflection in general. Far from interpreting the evaluative dimension as hypothetically normative maxims (to use the Kantian term), the interpretative dimension referring to agency and identity must be taken into consideration.

Thirdly, if narrative bioethics is, at least in part, a literary hermeneutical bioethics, then it needs to be conceived of methodologically as commentary and critique (Haker 1999). The commentary must comprehend and analyse a text as exactly as possible, keeping in mind that neither the story itself nor the analysis can capture the "inexpressible" and pre-reflective side of the experience or experiences, and thus is an interpretation. This is particularly the case with respect to non-fictional texts, which are subject to the criterion of authenticity. It is the prerogative of the critique, on the other hand, to confront the text or statements of the text with ethical theory and ethical knowledge. In this way, the aesthetic and the evaluative content of a text is taken seriously, and at the same time the argumentative content is transferred to the normative ethical discourse. Here, in a further stage of normative ethical reflection, it is to be examined with respect to its consistency and argumentative force.

Fourthly, a bioethical ethics of virtue – or perhaps rather an ethics of values and long-term convictions – cannot be conceived independently of reflection on the narrative hermeneutical foundation.[13]

Fifthly, it is not yet clear what role narrative bioethics can play in social ethics. However, socially and culturally mediated values, increasingly communicated by the media, have to be taken into account. They must nevertheless be examined critically with a view to the question of whether they do justice to respect for the individual human being and social practices.[14]

Bioethics as social ethics has to be conceived as an ethics of justice, and as such it will have to engage with structures and experiences of injustice rather than justice. But it must also participate in society's debate about the goals, for example, of health policy. Here, narrative bioethics is also certainly in its infancy, but I hope to have shown that there is good reason not to marginalize it within the bioethical reflection.

References

Arras, J.D. "Nice Story, But So What?" In *Stories and their Limits: Narrative Approaches to Bioethics*. Hilde Lindemann Nelson. New York: Routledge, 1997, 65–90.

Callahan, D. "The Social Sciences and the Task of Bioethics." *Daedalus* 1999, 128, 4, 275–294.

Charon, R. "Narrative Contributions to Medical Ethics: Recognition, Formulation, Interpretation, and Validation in the Practice of the Ethicist." In *A Matter of Principles? Ferment in U.S. Bioethics*. Edwin R. DuBose, Ronald P. Hamel and Laurence O'Connell (eds.). Pennsylvania: Trinity Press International, 1994, 260–283.

Charon, R. and Montello, Martha M. (eds.). *Stories matter: the role of narrative in medical ethics*. New York, 2002.

Childress, J.F. *Practical Reasoning in Bioethics*. Bloomington/Indianapolis: Indiana University Press, 1997.

DuBose, E.R.; Hamel, R.P.; O'Connell and Laurence J. (eds.). *A Matter of Principle? Ferment in U.S. Bioethics*. Pennsylvania: Trinity Press, 1994.

Evans, D. "Imagination and medical education." *Journal of Medical Ethics: Medical Humanities* 2001, 27, 30–34.

Frank, A.W. *The Wounded Storyteller: Body, Illness, and the human condition*. Chicago: University of Chicago Press, 1995.

Frank, A.W. "The Standpoint of the storyteller." *Health Research* 2000, 10, 3, 354–365.

Gadamer, H.G. *Über die Verborgenheit der Gesundheit*. Frankfurt am Main: Suhrkamp, 1993.

Haker, H. *Moralische Identität. Literarische Lebensgeschichten als Medium ethischer Reflexion. Mit einer Interpreation der 'Jahrestage' von Uwe Johnson*. Tübingen: Francke, 1999.

Haker, H. "Compassion als Weltprogramm des Christentums? Eine ethische Auseinandersetzung mit Johann Baptist Metz." *Concilium* 2001, 4, 436–450.

Haker, H. "'Wie die Ränder einer Wunde, die offenbleiben soll'. Ästhetik und Ethik der Existenz". In *Denken/Schreiben (in) der Krise – Existentialismus und Literatur*. Cornelia Blasberg and Franz Josef Deiters (eds.). 2004, 539–564.

Haker, H. "The fragility of the Moral Self." *Harvard Theological Review*, 1, 2005 (forthcoming).

Hudson Jones, A. "Narrative based Medicine: Narrative in medical ethics." *British Medical Journal* 1999, 318, 7178, 253–256.

Irrgang, B. *Praktische Ethik aus hermeneutischer Sicht*. Paderborn: Schönigh, 1998.

Jauss, H.-R. *Ästhetische Erfahrung und literarische Hermeneutik*. Frankfurt: Suhrkamp 1982.

Kleinman, A. *The Illness Narratives. Suffering, Healing, and the Human Condition*. New York: Basic Books, Inc. Publ., 1988.

Kleinman, A. "Moral Experience and Ethical Reflection: Can Ethnography Reconcile them? A Quandary for 'The New Bioethics'." *Daedalus* 1999, 128, 4, 69–97.

Leder, D. "Toward a Hermeneutical Bioethics." In *A Matter of Principles? Ferment in U.S. Bioethics*. Edwin R. DuBose, Ronald P. Hamel and Laurence J. O'Connell (eds.). Pennsylvania: Trinity Press, 1994, 240–259.

Lenk, H. *Philosophie und Interpretation. Vorlesungen zur Entwicklung konstruktionistischer Interpretationsansätze*. Frankfurt am Main: Suhrkamp, 1993.

Liebsch, B. *Geschichte als Antwort und Versprechen*. München: Alber, 1999.

Mead, George H. *Mind, Self, and Society: From a standpoint of a social behaviourist*. Chicago: Chicago University Press, 1967 [1934].

Mieth, D. *Moral und Erfahrung I. Grundlagen einer theologisch-ethischen Hermeneutik* (4. überarb. Aufl.). Freiburg i. Ue: Universitätsverlag Freiburg i.Ue./Herder, 1999.

Mieth, D. *Moral und Erfahrung II*. Studien zur Theologischen Ethik 76. Holderegger, A. and Bertouzoz, R.(eds.). Freiburg i. Ue: Universitätsverlag Fribourg/Herder, 1998.

Mieth, D. *Erzählen und Moral. Narrativität im Spannungsfeld von Ethik und Ästhetik*. Tübingen: Attempto, 2000.

Miller, J.H. *The Ethics of Reading: Kant, deMan, Eliot, Trollope, James, and Benjamin*. New York: Columbia University Press, 1987.

Nussbaum, M. *Love's Knowledge. Essays on Philosophy and Literature*. Oxford, 1990.

Nussbaum, M. *Poetic Justice. The literary imagination and public life*. Boston, 1995.

Ricœur, P. *Oneself as Another*. Chicago, 1992 [1990, Soi même comme un autre].

Ricœur, P. *Time and Narrative*. Chicago, 1984–88 [Temps et recit].

Schnädelbach, H. *Vernunft und Geschichte*. Frankfurt am Main: Suhrkamp, 1984.

Schapp, W. *In Geschichten verstrickt: Zum Sein von Mensch und Ding*. Frankfurt a.M., 1985.

Stoller, C. *Eine unvollkommene Schwangerschaft*. Zürich, 1996.

Seel, M. *Eine Ästhetik der Natur*. Frankfurt am Main: Suhrkamp, 1991.

Seel, M. *Ethisch-ästhetische Studien*. Frankfurt am Main: Suhrkamp, 1996.
Taylor, C. *Philosophical Arguments*. Boston: Harvard University Press, 1995.
Zaner, R.M. "Experience and Moral Life." In *A Matter of Principles? Ferment in U.S. Bioethics*. Edwin R. DuBose, Ronald P. Hamel and Laurence J. O'Connell (eds.). Pennsylvania: Trinity Press, 1994, 211–239.

---

1 I am grateful for the two persons who translated parts or a draft version of this paper: Glenn Patton and David Hester.

2 Transcendence of reality, in this sense, is immanent transcendence: it cannot go beyond the horizon of interpretation or the hermeneutically constructed meaning.

3 For a broader discussion of this point, including a discussion of J. Butler's *Kritik der ethischen Gewalt* (2003), see the author's *Wie die Ränder einer Wunde, die offen bleiben soll* (2004), and *The fragility of the moral self* (2005).

4 I am aware that literature is not to be identified with narrative literature. However, it could well be that in the process of reception, we transform, for example, a poetic form into a narrative. I will not deal with this question here, however.

5 In this approach, one might even find some parallels to the work of Stanley Hauerwas who introduced his narrative approach during the same period. The distinction between the two concepts lies in the way in which Mieth and Hauerwas treat texts. For Mieth narrative ethics was not a substitute for norms or normative ethics, and these remained important. Mieth, therefore, follows on a complementary model (see below). For a discussion of Mieth see the author's *Moralische Identität*, 1999.

6 Since her *Fragility of Goodness*, Nussbaum has changed her approach to an increasingly normative theory, and at present favours a concept of what I would call a theory of the good, although the "capacity approach" shows strong connections to human rights' theories and lately to the social contract theories. Nussbaum still remains fundamentally ambivalent.

7 After World War II, "Literature after Auschwitz" became the term for a literary genre that tried a) to recognize and respect the only possible answer to the mass-murder, which is silence and refrain from any kind of explanation, and b) to give an account of what happened by way of narratives. Even at this borderline between silence in the face of senseless and meaningless destruction of human beings and the recognized responsibility for giving an account of this history, it is true that literature takes part in the articulation of experiences.

8 William Faulkner is one of the most famous writers making use of this technique.

9 See, for example, S. Sontag: *Illness as Metaphor*, London, 1979.

10 Compare also: Charon 1994 and Charon/Montello 2002.

11 The following quotes are my translation of the German text.

12 For an alternative normative reasoning, which is not a-historical and still bound to truth claims, see H. Schnädelbach: *Vernunft und Geschichte*, 1984.

13 Could it be that Beauchamp's and Childress' approaches were better viewed as a kind of value ethics? In this case, the debate on "principlism" would have to begin at another point (see also Schöne-Seifert in this volume).

14 A key literary text for the development of a narrative social ethics may be Franz Kafka's *Castle*. Taking this text as a point of departure, one could subject social institutions to a rigorous critique, and their goals, methods and procedures,

efficiency and effects on social life could be examined against the negative foil of the *Castle*. The structure of experience in institutions can be experienced and analysed on the basis of this text in a unique way.

LIST OF CONTRIBUTORS

BEYLEVELD, Deryck

Professor of Jurisprudence at the University of Sheffield and Director of the Sheffield Institute of Biotechnological Law and Ethics (SIBLE), which he founded in 1994). He studied biochemistry (University of the Witwatersrand), philosophy and social and political sciences (University of Cambridge) and received his PhD on *Epistemological Foundations of Social Theory* (University of East Anglia). He held posts at the University of Cambridge Institute of Criminology and at Bradford College before joining the Faculty of Law (University of Sheffield) in 1978. He has a special interest in the regulation and ethics of biotechnology and medical science. He is the author of *A Bibliography on General Deterrence Research* and *The Dialectical Necessity of Morality*, and co-author (with Roger Brownsword) of *Law as a Moral Judgment, Mice Morality and Patents*, and *Human Dignity in Bioethics and Biolaw*, as well as of numerous articles in refereed journals and books.

CASTIGNONE, Silvana

Professor of Philosophy of Law, Faculty of Law, Genoa University, Italy since 1976. From 1993 to 1996 she directed the course in political philosophy at the Faculty of Law. She was head of the Department 'Giovanni Tarello' (Legal Culture) in 2000 and 2001. Editor of *Materiali per una storia della cultura giuridica* and numerous publications.

CORTINA, Adela

Professor of Ethics and Political Philosophy at the University of Valencia since 1987. She studied philosophy and held scholarships from the 'DAAD' in Munich and 'Alexander von Humboldt Stiftung research programmes' in Frankfurt am Main, and received her habilitation in 1980 in Madrid. Visiting Professor at the University of Louvain-la-Neuve, Vrije Universitet (Amsterdan) and Notre Dame University (USA). She is Director of the Foundation ÉTNOR (for business and organisational ethics), member of the 'Comisión Nacional de Reproducción humana asistida' and of the 'Comité Ético para la Investigación Científica y Tecnológica.' She participates in research and working groups in both Europe and Latin America. Her areas of research are: discourse ethics, applied ethics, theory of democracy, theories of citizenship, human rights. She is the author of many books, including *Ética mínima* (1986), *Ética aplicada y democracia radical* (1993), *Ética de la empresa* (1994), *Ciudadanos del mundo* (1997), *Alianza y Contrato* (2001), *Por una ética del consumo* (2002), *Razón pública y éticas aplicadas* (2003) and *Construir confianza* (2003).

DÜWELL, Marcus

Professor of Philosophical Ethics and Scientific Director of the Institute for Ethics at the University of Utrecht, the Netherlands. He studied philosophy at the Universities of Munich and Tübingen and received his PhD with a book on the *Relation of Ethics and Aesthetics*. From 1993 to 2001 he was scientific coordinator of the 'Interfakultäres Zentrum für Ethik in den Wissenschaften' at the University of Tübingen. His many publications include *Handbuch Ethik* (with Christoph Hübenthal and Micha H. Werner) and *Bioethik. Eine Einführung* (with Klaus Steigleider).

FEUILLET-LE MINTIER, Brigitte

Professor of Law in the Faculty of Law, Rennes University, France. Specialist in international law, family law and biomedical law. She is head of a research team, the 'West Center of Law Research', specializing in questions of bioethics and law. She leads the Masters program 'Law, Health and Social Welfare' and is a member of the editorial committees of several review boards. Publications: *L'embryon humain, approche multidisciplinaire* (Economica Paris 1996), *Philosophie, Ethique et droit de la médecine* (with J.F. Mattei and D. Folscheid, which won the prize *Le Dissez de Penanrum* of the Academy of Moral and Political Sciences in 1999), *Les lois bioéthique à l'épreuve des faits: réalités et perspectives* (Puff, Droit et Justice 1999), *Normativité et Biomédecine* (Economica 2002), *Santé, Argent et Ethique* (L'Harmattan 2005).

GRAUMANN, Sigrid

Senior Researcher at the Institute 'Mensch, Ethik und Wissenschaft' in Berlin since April 2002. Studied biology and philosophy at the University of Tübingen, receiving a Diploma in Biology and PhD from the Department for Human Genetics for the thesis *Somatic Gene Therapy for Monogenetic Diseases – Conceptual and Ethical Questions*. Scientific Coordinator of the European

Network for Biomedical Ethics (1997-1999, European Commission). She led the research project 'Ethical Questions Relating to *in vitro* Techniques at the Beginning of Human Life', German Science Foundation DFG. She was a member of the Enquete-Commission on 'Ethics and Law of Modern Medicine' of the German Bundestag 2000–2002 and 2003–2005, a member of the Central Ethics Commission of the German Medical Association, and board member of the Academy of Ethics in Medicine (Germany). She has many publications in the fields of bioethics and disability studies.

HAIMES, Erica

Professor of Sociology at the University of Newcastle, UK, and Executive Director and Director of Research, PEALS (Policy, Ethics and Life Sciences) Research Institute, a partnership between the universities of Newcastle and Durham and the International Centre for Life. She has written and researched extensively on socio-ethical aspects of genetic and reproductive technologies, ageing and nanobiotechnology. She has a long-standing interest in the sociology of identity and in the sociology of ethics.

HAKER, Hille

Professor of Moral Theology/Social Ethics at the Catholic Faculty of Frankfurt. She studied Catholic theology, German literature and philosophy at the Universities of Tübingen, Munich and Nijmegen (NL). From 1989-2003 she was staff member of the Center for Ethics in the Sciences and Humanities, and lecturer at the department of Ethics/Social Ethics at the Catholic Theological Faculty, both University of Tübingen. From 2003-2005, she was Associate Professor of Christian Ethics, Harvard University. Her main works are in several fields of ethics, above all in the theory of ethics, ethics and literature, and bioethics (with a focus on narrative bioethics). She is currently working on a new approach to a feminist ethics/gender ethics in the sciences. Her books include *Moralische Identität* (1999), *Ethik der genetischen Frühdiagnostik* (2002) and three co-edited volumes: *Ethics of Human Genome Analysis. European Perspectives* (1993), *The Ethics of Genetics in Human Procreation* (2000), and *Ethik-Geschlecht-Wissenschaften* (2005).

380

JONSEN, Albert R.

Emeritus Professor of Ethics in Medicine at the School of Medicine, University of Washington, where he was chairman of the Department of Medical History and Ethics from 1987–1999. He is now Co-director, Program in Medicine and Human Values, California Pacific Medical Center, San Francisco. From 1972 to 1987, he was chief of the Division of Medical Ethics, School of Medicine, University of California, San Francisco. Prior to that, he was President of the University of San Francisco. His latest book is *Bioethics Behind the Headlines* (Rowland-Littlefield 2005). He is also author of *A Short History of Medical Ethics* (Oxford University Press 2000) and *The Birth of Bioethics* (Oxford 1998), and many others.

KETTNER, Matthias

Dean of the Humanities Department at the University of Witten/Herdecke, He was Assistant Professor from 1994-2000 at the Institute for Advanced Studies in the Humanities in Essen, where his research focused on clinical ethics committees in Germany, communicative ethics, and the relationship of applied ethics, biopolitics and democracy. Prior to that, he collaborated with Karl-Otto Apel and Jürgen Habermas in research on discourse ethics at Frankfurt University where he had earned a PhD in philosophy and a diploma in psychology. He recently edited *Biomedizin und Menschenwürde* (Suhrkamp Verlag 2004).

KITTAY, Eva Feder

Professor of Philosophy at SUNY, Stony Brook. Her books include *Love's Labor: Essays on Women, Equality, and Dependency* (Routledge – Thinking Gender Series 1999), *Theoretical Perspectives on Dependency and Women* (with Ellen Feder: Rowman and Littlefield 2002), *Women and Moral Theory* (edited with D. T. Meyers: Rowman and Littlefield 1987), *Metaphor: Its Cognitive Force and Linguistic Structure* (Oxford University Press 1989), *Frames, Fields and Contrasts* (edited with A. Lehrer: Erlbaum 1992). She is currently working on the *Blackwell Guide to Feminist Philosophy* (with L. Alcoff). She has edited two Special Issues of *Hypatia: Feminism and Disability* (with A. Silvers and S. Wendell) and *Special Issue of Social Theory and Practice: Embodied Values: Philosophy and Disabilities* (with R. Gottlieb).

KRIZOVA, Eva

Senior Lecturer in Medical Sociology and Public Health at the Institute of Medical Ethics and Nursing of the Third Faculty of Medicine, Charles University, Prague since 1990. She graduated from the Faculty of Philosophy in Prague, Department of Sociology. She received her PhD for the thesis *Sociological Transformations of the Medical Profession*. She started her career as a clinical sociologist at the Teaching Hospital's Department of congenital heart defects. Her main topics of interest are comparative healthcare systems, patients' rights, equity in health, and alternative medicine. She is the author of many articles and monographs and participates in inter-national research networks (e. g. the 'Phoenix Thematic Network').

LESCH, Walter

Professor of Ethics at the Institute of Philosophy and the Faculty of Theology of the Catholic University of Louvain in Louvain-la-Neuve, Belgium, since 1999. He was formerly an assistant and researcher at the University of Fribourg in Switzerland. As a theologian and philosopher he has published widely on topics of fundamental moral philosophy, various areas of applied ethics and theories of religious and aesthetic experience.

MIETH, Dietmar

Professor of Theological Ethics and Social Ethics at the Catholic Faculty of the University of Tübingen, since 1981. His studies were in theology, philosophy and German philology and he received his PhD at the University of Würzburg, Germany, 1968. He was Professor for Moral Theology in Fribourg, Switzerland (1974–1981) and Founder and Director of the 'Interfakultäre Zentrum für Ethik in den Wissenschaften' in Tübingen (1990-2001). He was a member of the 'European Group on Ethics in Science and New Technologies' (1994–200), and of the Enquete-Kommission of the German Bundestag on 'Ethics and Law of Modern Medicine'. Selected publications: *Was wollen wir können? Ethik im Zeitalter der Biotechnik* (edited with Ch. Baumgartner: Freiburg i.Br. 2002), *Patente am Leben?* (Paderborn 2003), *Meister Eckhart: Mystik und Lebenskunst* (Düsseldorf 2004), *Kleine Ethikschule* (Freiburg i.Br. 2004).

MORDACCI, Roberto

Associate Professor of Moral Philosophy at the Faculty of Philosophy of San Raffaele University, Milan. His main research fields are bioethics, moral theories and metaethics, in particular the issue of moral reasons. Among his publications are "Health as an analogical concept" in the *Journal of Medicine and Philosophy* (20, 1995), "Medicine as a practice and the ethics of illness" in *Analecta Husserliana* (72, 2001), *Ethics and Genetics. A workbook for practitioners and students* (with G. de Wert, R. ter Meulen and M. Tallacchini: Berghahn Books 2003), *Una introduzione alle teorie morali. Confronto con la bioetica* (Feltrinelli 2003).

RAPP, Rayna

Professor of Anthropology at New York University. She received her Honors BS, MA, and PhD in cultural anthropology from the University of Michigan and taught for 28 years at the Graduate Faculty of the New School, New York. She serves on the University Committee on activities involving human subjects ('ethics commission'). Social movements for women's health and women's studies deeply influenced her scholarship. Her edited and co-edited books include: *Toward an Anthropology of Women, Promissory Notes: Women and the Transition to Socialism, Articulating Hidden Histories*, and *Conceiving the New World Order: the Global Politics of Reproduction*. Her *Testing Women, Testing the Fetus: the Social Impact of Amniocentesis in America* won four awards, and her current research with Faye Ginsburg investigates the cultural epidemic in learning disabilities in the US.

REHMANN-SUTTER, Christoph

Assistant Professor of Ethics in Biosciences and Biotechnology, Head of the Unit of Ethics in Biosciences, University of Basel, Switzerland. His academic training was in molecular biology, then philosophy and sociology in Basel, Freiburg i. Brsg. and Darmstadt. From 1997–1998 he was Research Fellow at the University of California, Berkeley. Since 2001 he has been President of the Swiss National Advisory Commission on Biomedical Ethics. Research interests: bioethics, natural philosophy, philosophical and ethical issues of gene therapy and genetic testing. His books include *Leben beschreiben* (Königshausen and Neumann 1996), *Partizipative Risikopolitik* (with Adrian Vatter and Hansjörg Seiler: Westdeutscher Verlag 1998), *Ethik und Gentherapie* (coedited with Hansjakob Müller: 2nd enlarged edition, Francke 2003), *Genes in Development* (with Eva Neumann-Held: Duke University Press 2005), and *Zwischen den Molekülen* (Francke 2005).

SCHÖNE-SEIFERT, Bettina

Chair in Medical Ethics at the University of Münster, Germany since 2003. She studied medicine and philosophy in Freiburg, Göttingen, Wien and Washington DC. Member of several committees and organisations, e.g. the 'National Ethics Council' in Germany. Her main research interests are in ethical problems of modern medicine (ethical theories and applied ethics).

SCULLY, Jackie Leach

Senior Researcher and Lecturer at the Unit of Ethics in Biosciences, University of Basel, Switzerland. She took a first degree in biochemistry from the University of Oxford and a PhD in molecular pathology from the University of Cambridge, and held postdoctoral research positions at the Swiss Institute for Experimental Cancer Research and the Institute for Physiology, University of Basel. Current research interests include empirical approaches to ethical evaluation, theories of ethics in embodiment and disability, feminist ethics, and applications of psychoanalytic theory to ethics. Recent publications include: "Admitting all variations? Postmodernism and genetic normality" in *Ethics of the Body: Postconventional Challenges* (eds. Shildrick and Mykitiuk: MIT Press 2005), *Quaker Approaches to Moral Issues in Genetics* (Mellen 2002), and "A postmodern disorder: moral encounters with molecular models of disability" in *Disability/ Postmodernity: Embodying Disability Theory* (eds. Shakespeare and Corker 2002).

SITTER-LIVER, Beat

Professor of Practical Philosophy at the University of Fribourg, Switzerland. He earned his PhD at the University of Berne after studying philosophy, German and English language and literature, law and theory of the state in Berne, London, Reykjavik, and Cologne. He lectured at the University of Berne, was a Visiting Professor at the University of Munich, and taught at the Swiss Federal Institute of Technology (Zürich), and the Universities of Lucerne, Basel and Bern. In 2003, the University of Lausanne conferred on him a *doctorat honoris causa* (social sciences). Honorary member of the Swiss Academy of Humanities and Social Sciences, member of the European Academy of Sciences and Arts, the Institute Grand' Ducal of Luxembourg, and of various national and European scientific bodies of ethics.

TISHCHENKO, Pavel

Associate Professor for Philosophy, Department of Philosophy, Moscow, Russia. Trained as a philosopher in Moscow, he held Fellowships in the Center For Philosophy and Health Care in Swansea (UK 1992), Utah Center for Human Genome Research (USA 1995–1996), the Hastings Center, (USA 1991 and 1995), and the Center for Ethics and Philosophy of Medicine, Tübingen University (Germany 1996). Selected books and papers include *Bio-power in the Epoch of Biotechnologies* (RASC Moscow 2001), "The Goals Of Moral Reflection" in *Advances in Bioethics* (ed. by R. Edwards and E.E. Bittar: Volume 4, 1998), "The Moral Status of Fetuses in Russia" in *Cambridge Quarterly of Healthcare Ethics* (6, 1, 1997), "Corruption: the Russian experience" in *Bulletin of Medical Ethics* (121, Sept. 1996).

WIESEMANN, Claudia

Director of the Institute for Medical Ethics and History of Medicine at Göttingen University, Germany. President of the German Academy for Ethics in Medicine and Secretary of the European Association for the History of Medicine and Health. Her focus of research lies on the history of medical ethics in the 20th century and on medical ethics from a cultural perspective. She has published on ethical problems of the brain death definition as well as the history of brain death. Together with T. Schlich she edited *Hirntod – Kulturgeschichte der Todesfeststellung.*

WILS, Jean-Pierre

Professor of Christian Ethics and Director of the Center for Ethics in Nijmegen (the Netherlands). Studied philosophy and theology in Leuven (Belgium) and Tübingen (Germany); received his PhD and his habilitation in Tübingen (1987 and 1990). Professor of Philosophy at the Humboldt-Zentrum of the University of Ulm, Germany (1992-93); held a ,Werner-Heisenberg-Stipendiat' from the German Research Foundation DFG (1993-1996). Selected publications: *Subjektivität und Sittlichkeit* (1987), *Ästhetische Güte* (1990), *Verletzte Natur* (1992), *Die große Erschöpfung* (1994), *Die Moral der Sinne* (1999), *Euthanasie* (1999), *Handlungen und Bedeutungen* (2002), *Versuche über Ethik* (2004), *Nachsicht* (2005).

International Library of Ethics, Law, and the New Medicine

1. L. Nordenfelt: *Action, Ability and Health*. Essays in the Philosophy of Action and Welfare. 2000   ISBN 0-7923-6206-3
2. J. Bergsma and D.C. Thomasma: *Autonomy and Clinical Medicine*. Renewing the Health Professional Relation with the Patient. 2000   ISBN 0-7923-6207-1
3. S. Rinken: *The AIDS Crisis and the Modern Self*. Biographical Self-Construction in the Awareness of Finitude. 2000   ISBN 0-7923-6371-X
4. M. Verweij: *Preventive Medicine Between Obligation and Aspiration*. 2000
                                                                             ISBN 0-7923-6691-3
5. F. Svenaeus: *The Hermeneutics of Medicine and the Phenomenology of Health*. Steps Towards a Philosophy of Medical Practice. 2001   ISBN 0-7923-6757-X
6. D.M. Vukadinovich and S.L. Krinsky: *Ethics and Law in Modern Medicine*. Hypothetical Case Studies. 2001   ISBN 1-4020-0088-X
7. D.C. Thomasma, D.N. Weisstub and C. Hervé (eds.): *Personhood and Health Care*. 2001
                                                                             ISBN 1-4020-0098-7
8. H. ten Have and B. Gordijn (eds.): *Bioethics in a European Perspective*. 2001
                                                                             ISBN 1-4020-0126-6
9. P.-A. Tengland: *Mental Health*. A Philosophical Analysis. 2001   ISBN 1-4020-0179-7
10. D.N. Weisstub, D.C. Thomasma, S. Gauthier and G.F. Tomossy (eds.) : *Aging: Culture, Health, and Social Change*. 2001   ISBN 1-4020-0180-0
11. D.N. Weisstub, D.C. Thomasma, S. Gauthier and G.F. Tomossy (eds.) : *Aging: Caring for our Elders*. 2001   ISBN 1-4020-0181-9
12. D.N. Weisstub, D.C. Thomasma, S. Gauthier and G.F. Tomossy (eds.) : *Aging: Decisions at the End of Life*. 2001   ISBN 1-4020-0182-7
                                            (Set ISBN for Vols. 10-12: 1-4020-0183-5)
13. M.J. Commers: *Determinants of Health: Theory, Understanding, Portrayal, Policy*. 2002
                                                                             ISBN 1-4020-0809-0
14. I.N. Olver: *Is Death Ever Preferable to Life?* 2002   ISBN 1-4020-1029-X
15. C. Kopp: *The New Era of AIDS*. HIV and Medicine in Times of Transition. 2003
                                                                             ISBN 1-4020-1048-6
16. R.L. Sturman: *Six Lives in Jerusalem*. End-of-Life Decisions in Jerusalem - Cultural, Medical, Ethical and Legal Considerations. 2003   ISBN 1-4020-1725-1
17. D.C. Wertz and J.C. Fletcher: *Genetics and Ethics in Global Perspective*. 2004
                                                                             ISBN 1-4020-1768-5
18. J.B.R. Gaie: *The Ethics of Medical Involvement in Capital Punishment*. A Philosophical Discussion. 2004   ISBN 1-4020-1764-2
19. M. Boylan (ed.): *Public Health Policy and Ethics*. 2004
                                            ISBN 1-4020-1762-6; Pb 1-4020-1763-4
20. R. Cohen-Almagor: *Euthanasia in the Netherlands*. The Policy and Practice of Mercy Killing. 2004   ISBN 1-4020-2250-6
21. D.C. Thomasma and D.N. Weisstub (eds.): *The Variables of Moral Capacity*. 2004
                                                                             ISBN 1-4020-2551-3
22. D.R. Waring: *Medical Benefit and the Human Lottery*. An Egalitarian Approach. 2004
                                                                             ISBN 1-4020-2970-5
23. P. McCullagh: *Conscious in a Vegetative State? A Critique of the PVS Concept*. 2004
                                                                             ISBN 1-4020-2629-3

International Library of Ethics, Law, and the New Medicine

24. L. Romanucci-Ross and L.R. Tancredi: *When Law and Medicine Meet: A Cultural View*. 2004
                                                                      ISBN 1-4020-2756-7
25. G.P. Smith II: *The Christian Religion and Biotechnology*. A Search for Principled Decision-
     making. 2005   ISBN 1-4020-3146-7
26. C. Viafora (ed.): *Clinical Bioethics*. A Search for the Foundations. 2005 ISBN 1-4020-3592-6
27.
28. C. Rehmann-Sutter, M. Düwell and D. Mieth (eds.): *Bioethics in Cultural Contexts*. Reflections
     on Methods and Finitude. 2006   ISBN 1-4020-4240-X

Printed in the United States
86152LV00002BA/82/A